Developing Safe Systems

Related Titles

Developments in Risk-based Approaches to Safety
Proceedings of the Fourteenth Safety-critical Systems Symposium, Bristol, UK, 2006
Redmill and Anderson (Eds)
1-84628-333-7

The Safety of Systems
Proceedings of the Fifteenth Safety-critical Systems Symposium, Bristol, UK, 2007
Redmill and Anderson (Eds)
978-1-84628-805-0

Improvements in System Safety
Proceedings of the Sixteenth Safety-critical Systems Symposium, Bristol, UK, 2008
Redmill and Anderson (Eds)
978-1-84800-099-5

Safety-Critical Systems: Problems, Process and Practice
Proceedings of the Seventeenth Safety-critical Systems Symposium, Brighton, UK, 2009
Dale and Anderson (Eds)
978-1-84882-348-8

Making Systems Safer
Proceedings of the Eighteenth Safety-critical Systems Symposium, Bristol, UK, 2010
Dale and Anderson (Eds)
978-1-84996-085-4

Advances in Systems Safety
Proceedings of the Nineteenth Safety-critical Systems Symposium, Southampton, UK, 2011
Dale and Anderson (Eds)
978-0-85729-132-5

Achieving Systems Safety
Proceedings of the Twentieth Safety-critical Systems Symposium, Bristol, UK, 2012
Dale and Anderson (Eds)
978-1-4471-2493-1

Assuring the Safety of Systems
Proceedings of the Twenty-first Safety-critical Systems Symposium, Bristol, UK, 2013
Dale and Anderson (Eds)
978-1481018647

Addressing Systems Safety Challenges
Proceedings of the Twenty-second Safety-critical Systems Symposium, Brighton, UK, 2014
Dale and Anderson (Eds)
978-1491263648

Engineering Systems for Safety
Proceedings of the Twenty-third Safety-critical Systems Symposium, Bristol, UK, 2015
Parsons and Anderson (Eds)
978-1505689082

Mike Parsons • Tom Anderson
Editors

Developing Safe Systems

Proceedings of the Twenty-fourth
Safety-critical Systems Symposium,
Brighton, UK, 2nd-4th February 2016

Safety-Critical
Systems Club

The publication of these proceedings is
sponsored by BAE Systems plc

BAE SYSTEMS

Editors

Mike Parsons
NATS CTC
4000 Parkway
Whiteley, Fareham
PO15 7FL
United Kingdom

Tom Anderson
Centre for Software Reliability
Newcastle University
Newcastle upon Tyne
NE1 7RU
United Kingdom

ISBN-13: 978-1519420077
ISBN-10: 1519420072

Preface

This book contains the papers presented at the twenty-fourth Safety-critical Systems Symposium (SSS'16). This year's authors have produced interesting, informative and important material covering topics that are of concern to safety-critical systems practitioners; we are very grateful to them for their contributions.

The first day themes were New Views and Lessons Learnt. Martyn Thomas opened the symposium with a keynote presentation explaining how cybersecurity should be integrated into the safety lifecycle which is an issue that has relevance for very many critical systems. Talks on resilience and smart safety assessement concluded the first morning. The afternoon featured presentations on data safety and on new ways forward, including better aviation approvals processes and use of smart devices in nuclear installations. Rod Chapman then explained his interesting views on Agile development for high integrity software and Peter Ladkin tackled the big issue of statistical evaluation of critical software. The day finished with a stimulating keynote talk from Chris Johnson on how techniques from both safety and security can fail when used in each other's domain.

The two themes of the second day were The Human Factor and Vehicles and Autonomy. Harold Thimbleby started the day with a keynote talk looking at the errors people make in programming. Papers on competency and confirmation bias in safety arguments and safety culture completed the morning. In the afternoon the topical issue of automation in road vehicles was considered with views from manufacturers, risk assessors and also insurers. The day finished with a lively "Practitioners' Question Time" where eminent panellists responded to questions.

Formal Models and New Techniques were the themes of the final day. The morning's papers considered the underlying models used in safety development, including model-based risk management and creation of an ontological model for new guidance in data safety. The following talks considered new methods for testability, detecting memory corruption, and guidance for a new domain, sub-sea gliders. The symposium concluded with a keynote address by Graham Braithwaite which considered the way aircraft accident investigations are conducted.

We are grateful to our sponsors for their valuable support and to the exhibitors at the Symposium's tools and services fair for their participation. And we thank Joan Atkinson and her team at Newcastle for laying the event's foundation through their exemplary planning and organisation.

MP & TA

A message from the sponsors

BAE Systems is pleased to support the publication of these proceedings. We recognise the benefit of the Safety-Critical Systems Club in promoting safety engineering in the UK and value the opportunities provided for continued professional development and the recognition and sharing of good practice. The safety of our employees, those using our products and the general public is critical to our business and is recognised as an important social responsibility.

The Safety-Critical Systems Club

organiser of the

Safety-critical Systems Symposium

Safety-critical systems and the accidents that don't happen

When an aircraft crashes, it makes headlines. That hundreds of thousands of flights each week do not crash is accepted as routine. Airliners, air traffic control systems, railway signalling, car braking systems, defence systems, nuclear power stations and medical equipment are some of the critical systems in use, on which life and property depend. New autonomous systems that will affect our daily life are coming on-stream soon, including self-navigating drones and self-driving road vehicles. That safety-critical systems do work well is because of the expertise and diligence of professional systems safety engineers, regulators and other practitioners who work to minimise both the likelihood that accidents will occur, and the consequences of those that do. Their efforts prevent untold deaths and injuries every year. The Safety-Critical Systems Club (SCSC) has been actively engaged for twenty-five years to help to ensure that this continues to be the case.

What is the Safety-Critical Systems Club?

The SCSC is the UK's professional network and community for sharing knowledge about safety-critical systems. It brings together engineers and specialists from a range of disciplines working on safety-critical systems in a wide variety of industries, academics researching in the field, providers of the tools and services that help develop the systems, and the regulators who oversee safety. It provides, through publications, seminars, workshops, tutorials, a web site and, most importantly, at the annual Safety-critical Systems Symposium, opportunities for them to network and benefit from each other's experience in working hard at the accidents that don't happen. It focuses on current and emerging practices in safety engineering, software engineering, and product and process safety standards.

What does the SCSC do?

The SCSC maintains a website (scsc.org.uk), which includes directories of tools and services that assist in the development of safety-critical systems. It publishes a regular newsletter, Safety Systems, three times a year. It organises seminars, workshops and training on general matters or specific subjects of current concern, which are prepared and led by world experts. Since 1993 it has organised the annual Safety-critical Systems Symposium (SSS) where leaders in different aspects of safety, from different industries, including consultants, regulators and academics, meet to exchange information and experience, with the papers published in a proceedings volume. From time to time, the SCSC supports relevant initiatives, such as the current Data Safety Initiative that is addressing concerns raised about data in safety-related systems. The SCSC carries out all these activities to support its mission:

> ... to raise awareness and facilitate technology transfer in the field of safety-critical systems ...

History

The SCSC began its work in 1991, supported by the Department of Trade and Industry and the Engineering and Physical Sciences Research Council. The Club has been self-sufficient since 1994, but enjoys the active support of the Health and Safety Executive, the Institution of Engineering and Technology, and BCS, The Chartered Institute for IT; all are represented on the SCSC Steering Group.

Membership

Membership may be either corporate or individual. Individual membership, which costs £95 a year, entitles the member to Safety Systems three times a year, other mailings, and discounted entry to seminars, workshops and the annual Symposium. Frequently individual membership is paid by the employer.

Corporate membership is for organisations that would like several employees to take advantage of the benefits of SCSC programmes. The amount charged is tailored to the needs of the organisation.

For more information about membership, or to join, please:
call Joan Atkinson on +44 191 221 2222 (email Joan.Atkinson@ncl.ac.uk) or email mike.parsons@scsc.org.uk

The Safety Community and Safety Club for everyone working with safety systems

Contents

Cybersecurity in the Safety Life-cycle

Martyn Thomas

Gresham College

London, UK[1]

Abstract *Computer-based systems may fail catastrophically for a variety of reasons. Assurance processes for safety-related systems have focused primarily on the risks from random hardware faults, design faults, and operator error. Where failures are triggered by unpredictable events such as an unexpected combination of inputs, or one or more operator errors, safety analysts may assume such events are independent and/or stochastic. That assumption cannot be sustained if there is a credible threat of any form of cyber attack, because an attacker might be able to create any desired pattern of unlikely events. Consequently, no safety-related system can be considered adequately safe unless it is also adequately secure against cyber-attack, and this raises issues that need to be considered throughout the safety life-cycle.*

1 Introduction

I first attended a conference on safety-related systems (SRS) in Jersey in the Channel Islands in the mid 1980s. I had assumed that this would be where developers talked seriously about rigorous software engineering and quality management; instead I found the conference dominated by discussions about random hardware failure and its consequences. When the issue of software failure was mentioned in passing, the audience laughed – there was a view that software

---

[1] Livery Company Professor of Information Technology at Gresham College, Barnard's Inn Hall, London UK

didn't *fail* in the same sense as hardware. I recall one discussion about the feasibility of assessing the probability of failure of basic software components such as the assignment statement – analogously to assessing the failure probability of a hardware switch or relay.

A lot of progress has been made since then. Perhaps most notable is the work on the safety of Programmable Electronic Systems, much of which was led by Ron Bell and John Brazendale of the UK Health & Safety Executive (HSE) and which resulted in the standard IEC 61508 *Functional safety of electrical/electronic/programmable electronic safety-related systems* (IEC). There are probably several people at this conference or reading this who contributed to that work.

Those were innocent days, in the 1980s and 1990s. We discussed the difference between systemic and random faults and whether or not it was reasonable to talk about software failing randomly, since the faults that led to the failure would do so whenever the same system state was reached with the same inputs. After a while it became accepted that software did fail randomly because the randomness of possible states and inputs was such that resultant failures could legitimately be treated as stochastic and independent (though this debate still raises its head periodically on the software safety mailing lists). Independence was a key issue, because it allowed the probability of two failures to be combined by multiplication if both failures were required before a hazardous state would be created. Two unlikely events could then be treated as extremely unlikely, and three as incredible.

The threat that someone would deliberately create the circumstances that would lead to failure was rarely part of the analysis. This could be for good reasons – or reasons that seemed good - for example that the system could only be accessed by trusted staff and that it was inaccessible to outsiders even if someone had malicious intent.

Of course, that assumption could be false even in the 1980s. During some work that my company *Praxis plc* did for a large international airport, we discovered that the system that controlled the taxiway guidance lighting for ground movement of aircraft and vehicles when there was low visibility was connected to an unprotected dial-up modem, to provide remote access for the contractor who designed and maintained it. I later learnt that it was not uncommon for industrial control systems to be monitored and controlled in this way, sometimes from a distance of several hundred or thousands of miles. Gradually, dial-up modems and dedicated lines were replaced by internet connections, usually for convenience and economic reasons, but with the consequence that computer-based systems became far easier to discover, explore and penetrate.

In the quarter-century since Robert Morris released the first major internet worm, malware and computer misuse has become an industry.

First there were the hackers – clever programmers who mostly wanted to experiment and to see what they could find and what they could do on remote computers.

Then some of their methods and the tools the hackers had developed were made available on the internet and a generation of *script kiddies* started using them, with less knowledge and far less self control.

Now, in 2016, we have to protect our systems against a spectrum of negligent or malevolent intruders, ranging from school students through to serious organised criminal gangs and nation states. The pace of change is high and technologies that five years ago would have been thought to require the resources of a nation state are now being used by criminals. By 2020 the script kiddies will probably have the tools to build Stuxnet (Wired 2014).

Edition 2 of IEC61508 requires that a hazard analysis should be carried out that includes security threats to safety and that (Section 7.4) where the hazard analysis identifies that a security threat is reasonably foreseeable, a security threat analysis should be carried out. But cyber threats are now endemic and it would be hard to justify a claim that a security threat is *not* reasonably foreseeable.

Cyber threats are part of the environment that our systems inhabit. It is no longer enough to consider that no-one has the motive or the capability to attack our systems, because motives and capabilities may change considerably during the service lifetime of any system developed today, so any threat analysis must consider threats over the whole service lifetime. Cybersecurity vulnerabilities must be part of every hazard analysis and risk register. No *safety* assurance can be adequate unless it fully incorporates *cybersecurity* assurance. In the presence of a cybersecurity threat, it is much harder to justify treating potential failures as independent.

These considerations have far-reaching implications for the way that safety-related systems should be developed and for the evidence that should be presented in safety cases.

2 Risks and Properties

In 2004 the UK Ministry of Defence commissioned Praxis High Integrity Systems (now part of Altran) to provide an integrated methodology for safety and security certification for avionics. The resulting *SafSec* methodology and guidance material is available online (Altran 2004).

> One of the SafSec Methodology principles came from the realisation that both fields work from a base concept of Risk assessment and management. Identifying and mitigating risk is an essential driver for development processes. However, this concept is rarely given a central role in current certification approaches based on procedural frameworks.
> Another emerging principle was the need to consider Properties of systems alongside the functions they perform, if safety and security are of concern. This leads to the idea that properties (expressed as objectives and assurance requirements (Hawes and Steinacker 1997)) should be central to the design process as well as function. Although this approach of focusing on properties can be

applied to other non-functional aspects of design, SafSec has restricted its attention to just safety and security aspects.

The three technical areas which are central to the Methodology are those concerned with the processes of risk management, argument and evidence production during development, and modular certification. (Altieri et al, 2005)

I shall refer to the central role of risks and system properties throughout the rest of this paper.

2.1 Full-spectrum cybersecurity

Cybersecurity threats exist across the electromagnetic spectrum. Successful attacks have been demonstrated against digital radio receivers, Global Navigation Satellite Systems such as GPS (RAEng 2011), and LIDAR sensors in automobiles. Every source of data that is external to the system should be viewed as a potential route for a cyberattack.

The insider threat must also be considered. Most systems require human interactions, to operate, maintain, support, archive and more. Even the staff who clean the rooms housing computer equipment are likely to have some unsupervised access to those systems. Anyone with access could be influenced to compromise the system, through threats, blackmail or bribery for example. They can also be deceived into taking incorrect action (phishing emails being just one of the possible mechanisms, but one that has become increasingly sophisticated and successful—in part because so many staff with privileged access to systems advertise this fact and other personal details through their profiles on social media such as LinkedIn).

3 Specification, architecture and design

Safety and security are related and should be considered together in all phases of the system lifecycle. When **specifying** the system properties that are important for safety, it is necessary to specify the security properties that the system must have if the safety properties are to be achieved, *and how these security properties will be assured.*

The system **architecture** will be strongly affected by security considerations. It will rarely be enough to protect the perimeter of a system in the hope of preventing intrusion. The Stuxnet attack on Iranian nuclear centrifuges showed that even an air-gap can be crossed by a determined attacker, and there have been examples of systems being compromised through the well-meaning thoughtlessness of a staff member who picks up a USB stick in the car-park and puts it into a company computer to see who it belongs to so that it can be returned (or because it was labeled "staff salaries"!).

Increasingly, companies are allowing or encouraging staff to bring their own computers, tablets and phones to work and to connect them to company networks, or to use company devices for private purposes, increasing the opportunities for malware infection that could exfiltrate confidential data or compromise operational software, firmware, configuration data or development tools.

If a company wi-fi network has a meaningful name, then any device that connects to it and that remembers the network SSID will broadcast that SSID to anyone listening, making it easy to identify people who have access inside the company. Even networked wearables, such as watches and fitness devices, are a potential security weakness. Mitigation of all these vulnerabilities should be part of every safety case.

There are so many ways that an attacker can gain access that it is wise to assume that some of your systems have already been compromised. The architectures of corporate systems, development systems and operational systems should be designed to restrict the ability for an attacker to move between systems and subsystems (or at a minimum to detect when a user performs actions beyond their usual role without appropriate authorization, as Edward Snowden did in the NSA).

A recent report by the US Senate Committee on Commerce, Science and Transportation (CCST 2014) shows how a small security vulnerability can turn into a major breach.

In November and December 2013, cyber thieves executed a successful cyber attack against Target, one of the largest retail companies in the United States. The attackers surreptitiously gained access to Target's computer network, stole the financial and personal information of as many as 110 million Target customers, and then removed this sensitive information from Target's network to a server in Eastern Europe.

Target gave network access to a third-party vendor, a small Pennsylvania HVAC [heating, ventilation and air-conditioning] company, which did not appear to follow broadly accepted information security practices. The vendor's weak security allowed the attackers to gain a foothold in Target's network.

The attackers were able to use the limited credentials that they had gained to move into more sensitive areas of Target's systems, finally gaining the ability to download modified software or firmware onto the retail chain's point-of-sale terminals. This software included RAM-scraping code that captured the financial and personal data of more than 100 million customers.

It may be possible to detect that a system has been compromised by monitoring its behaviour – for example, monitoring all network ports to detect any connections to unauthorised IP addresses. Of course, this requires the existence of a verified and up to date list of authorized connections, and this list must itself be protected against compromise.

4 Soup and nuts

Almost all safety-related systems contain some software that has not been developed by the most recent or current development teams. This software may be part of a previous system, considered good enough, or too difficult to re-write. It may be a commercial operating system or open-source software. It may even be software copied from the internet, by taking the source of a web-page or code from some library of routines made available for public use (in which case, even the technical lead on the project may be unaware that the software has not been written line-by-line by their programming team).

In all these cases, it is unlikely that the current developers fully understand the security vulnerabilities in the software they are working with. Even widely used and small open-source components have been discovered to contain security problems that have remained unreported for many years; an example is the Heartbleed bug in OpenSSL. (Heartbleed 2014).

So the use of Software of Uncertain Provenance (SOUP) and the behaviour of Not Uniformly Trustworthy Staff (NUTS) both have to be part of the security analyses and safety cases.

5 Data

Most software failures are triggered by specific data or combinations of data and safety analyses should therefore already include data analyses. Unfortunately, it is still rare for safety analyses to include comprehensive static analyses to detect data values that could trigger failures, and testing can only explore a minuscule percentage of possible data combinations.

The *security* analysis of data vulnerabilities should consider all the data in, or used by, the system throughout its lifecycle – including all the development resources and maintenance data.

For each class of data, the analysis might need to consider:

- What could be at what risk if these data are leaked or compromised?
- Should the system holding these data be connected to a network or other systems at all?
- If it must be connected, should any connection be one-way only (so that this system cannot be controlled across the network, for example)?
- Should the data be protected cryptographically? At rest? In transit across the network?
- How important is data availability?
- How important is data integrity?
- How important is data confidentiality?

- How confident do we need to be that the data are protected against these risks?
- How will we achieve that degree of confidence?
- How will we demonstrate the confidence level that has been achieved?

6 Assurance

Safety-related applications often require a high degree of assurance of the properties that affect safety – and this assurance requirement extends to security for the reasons discussed above.

High assurance requires strong evidence. This must start with a clear and complete statement of the required properties; there must then be a detailed justification for the belief that the stated properties are adequate, followed later by evidence that the required properties have been achieved. Creating the necessary evidence and showing that it is adequate will inevitably impose constraints on the development processes that are followed; for example, if it is necessary to show that the system cannot run out of stack space, it may be necessary to use a memory safe language.

Testing can provide conclusive evidence that the required properties have *not* been achieved, but rarely can testing show that they *have*. Strong evidence for general properties (such as the absence of buffer overflow for all possible inputs) is likely to involve automated analysis of the program which, in turn, may require the use of a strongly-typed language with sound semantics. There are some powerful checkers available for C and C++ and these should certainly be used where possible, but the degree of assurance they can provide will need to be understood and stated clearly in any safety case that depends on their evidence.

To perform effective security assurance you need to think like an attacker. Every source of an input should be seen as a possible attack vector. How will you detect that an attack is in progress or that it has been successful? When your system is running, will you monitor its behaviour? Do you know accurately enough what normal behaviour looks like? Can you be confident that you can detect deviations from normal behaviour without too many false alarms? What fall-back or resilience will you incorporate so that you have a tactic to employ when a problem is suspected?

7 Maintenance and support

Most safety-related systems will need to undergo maintenance or support, to correct errors or to change or add functionality. This may involve program changes or data changes and every change will offer an opportunity for a cyberattack. Even the development tools may need cryptographic security so that they cannot be changed to incorporate Trojan code, as the XcodeGhost attack demonstrated (Register 2015).

If the system contains open-source or COTS software, it will be important to recognise that security updates are likely to be released and that the underlying vulnerabilities will become well known, exposing any unpatched system to increasing risk. If the decision is taken to trust updates by a software vendor, this becomes a further vulnerability.

8 Disposal

Cybersecurity must also be considered when the time comes to dispose of part or all of the system, because it may contain confidential data, or algorithms (or even passwords!) that could expose vulnerabilities in other versions of the same system that are still live or that will be implemented in future.

9 Conclusion

Safety specialists are used to contending with Murphy's Law – that anything that could go wrong, will go wrong – and that has been hard enough! The threat of cyberattack must now part of every safety-related system's environment, and this raises new difficulties because it is very difficult to assign a probability to a cyberattack. In the presence of a cyber threat, statistical testing is uninformative. Indeed, as Volkswagen have demonstrated with the emissions from certain diesel vehicles, test results can be misleading when a programmer seeks to make them so (VW 2015) so the job of the independent safety assessor is difficult indeed.

The job of safety assurance has become very much harder. Cybersecurity must be considered throughout the system lifecycle.

References

Altieri et al, (2005). SafSec: Commonalities Between Safety and Security Assurance. SSS 05, February 2005. http://intelligentsystems.altran.com/fileadmin/medias/0.commons /documents /Technology_documents/SafSecCommonalities.pdf. Accessed 28 September 2015.
Altran (2004) SafSec standards. http://intelligent-systems.altran.com/technologies/security /safsec/safsec-standards.html. Accessed 28 September 2015.
CCST (2014) http://www.commerce.senate.gov/public/?a=Files.Serve&File_id=24d3c229-4f2f-405d-b8db-a3a67f183883. Accessed 30 September 2015.
Hawes and Steinacker (1997). Combining Assessment Techniques from Security and Safety to Assure IT System Dependability—The SQUALE Approach, VIS97 security conference, Freiburg, Germany.
Heartbleed (2014) http://heartbleed.com/. Accessed 30 September 2015.

IEC http://www.iec.ch/functionalsafety/. Accessed 13 November 2015.

RAEng (2011) Global navigation space systems: reliance and vulnerabilities. Royal Academy of Engineering. http://www.raeng.org.uk/gnss. Accessed 31 August 2011.

Register (2015) http://www.theregister.co.uk/2015/09/21/xcodeghost _apple_ios_store_malware_zapped/ Accessed 1 October 2015.

VW (2015) http://www.volkswagen.co.uk/about-us/news/686. Accessed 1 October 2015.

Wired (2014) An Unprecedented Look at Stuxnet, the World's First Digital Weapon. http://www.wired.com/2014/11/countdown-to-zero-day-stuxnet/. Accessed 28 September 2015.

Resilience is an Emergent System Property: A Partial Argument

Peter Bernard Ladkin, Bernd Sieker

Causalis Limited & Causalis IngenieurGmbH

Bielefeld, Germany

Abstract   *Systems are collections of objects exhibiting joint behaviour. Sometimes this behaviour is anticipated, sometimes not. We have studied a number of types of complex systems and their failures, including electricity supply grids, motorways, the financial system, and air traffic control. We argue that the resilience properties of such systems are largely emergent. We illustrate the thesis through analysis of three electricity blackout events. We consider one event in detail and two others summarily.*

1 Systems

In order to talk about properties of systems, including emergent properties, we first need some definitions. We have been using the following definitions for over a decade (Ladkin 2001, Chapter 3). We will not use here the formal properties of these definitions, but we judge it is well to state the vocabulary and its meaning to us. We illustrate the definitions below.

- A system is a collection of agents with joint behaviour
- An agent is an object with behaviour
- Agents have properties (attributes)
- Multiple agents have relations
- Behaviour is considered as: change in properties and relations over time

Systems have boundaries: some agents and other objects belong in a system; others are outside. Natural system boundaries are often drawn to satisfy the following criterion (note that this is just one criterion; there are others): relations/joint behaviour of objects that "crosses the boundary", that is, some objects in the joint behaviour are in the system and others out, are relatively sparse, whereas the relations/joint behaviour of objects, all of whom are in the system, are relatively abundant.

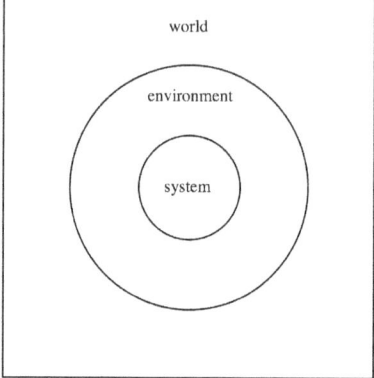

Fig. 1: A Venn Diagram of a System, Its Environment, and the World Outside (Ladkin 2001)

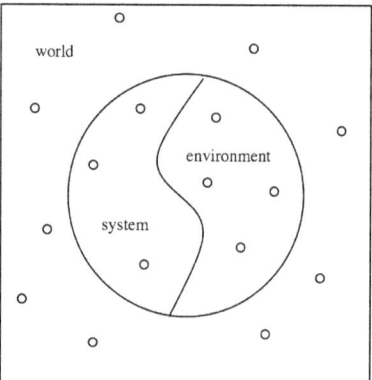

Fig. 2: A System May Interact with Objects Not in the Understood Environment (Ladkin 2001)

When a system is conceived, and its boundary is drawn, the system designers try to understand the environment as all objects outside the system with which the system interacts; which the system influences or vice versa. The ideal is shown in a Venn Diagram in Figure 1. However, mistakes may well be made. There may be parameters simply missed; interactions not foreseen or understood. The reality is more often as in the Venn Diagram in Figure 2.

When a system is conceived and designed, various parameters, properties and relations of system objects and environment, are laid down and the behaviours specified. This may be formal or, more usually, informal. A state of a simple system, formally described, is shown in Figure 3.

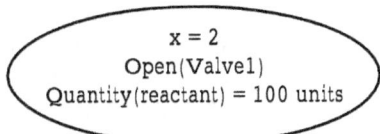

Fig. 3: A State of a System, Formally Described (Ladkin 2001)

A behaviour, a change in system state, is shown in Figure 4: Valve1 is opened; nothing else is changed - parameter x and the quantity of reactant retain their values.

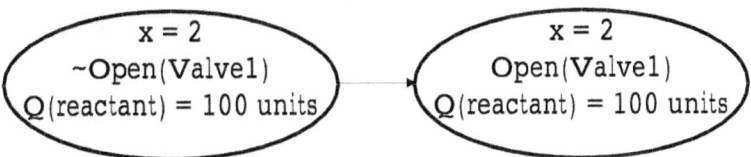

Fig. 4: A Behaviour (Ladkin 2001)

We shall use the following conception of *emergent property*. Emergent properties of a system are properties which are not base; that is, they are not defined in terms of the properties and relations of the objects constituting the system, as designed or conceived.

Emergent behaviour is joint behavior of objects which is/are (behaviour/objects) not initially considered. One kind of system of systems which is currently attracting a lot of attention consists of swarms of small, simple aircraft. Swarm behaviour cannot be completely described using the individual properties and relations of a single agent or its immediate interactors. A new vocabulary is often needed; the swarm behaviour is thereby emergent. An example is murmurations of starlings, shown in Figure 5. Describing the density and movementof the mass of starlings, it may well be that the vocabulary of fluid mechanics would be useful, which is far away from any vocabulary useful for describing individual starlings or their interaction with neighbors.

One often overlooked way in which properties can turn out to be emergent is when there is a failure scenario of a type which had been unanticipated. This very often happens when the hazard analysis of the system is incomplete. Here is an example. Hazard analyses are often conducted with the help of techniques such as FMEA1. The FMEA conducted on the Boeing 787 Lithium-ion-type main and auxiliary batteries considered the phenomenon of thermal runaway. The Boeing analysis identified overcharging as the only event which would result in (smoke and) fire. An FMEA by the battery manufacturer GS Yuasa suggested that an internal short-circuit would only result in smoke production (NTSB 2014, pp50-52). We understand the analysis was not revised when a battery under test underwent thermal runaway and burnt down a building in November 2006 (*op.cit.* p43). Batteries ignited twice in 2013 on Boeing 787 aircraft in line service and the probable cause of one of the incidents was "*internal short circuit within a cell of the auxiliary power unit (APU) lithium-ion battery, which led to thermal runaway that cascaded to adjacent cells, resulting in the release of smoke and fire*" (*op.cit.* p79).

Fig. 5: A Murmuration of Starlings
(© Walter Baxter, reused under a Creative Commons Licence)

We consider the resilience of systems of electricity supply through the grid. In other work, we consider collision avoidance in air traffic operations, collision avoidance in rail operation, auffahr accidents on motorways, and asset protection and enhancement in banking and company activity.

We call systems of the sort we consider *teleological systems*. They are systems built (by people or other animals) for a purpose. Teleological systems are distinguished from such naturally-occurring systems such as predators and their prey, or other ecosystems, which have system characteristics but no overt purpose intended by any conscious entity. Most engineered systems are teleological. Some systems are not. For example John Conway's Game of Life (Conway 1970) is a mathematical system whose original purpose was, if anything, for its creator and others to have fun. One emergent property of the Game of Life is its usefulness in illustrating the talk accompanying this paper.

We need a definition of resilience. The convenor of the EU ReSIST project[1], Jean-Claude Laprie, defined it as *"The persistence of service delivery that can be justifiably be trusted, when facing changes"*, cited in (Meyer 2009). Meyer also considers the definition *"the ability of a system to deliver service under conditions that lie beyond its normal domain of operation,"* as well as others, such as that of Woods: *"how well can a system handle disruptions and variations that fall outside of the base mechanisms/model for being adaptive as defined in that system."* (*op. cit.*)

2 Electricity Blackouts

To understand how blackouts may happen, it is necessary to understand some qualitative physics of grid supply. Except for a few direct-current (DC) lines, almost the entire European grid is a synchronised, in-phase alternating-current (AC) grid. The North American grid system is a set of three grids, with some inter-grid connection through high-voltage DC lines. The main reasons for using AC are historical, as alternating voltage conversions are technically trivial using transformers.

The downside of AC is that the entire connected network must be at exactly the same frequency and in phase to avoid large energy losses. The frequency may change as a result of the mechanics of energy – see below. Such change must be actively managed.

In most power sources (power stations), mechanical energy is converted into electrical energy. In some, nuclear energy is converted into heat and via steam and mechanical energy into electrical energy. Large centralised power stations include nuclear, coal-fired and gas-fired stations, and hydroelectric stations. Decentralised power generation includes wind turbines (mechanical into electricity) and photovoltaic installations (light into electricity).

AC electricity supply divides into active power and reactive power. Instantaneous power is, as in other areas of physics, the product of voltage and current at an instant. These quantities vary sinusoidally with time in AC supply. *"Sinusoid"* means the following, illustrated in Figures 6 and 7. Suppose there is a circle with centre at (0,0) on a two-dimensional surface, and a radius of that circle which is rotating at constant angular speed. Then the sinusoid quantity is the y value of the (x,y) values traced out by the tip of the radius. The angle which the radius makes with the x axis is called the *"phase"* of the sinusoid. In Figure 6 the radius is given by the blue line labelled U_C.

---

[1]   Full Disclosure: The first author was a formal reviewer of the ReSIST project.

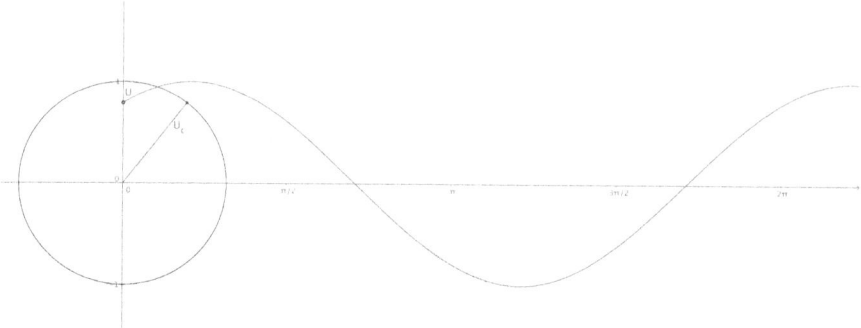

Fig. 6: A Sinusoid Curve generated by Voltage over time

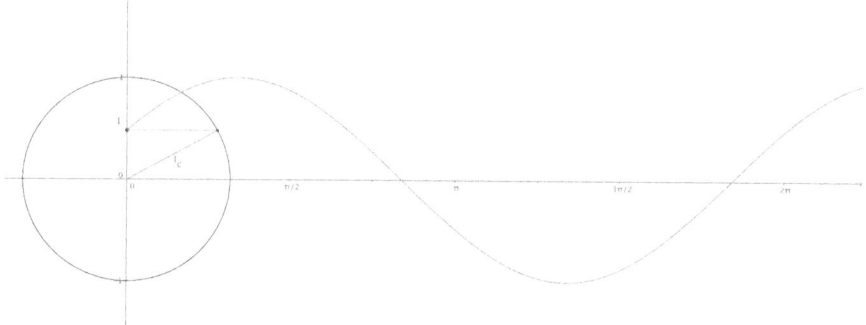

Fig. 7: A Sinusoid Curve generated by Current over time

Voltage and current are two parameters of electricity flow which result in sinusoid fashion from AC generators. Their respective values given by the sinusoid curves may not cohere.

Power is the product of voltage and current. When voltage and current are "*in phase*", then their values are always either both zero or both positive or both negative, so their product is zero or positive and so the power delivered over one cycle (= rotation of the radius) is the integral of that and is positive. That power may be used to do work in a recipient device and is called "*active power*". This is illustrated in Figure 8. Instantaneous power is shown by the brown curve, and its mean over a cycle by the area under this curve, here normed to 1.

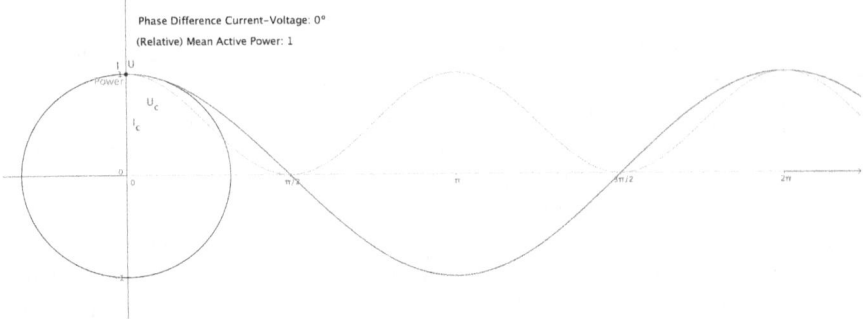

Fig. 8: "In phase" voltage and current – positive average power

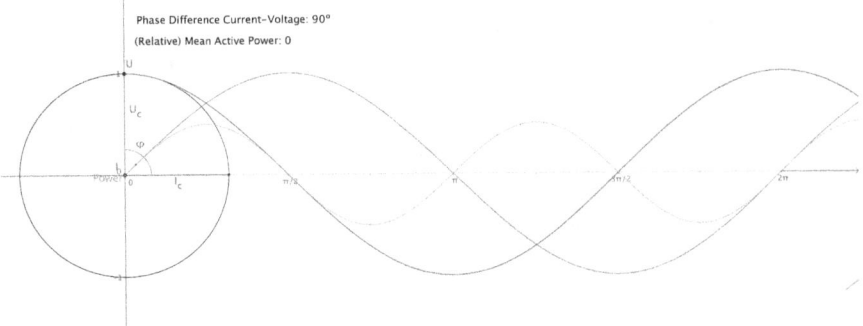

Fig. 9: 90° "out of phase" voltage and current

If the voltage and current are 90° "*out of phase*", then the integral is zero over one cycle. This means that the average power delivered over a cycle is zero. Such an out-of-phase current is called "*reactive power*". This situation is shown in Figure 9. The general situation is given in Figure 10.

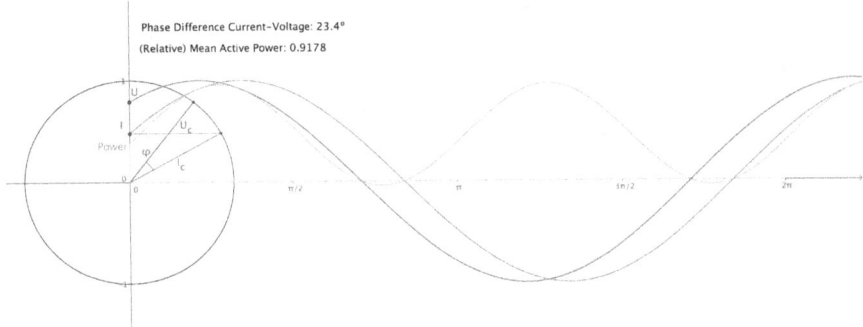

Fig. 10: General case of "out of phase" voltage and current

Any phase is mathematically given by a vector sum of orthogonal components. It follows that any AC power, as in Figure 10, can be considered as a sum of active and reactive power. Relative phase shifts, and thereby the reactive power component, is caused by capacitive and inductive elements in the grid. Reactive power has a large influence on voltage levels and creates additional losses on the transmission lines. It must therefore be carefully controlled.

Energy is also stored in the electrical grid in other ways. For example, surplus power in the grid can be absorbed by generators if their drives are disconnected – they turn into motors and electrical energy can be turned into momentum of the rotating armature. When this happens, the armature will speed up, so that when the item is reconnected as a generator its frequency is a little higher[1]. Conversely, the spinning armature can temporarily supply additional energy into the grid, by transferring some of its momentum into electric power, thereby slowing down. Such frequency fluctuations (of the nature of mHz to, say, cHz) are appropriate over a short period of time, provided they are counteracted before the phases difference to adjacent parts of the net becomes too large.

The reliability of a grid-based electricity supply means providing electrical energy at the required voltage and in the demanded amount to consumers. "*Maintaining reliability is a complex enterprise that requires trained and skilled operators, sophisticated computers and communications, and careful planning and design.*" (US-Canada Task Force 2004).

2.1 The 2003 North-Eastern North American Blackout

We consider first the August 2003 blackout of large portions of the Eastern Interconnection in the USA and Canada. The North American grid is divided into three, each component called an "Interconnection". An interconnection is a more or less open network in which the flow of electricity is physically determined by supply and demand, operating according to basic laws of physics. Flow can be controlled only be regulating supply and demand. Within an interconnection, there are usually many pathways available to satisfy a demand, and this yields a certain resilience. Connections between the interconnections are often established by DC lines. The North American interconnections are shown in Figure 11.

---

[1]   Large power stations usually employ synchronous generators where the AC frequency is tied to the rotational speed of the armature.

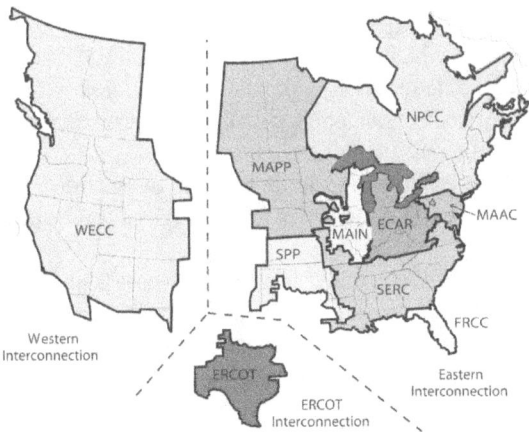

Fig. 11: The North American grid interconnections (NAERC 2014, under fair use)

August 14, 2003 was not a particularly hot day in the U.S. Midwest, with temperatures in the mid-80's Fahrenheit (28-30°C). Electric air conditioning systems were operating in many homes, but the electricity distribution system (the "grid") had dealt with 100°F (38°C) temperatures a year before without problems.

Within each interconnection, power supply and demand must be matched, else frequency fluctuations occur which can damage equipment. Reactive power must be balanced to maintain acceptable voltages; voltage fluctuations can cause a collapse in supply when low, and can damage equipment and result in arcing when high. Electricity flow over transmission lines heats up the lines and must be controlled to ensure that thermal limits are maintained; hot lines expand and sag, and clearances from other objects must be maintained. Furthermore, flows must be managed to absorb "contingency events", such as a generator going off-line or a transmission line "tripping" (shutting down). External insults such as contact with trees or physical damage to lines are usually handled through tripping.

The North American Electric Reliability Corporation is a voluntary organisation whose mission is to assure the reliability of electrical power supply in North America. Its members are ten regional reliability organisations. The August 14 blackout affected three of the ten regions. There are 140 "control areas" in the US. A control area has one entity, either an "independent system operator" (ISO) or "regional transmission organisation" (RTO), responsible for balancing generation and loads in real time to maintain stability (one of two primary functions determined by legislation). They also control generation directly, to support interchange schedules with other control areas, and operate collectively to maintain stability of their interconnection. Control area dispatch centers monitor and control electricity generation and flow and are staffed continuously.

The initiating events of the blackout involved two control areas, FirstEnergy (FE) and American Electric Power (AEP), and their reliability coordinators, Midwest Independent System Operator (MISO) and PJM Interconnection (PJM). FE operates a control area in northern Ohio. FE consists of seven electric utility operating companies, four of which operate in the NAERC ECAR region, with MISO as their reliability coordinator. AEP operates a control area in Ohio just south of FE. AEP is both a transmission operator and a control area operator. PJM is AEP's reliability coordinator.

The area of blackout is shown in Figure 12. The two-nation Task Force Report (U.S.-Canada Task Force 2004) identifies four classes of causes of the blackout:

1. FirstEnergy and ECAR failed in general to assess and understand the inadequacies of FE's system, particularly concerning voltage instability and the vulnerability of the Cleveland-Akron area. FE did not operate its system with appropriate voltage criteria.
2. There was continuing inadequate situational awareness at FirstEnergy. FE did not recognize or understand the deteriorating condition of its system.
3. FE failed to manage adequately tree growth in its transmission rights-of-way. Tree contact was the cause of the outage of three FE 345-kV transmission lines and one 138-kV line during the incident.
4. The ISO/RTOs failed to provide effective real-time diagnostic support.

It is interesting that this enumeration of causes, as well as a similar but lengthier list in the NAERC report, concerns exclusively human or organisational failures. The physical causes of the event, what a physicist or scientist would say were the causes, are omitted, as well as some system characteristics which appear to us to be crucial in explaining system behaviour.

The detailed history of events (NAERC 2004), shown in general area form in Figure 13, enables further observations about how the grid system functions.

- As we have noted, an interconnection is a network flow operating under purely physical laws. Stability of flow is maintained through partly automatic and partly human intervention. Specific loading (energy consumption by consumers) varies according to circumstances generally not under network control. Network controllers can moderate active power flow; also reactive power flow in order to equilibrate its generation inside the network. Methods for moderation include supplementing generating capacity (increasing power output from generation plants, or bring off-line generation equipment online). In principle, transmission lines may also be taken out of service, but this did not happen in this event. Transmission lines took themselves out of service ("tripped", then "locked out") for a variety of reasons.
- A, maybe *the*, key event in this blackout process was the tripping of the Sammis-Star 345kV transmission line at 16:05. This is shown in Map 3 of Figure 13. This "*completely severed*" (*op. cit.* p55) the 345kV transmission path from South-eastern Ohio into Northern Ohio (Cleveland-Akron, on the southern shore of Lake Erie, and Toledo on the western shore), which had

significant net import of power at the time due to the demand. Three pathways were still available, namely from northwestern Pennsylvania to northern Ohio around the south shore of Lake Erie, from southern Ohio, from eastern Michigan and Ontario. However, *"no events, actions, or failures to take action after the Sammis-Star trip can be deemed to have caused the blackout."* (*op. cit.*, p 56). In other words, in the demand-supply circumstances prevailing at the time, the Sammis-Star line trip was a single point of system failure. At the point of tripping, the reactive power carried on the line was ten times as high as earlier in the day.

- Previous to the Sammis-Star line trip, the NERC report suggests that operator load-shedding may have been appropriate to maintain stability of the system but that afterwards "only automatic protection systems" would have mitigated the consequences (*op. cit.*, p57).

- The cascade developed into a blackout for "three principal reasons" (*op. cit.*, p58):
  - Loss of the Sammis-Star line triggered many subsequent line trips;
  - Many lines operated with so-called "zone 3" impedance relays, as did Sammis-Star, which respond to overloads rather than line faults;
  - Relay protection settings for line, generators, and under-frequency load shedding may not be sufficient to reduce the likelihood and consequences (= risk) of a cascade, *"nor were they intended to do so"* (*op. cit.*, p58).

The blackout happened within about seven minutes after the Sammis-Star line trip. There were large power surges, for example a 3,700 MW flow from Michigan to Canada turned into a 2,100 MW flow in the other direction within one second, a 5,800 MW flow reversal. The events caused a large electrical island separated from the rest of the Eastern Interconnection. The region had been importing power and did not have enough generational capacity within to satisfy the demand. However, pockets within this island did stabilise, and recover. Phase and synchronisation mismatches often hinder facility resetting after trips. Some pockets took a long time to recover.

The reason for the Sammis-Star 345kV line trip was that a protective relay sensed *"low apparent impedance"*, that is, low voltage and high current (*op. cit.*, p57). There was in fact no fault. The protective relay cannot physically distinguish a fault from high load, and the line was operating at 130% of nominal capacity and the voltage was lowering. As mentioned above, before this event operator load-shedding could have reduced load, thereby avoiding the trip, but after the event *"only automatic protection systems would have mitigated the cascade"* (*op. cit.*, p57) and there were none, or insufficient, in place.

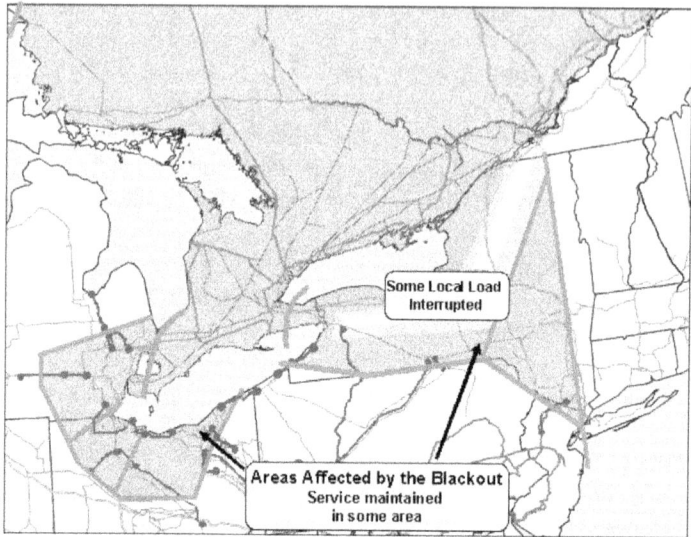

Fig. 12: Areas affected by the 14 August 2003 blackout
(NAERC 2004, reproduced under fair use)

The above observations concerning the Sammis-Star trip can be summarised as:

- The Sammis-Star trip led inexorably, under the supply-demand conditions prevailing at the time, to blackout;
- The Sammis-Star trip was a normal, designed reaction to the conditions on the line, namely a condition of "low apparent impedance".

Two conclusions follow directly from these observations:

- Given the conditions prevailing at the time, the Sammis-Star line was a single point of failure of the grid;
- The blackout as a consequence of the Sammis-Star trip was a "normal accident" in the sense of Perrow (Perrow 1984). It was an undesired event which followed as a physical consequence of the Sammis-Star trip, which itself was a correct functioning of the system as designed to the physical grid state prevailing at the time.

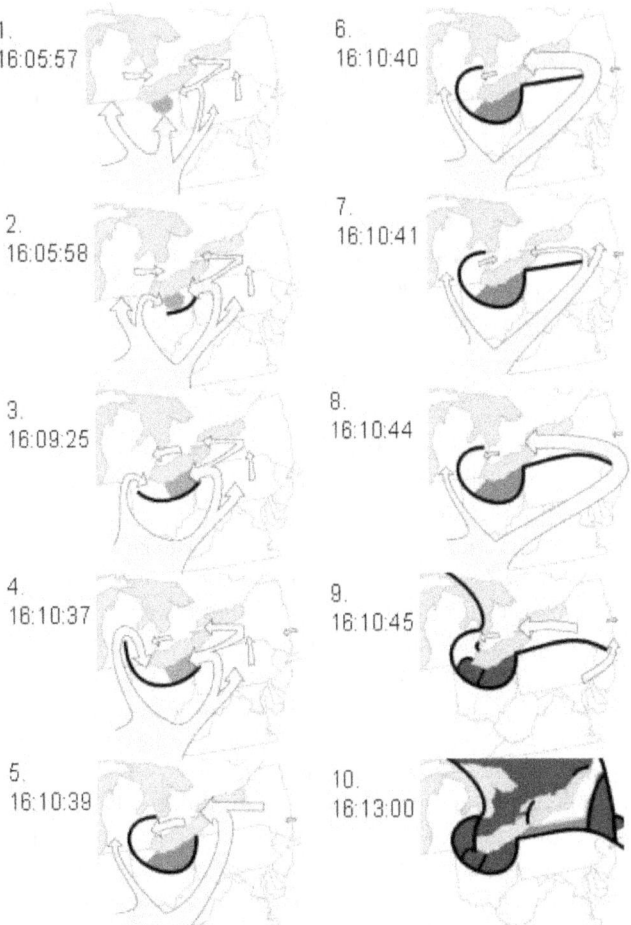

Fig. 13: Power flows from immediately before the Sammis-Star trip (shown in diagram 1; Sammis-Star trip is in diagram 2) to complete system collapse
(NAERC 2004, reproduced under fair use)

The trigger circumstances of normal accidents is often an emergent condition – it is rarely the case that such are known in advance, because, when they are, they would have been classified as a hazard and normal practice would have required mitigation or avoidance measures to be taken. During this incident, it appears also that the status of being a single point of failure under given grid conditions is also emergent.

2.2 History of the 2003 Event: Computer Problems Hindering In-formation Flow

It follows from the above conclusion in the report (NAERC 2004) that the most causally-significant events, namely those causal events where intervention was both possible and could have been effective in avoiding the blackout, occurred previous to the Sammis-Star trip. We recount the significant precursor events, illustrated in Figures 8 and 9 after the text.

- During the day, voltages were sagging in the Cleveland-Akron region and the system was judged retrospectively by NERC to have been approaching voltage collapse. However, this condition was not causal to the blackout
- Power transfers were *"high, but within studied limits and less than historical values"* (*op. cit.*, p12)
- At around 12:00, several lines in SE Indiana tripped
- At 13:31, Eastlake Unit 5, a generating station on the south shore of Lake Erie, tripped (Figure 14)
- At 14:02, the Stuart-Atlanta 345kV line tripped (Figure 14). This line was not in the MISO area, so MISO had no information on it, but its non-function caused the MISO state estimator to operate incorrectly.
- FE was during the entire time a major importer of power (*op. cit.*, p20). In the metropolitan area of south Lake Erie, air-conditioning loads were "consuming" reactive power, of which Northern Ohio was then a net importer. The system was not "reliable" with respect to reactive power, but this state was not causal in the blackout.
- At 14:14, the FE operators lost the "alarm function", on the computerised Energy Management System (EMS). The alarm function is an audible and visible annunciation of problematic status of some piece of kit (line, generator, capacitor bank, and so on). Operators remained unaware of this loss until the failure of the second EMS at 14:54, and did not realise that they had in fact lost alarm function 40 minutes earlier. There was no technical indication of loss of function. There had been calls from other operators, which hinted that the system state was not fully understood at FE, but these interactions apparently had "little effect". The EMS continued to exercise supervisory control and send correct status updates to other entities, including MISO and AEP.
- Although there had been partial losses of the alarm function before, this was the first time that total loss of function occurred.
- Between 14:20 and 14:25, various remote control terminals in substations ceased to function. This was noticed only at 14:36 through on-site inspection at a substation.

- At 14:27, the Star-South Canton 345kV line tripped, and then reclosed, at 54% nominal load (Figure 14). At this point, the FE operators had begun to lose situational awareness.
- At 14:41, the primary EMS server failed. The server function was taken over by a "hot stand-by", but, because the alarm process was stalled, this transfer caused this backup system to fail at 14:54 (it is not explained how a stalled process caused the EMS server to fail).
- The failure of two EMS servers apparently caused the refresh rate on operators' screens to slow to almost a minute, compared with the usual refresh rate of 1-3 seconds. There was a "warm reboot" at 15:08, but a warm reboot does/did not restart the alarm function. When FE operators became aware of the alarm-function problem at 15:42, another warm reboot was attempted between 15:46 and 15:59. Operators were not, however, aware that this action would not restart the alarm function.
- The MISO state estimator normally runs automatically every five minutes. A real-time contingency analysis is also performed, less frequently. The state estimator takes real-time telemetry data and constructs a "best-fit" power-flow model from that data. A contingency analysis is used to alert operators if the system is running "insecurely" [sic]. The state estimator sometimes may not resolve, if information is inaccurate, or may also report a high degree of error (presumably, an estimate). Both tools were said to have been under development and "not fully mature" (op.cit., p36).
- At 12:15, the state estimator reported results "out of tolerance" (op.cit., p36), due to a line in Indiana which had tripped but which the state estimator recorded as still in service. This information was updated manually; a correct update followed at 13:00 at which point the state estimator resolved acceptably. However, to troubleshoot the problem, the MISO operator had disabled the automatic five-minute state estimation regime. He left his position. The fact that the state estimator was not running regularly was discovered at 14:40. When the state estimator was rerun, it failed to resolve.
- The likely cause of the non-resolution of the MISO state estimator at 14:40 was the tripped Stuart-Atlanta 345kV line. This line is outside MISO's area of responsibility and its status is not automatically linked to the MISO state estimator. There was a repeated failure to resolve until the MISO operator called PJM at 15:29 to determine the line's status. After updated to the correct status (tripped), the MISO state estimator then resolved. Contingency analysis was run manually and resolved at 15:41.
- The MISO state estimator and contingency analysis were "back under full automatic operation and solving effectively" by 16:04. (op.cit., p37). However, this was only a couple of minutes before the Sammis-Star trip and the start of the cascade.

- The MISO state estimator and contingency analysis were thus "*effectively out of service*" between 12:15 and 15:41 (*op.cit.*, p37). The report concludes, reasonably, that the lack of MISO diagnostic support contributed to the lack of situational awareness at FE.
- At 15:05, the Chamberlin-Harding 345kV line tripped and then locked out (Figure 15).
- At 15:32. the Hanna-Juniper 345kV line tripped and then locked out (Figure 15).
- At 15:41, the Star-South Canton 345kV line crossing the FE/AEP boundary tripped and locked out (Figure 15).
- The first two of these trips were not recognised by FE because of the loss of alarm function
- These trips obviously degraded the condition of the system.
- Between 15:39 and 16:08 there was a localised cascade of tripped 138kV lines in Northeastern Ohio
  - Seven lines tripped
  - Then the Dale-West Canton line, whose tripping caused the Sammis-Star 345kV line to overload, which initiated the blackout cascade irreversibly.
  - Then three more

Fig. 14: Initial line and plant trips (NAERC 2004, under fair use)

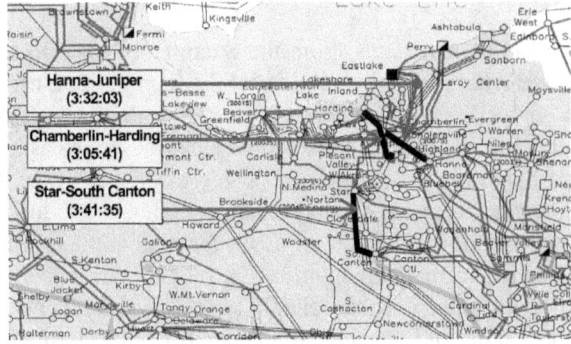

Fig. 15: Three 345kV line trips (NERC 2004, under fair use)

The Sammis-Star trip at 16:05:57 EDT and the resulting alteration of network flows into the Cleveland-Akron metropolitan area are shown in Figures 10 and 11. This event is considered in the report to be key, in that the consequences could not have been avoided within the sociotechnical control system as it was at that time (*op.cit.*).

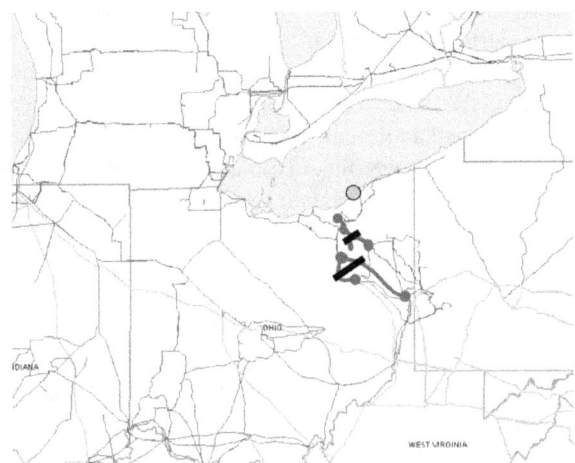

Fig. 16: Cleveland and Akron supply cuts, through Sammis-Star Trip
(NAERC 2004, under fair use)

The sequence of events after the Sammis-Star trip is detailed, and to our minds interesting, in particular how islands formed in sequence. A detailed set of maps showing the progression may be found in (*op.cit.*), also (Ladkin 2015), along with a synopsis of the full sequence of events from the NAERC report. We do not have space to consider it here. It contributes to our conclusions only through the observation of the rapidity of the decline to blackout. The entire sequence, covering multiple U.S: and Canadian states, from the Sammis-Star trip to the blackout in Figure 12, occurred within very few minutes of the Sammis-Star trip.

To a system analyst, the sequence of events before the Sammis-Star trip as related above are notable for their contrast with the causal factors as identified in the report. Effective human load management is key to the resilience of the system. Active management at times of high and variable load is key to load management. It is apparent that operators relied on the EMS for system management to the extent that loss of an alarm function significantly reduced the effectiveness of their management. It is also apparent that reliable state estimation is also key to effective management at times of high and variable load. The EMS, and the MISO state estimator, are computational systems whose functions are critical to resilience at times of high and variable load. However, it appears these systems were neither identified nor treated as mission-critical:

- Loss of an EMS function on which operators had ostensively come to rely, the alarm function, was not annunciated. Had the EMS system been identified as mission-critical and the usual design criteria applied, such a loss would have been annunciated.
- The MISO state estimator was temporally skewed for up to two and a half hours during the course of the incident, and did not reflect the true system state when skewed. This condition was likewise not annunciated, except indirectly, when the system failed to resolve, likely because a "hidden input" had a different value (the Stuart-Atlanta line). However, it was not annunciated over this two-plus hours that the estimate was static , a time when active management would have been required to ensure resilience.

It follows that the actual resilience of the Northeastern Interconnection is reliant upon the resilience of computerised systems, the FE EMS and the MISO state estimator and presumably equivalent kit at other operators and reliability organisations, which were not at the time either identified or managed as mission-critical. It is curious that this dependence was not addressed in the incident reports and analyses (*op. cit.*)

2.5 The November 2006 European Blackout

Except for a few DC lines, almost the entire European grid is a synchronised, in-phase AC grid. As in North America, the control stations of the network operators in Europe can usually see (part of) the states of the different components of the network, such as current load in megawatts (MW) and/or amperes (A), as well as the load limits for these lines. They will typically also be able to see how much is produced and consumed where, and also the states of the switches. High-voltage lines have automatic circuit breakers which will disconnect the line in case of overload. In contrast to the event we have just considered, in the European incident there do not appear to have been any computer anomalies causing misinformation. But not all the available information was used, and some important and faulty decisions were not checked.

At any point in time, the so-called *N-1 criterion* must hold in the European electric energy distribution grid. It means that "*any single loss of transmission or generation element should not jeopardize the secure operation of the interconnected network*" (BNA 2007). Closely related is the notion which we might call "*N-1 resilience*", that grid remains resilient when one element is lost. The term "*N-1*" comes from a single loss. N-2 resilient would be that the system remains resilient when two elements are lost. N-2 would be a much stronger criterion. The November 2006 incident showed that the grid at that point in time was N-1 resilient but not N-2 resilient.

All it took was a tall ship. On the evening of November 4 2006, a large cruise ship that had been built at the Meyer shipyards in Papenburg, on the River Ems near the north-east German coast, was scheduled to be conveyed along the river towards the North Sea.

There are several high-voltage lines passing across the River Ems, underneath which ships built in Papenburg have to pass on their way to the open sea. Most of the lines have been raised to allow safe passage, but for some ships the clearance of some lines still is not sufficient. Long-standing practice was to turn off some of these lines during a launch to allow the ship to pass. The disconnection had been requested for this ship some time in advance, and had been tentatively agreed by E.ON, the operator of the line in question. E.ON also informed the operators of neighboring network sections of the event.

On the day of passage, the shipyard requested shutdown of the line three hours earlier than originally planned. The earlier time was deemed more favourable and, after co-ordination with the neighboring operators RWE and TenneT, the request was granted by E.ON. RWE and TenneT checked for the fulfilment of the N-1 criterion prior to giving their approval to the disconnection, but E.ON did not do so (BNA 2007). In communications with RWE, it was found that, as a consequence of the disconnection, another high-voltage line, Landesbergen-Wehrendorf, which connected the networks operated by E.ON and RWE, was close to its load limit. In an attempt to reduce the load on that line, E.ON operators coupled busbars at a switching station. Because of the urgency of the situation, this action was not coordinated with RWE. Instead of reducing, the load on the line rose, and two seconds later the overloaded line tripped.

Because of the rapidly-changing distribution of the current flow, other lines in Germany and other parts of Europe tripped in quick succession, illustrated in Figure 17, which caused the transmission network to be split into three parts.

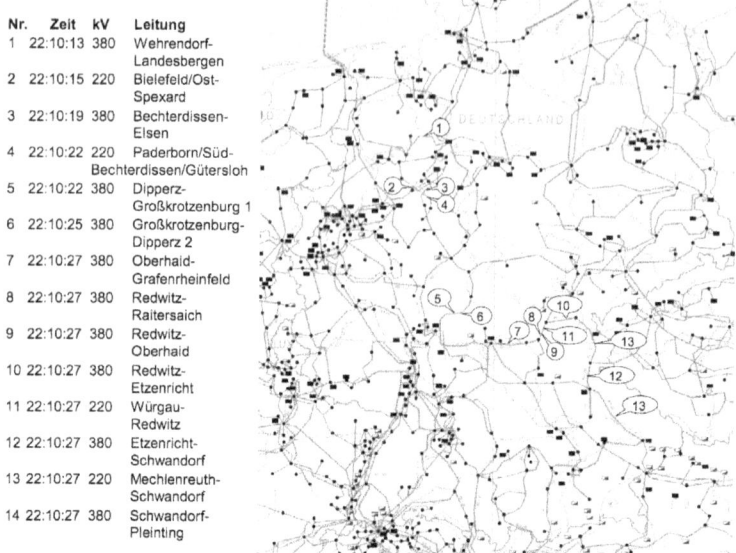

Nr.	Zeit	kV	Leitung
1	22:10:13	380	Wehrendorf-Landesbergen
2	22:10:15	220	Bielefeld/Ost-Spexard
3	22:10:19	380	Bechterdissen-Elsen
4	22:10:22	220	Paderborn/Süd-Bechterdissen/Gütersloh
5	22:10:22	380	Dipperz-Großkrotzenburg 1
6	22:10:25	380	Großkrotzenburg-Dipperz 2
7	22:10:27	380	Oberhaid-Grafenrheinfeld
8	22:10:27	380	Redwitz-Raitersaich
9	22:10:27	380	Redwitz-Oberhaid
10	22:10:27	380	Redwitz-Etzenricht
11	22:10:27	220	Würgau-Redwitz
12	22:10:27	380	Etzenricht-Schwandorf
13	22:10:27	220	Mechlenreuth-Schwandorf
14	22:10:27	380	Schwandorf-Pleinting

Fig. 17: The first 14 lines tripped within 14 seconds (EON 2006)

Due to a discrepancy between production and consumption, frequencies in these areas began to drift apart, making a quick reconnection impossible. Generators were shut down and consumers had to be disconnected. More than 15 million people in Europe were affected by the blackout.

At more than one point, E.ON operators did not carry out a computer-assisted flow analysis before performing actions which altered the load distribution in their network. It appears that they relied instead on their experience to assess the state and security of the grid. The first instance was when the approval was given to disconnect the line over the River Ems. The second instance was when they decided to couple the busbars to alleviate the load on the Landesbergen-Wehrendorf line. However, even experienced operators cannot judge the behaviour of highly-complex interconnected systems intuitively.

A Why-Because Analysis of all the causal factors is available (Sieker 2008).

2.6 Total Power Blackout in the Swiss Federal Railways (SBB) Network on June 22, 2005

On June 22, 2005, the Swiss Federal Railways suffered a total power blackout. In contrast to the other two blackout events above, the initial events in the Swiss incident happened rapidly, within a few seconds, leaving operators almost no time for intellectual analysis but maybe just time enough to react. However, an aspect of system design leading to an "alarm flood", with alarms required to be manually discharged before any action could be taken, contributed to the severity of the event.

In most electric-railway power grids, it is possible for the trains both to draw power for operation and to feed power back into the grid during braking. In order to avoid overloading the lines, the voltage of the line is measured, and a decision is made whether or not feeding power back in would be safe or not. In normal operation, this energy recuperation during braking both saves electrical power and reduces wear on the mechanical brake systems.

The sequence of events is elaborated in the report (SBB 2006). Two out of three power lines between two regions of the Swiss railway power network were shut down according to schedule due to construction work. The one remaining line tripped at 17:08h because of overload; there was no power connection between the Gotthard region and Central Switzerland. The railway power grid was separated into two islands, "North" and "South". The South island was overproducing electricity, and an attempt at transferring power into the 50-Hz-network failed. Most generators were shut down automatically within seconds. All SBB railway operations in the canton of Ticino and at the Gotthard ceased.

In German-speaking and Western Switzerland, production in the powerplants Chatelard, Vernayaz and Etzel was increased. In concert with transfer from the Deutsche Bahn, the underproduction could be temporarily compensated. At 17:35, the coupling to the network of Deutsche Bahn was shut down. Remaining power stations in German-speaking and Western Switzerland further increased their power output, but ceased operations shortly after 18:00. Railway operations stopped in the North island as well. The islands are shown in Figure 18.

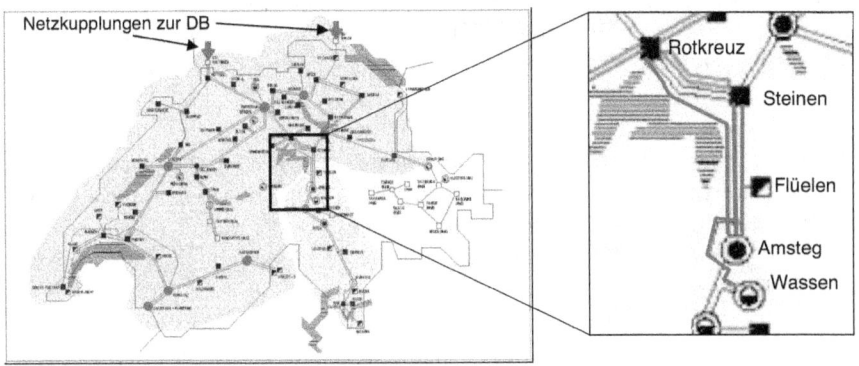

Fig. 18: Grid islands which formed during the June 2005 SBB blackout (SBB 2005)

There were three main causes identified by SBB's analysis (SBB 2006).

1. *Inappropriate risk estimate due to incorrect parameter values.*
   Wrong device parameters were a causal factor for an inaccurate risk

analysis. The control centre assumed that the high-voltage line Amsteg-Rotkreuz, which subsequently tripped, had a capacity of 240 MW. Although the line itself did in fact have a thermal capacity of 240 MW, the circuit breaker was set to 211.2 MW, limiting the usable capacity to this lower value. This latest current information was not available.

2. *Impossibility of timely and accurate assessment due to alarm flooding.*
There were four individual alarm messages about the overload of the couplings to the network of Deutsche Bahn, but these were inundated under the flood of other alarm messages. In the first 60 minutes after the first line failed, 18,000 messages, including 3,400 critical messages accumulated in the control centre (SBB 2006). A filtering of messages was not possible, and each message had to be acknowledged manually, individually, before the status display of all network components in the control centre was updated. An early recognition of the alarms about overload of the couplings to DB would have allowed the timely reversal of the energy flow through the frequency converters from the civil 50-Hz energy grid to augment the missing power in the railway network. This reversal could have been completed in a few seconds. Instead, the transformers continued operating in "rigid" mode, supplying railway power into the 50-Hz grid.

3. *A Scenario like this had never been considered.*
The possibility of a complete country-wide blackout of the railway power supply had never been considered prior to this incident, and was never included in operative risk management. Consequently, no contingency plans had been in place to prevent such an occurrence, or to minimize its consequences. Existing documentation about the prevention of (partial) blackouts proved unhelpful, because they were not tailored to the magnitude of this incident.

Of particular interest here is also the role of the N-1 criterion defined above. The N-1 criterion was knowingly disregarded, partly due to economic considerations (SBB 2006). When two of the three lines in Reusstal were shut down, the criterion was clearly violated, although continuing stable operation in both island networks would have been technically possible.

3 Conclusions Concerning Resilience

We wish to draw some straightforward conclusions. First, three observations.

1. In all three incidents, information was available to operators which, had it been acted upon, would have averted the blackout or mitigated its severity.

2. In two incidents, misleading information was displayed to operators and acted upon (unhelpful actions were taken; helpful actions were not taken). In one incident, some operators did not check available information but rather acted on an assumption which turned out not to hold.

3. In all three incidents, the generation and presentation of critical information was not subject to what critical-systems engineers would regard as appropriate assurance of dependability. In one incident, a design feature of the system inhibited timely action (the "alarm flood").

First, it follows from Observation 1 that the grid system considered as a physical system is theoretically resilient. All three incidents could have been avoided or mitigated through appropriate use of available information by operators. Second, considered as a sociotechnical system in which the actual behaviour rather than some idealised behaviour of human operators is taken into account, the system is manifestly less resilient than theory suggests. Third, the actual resilience of the sociotechnical systems could be significantly improved by routine critical-system engineering: identifying mission-critical functions in system components and ensuring their availability, or at least that their unavailability is not masked.

As things stand, the resilience properties of electricity grids are emergent. Considered as a physical system with ideal operator behaviour, say during design-time analysis, a grid appears adequately resilient. As an actual sociotechnical system, we have seen three cases in which it is in fact less resilient than supposed.

This situation in which presumed resilience is affected by actual implementation is also seen in other engineering domains. For example, air traffic control ground communications are effected though dedicated services. These services are often contracted out to telecommunications service providers, which run the dedicated services along with public telephone service and other services through non-dedicated equipment. A failure of this equipment, which may not be considered critical by the service provider, also causes the dedicated critical ATC services to fail (Neumann 1991). A contrasting case is that of motorway auffahr-accidents. The first author has shown, using Rational Cognitive Model checking, that auffahr-accidents are an emergent property of the motorway system-of-systems itself (Ladkin 2011).

We conclude from the three examples we have considered that the actual resilience of some sociotechnical systems is lower than a theoretical analysis might have led engineers to believe. We have also observed that conditions which causally lead to a normal accident are often emergent. Finally, we have observed that being a single point of failure under certain conditions is often an emergent property. It follows that the resilience properties of these systems are in large part emergent.

Acknowledgements A talk based on a version of part of this paper, as well as events from motorways and finance, was given at an SCSC seminar in London in April 2015. The first author thanks Michael Parsons for the invitation to present there, which gave the impetus to prepare a preliminary version of this paper. Both authors thank him for his helpful comments on the first version of this paper.

References

Berlekamp E R, Conway J H, Guy R K, (1982) Winning Ways for Your Mathematical Plays, Academic Press, 2nd edition A K Peters, 2001-4.

Bundesnetzagentur (2007) Report by the Federal Network Agency for Electricity, Gas, Telecommunications, Post and Railways on the disturbance in the German and European power system on the 4th of November 2006. Technical report, Bundesnetzagentur

Conway J H (1970) Life (game). Explained in (Berlekamp et. al. 1982). Also see https://en.wikipedia.org/wiki/Conway%27s_Game_of_Life , accessed 2015-11-14.

E.ON Netz GmbH (2006) Bericht über den Stand der Untersuchungen zu Hergang und Ursachen der Störung des kontinentaleuropäischen Stromnetzes am Samstag, 4. November 2006 nach 22:10 Uhr. Technical report, E.ON Netz GmbH.

Ladkin P B (2001) Causal System Analysis, e-book available from http://www.rvs.uni-bielefeld.de/publications/books/CausalSystemAnalysis/index.html , RVS Group, University of Bielefeld

Ladkin P B (2011) The Assurance of Cyber-Physical Systems: Auffahr Accidents and Rational Cognitive Model Checking, RVS Group, University of Bielefeld, December 2011. Available from http://www.rvs.uni-bielefeld.de/publications/Papers/20111230CPSV2.pdf , accessed 2015-11-15.

Ladkin P B (2015) Synopsis of the 2003 North American Blackout, available from http://www.rvs.uni-bielefeld.de/publications/Reports/LadkinSynopsis2003Blackout.pdf , RVS Group, University of Bielefeld

Meyer J (2009) Defining and Evaluating Resilience: A Performability Perspective, slides from a talk at PMCSS9, September 17, 2009. Available from http://web.eecs.umich.edu/people/jfm/PMCCS-9_Slides.pdf , accessed 2015-11-15.

U.S. National Transportation Safety Board (2014) Auxiliary Power Unit Battery Fire, Japan Airlines Boeing 787-8, JA829J Boston, Massachusetts January 7, 2013, Report AIR-14-01, NTSB 2014.

Neumann P G (1991) AT&T phone failure downs three New York airports for four hours, Risks Forum 12.36, September 1991. Accessable from http://catless.ncl.ac.uk/Risks/12.36.html#subj1.1 accessed 2015-11-19.

North American Electric Reliability Council (2004) Technical Analysis of the August 14, 2003, Blackout: What Happened, Why, and What Did We Learn?, Report to the NERC Board of Trustees by the NERC Steering Group, July 13, 2004. Available from http://www.nerc.com/docs/docs/blackout/NERC_Final_Blackout_Report_07_13_04.pdf)

Perrow C (1984) Normal Accidents: Living with High-Risk Technologies, Basic Books, 1984. Updated edition, Princeton University Press, 1999.

Schweizerische Bundesbahnen (SBB) (2005) Strompanne der SBB vom 22. Juni 2005. Technical report, Schweizerische Bundesbahnen

Schweizerische Bundesbahnen (SBB) Zentralbereich Revision (2006) Second Opinion zur Strompanne der SBB. Technical report, Schweizerische Bundesbahnen

Sieker B M (2008) European Electricity Blackout, November 2006. Causalis Technical Report, 2008. Available from http://www.causalis.com/90-publications/EuropeanElectricityBlackout.pdf , accessed 2015-11-15.

U.S.-Canada Power System Outage Task Force (2004) Final Report on the August 14, 2003 Blackout in the United States and Canada: Causes and Recommendations

Smart Safety Assessment, SSA

A G Hessami

Vega Systems

London, UK

Abstract *The commonly practiced approach to the safety assessment of complex technology, including those in the High Speed Railways[1] lacks a supporting theoretical foundation as a guiding and supporting backbone. In practice, this results in a confused, poorly conceived and often inadequate application of a mixed bag of methodologies, rules and standards that, due to the effort-intensive nature give a semblance of adequacy and completeness. In this uncharted and poorly structured landscape, demonstration of compliance with a given rule or standard is broadly regarded as adequate input to the safety assessment, potentially missing other analysis, effort and evidence. The key aim of the research outcomes presented in this paper is to give an overview of the principal requirements and qualities for robust, credible and systematic assessment to be supported by a host of relevant processes, rules, tools, codes of practice and standards.*

1 Introduction

Only in the light of a theoretical backbone and roadmap, can we make progress in arriving at a credible and required degree of confidence when applying scientific knowledge, methodologies, reviews, tests and evidence. It will guide us in our judgement about how much confidence we have or need to have in the safety of a product, process, undertaking or service and how best to arrive at this.

In this endeavor, we revisit the current practice from two complementary but critical perspectives comprising the process and the dominant culture that pervades the relationship between the stakeholders. The theoretical architecture/framework will be reviewed, enhanced and deployed to carefully examine and critique the current best practice in safety assessment. This will identify processes and activities that can be optimized to bring genuine real value to the cause of safety globally, whilst reducing wasteful effort, time scale, cost and scarce re-

---

[1] Keywords: Assessment, safety assessment, high speed railway, safety critical systems

sources. We also attempt to address the dominant adversarial culture in safety that divides the key stakeholder classes namely the producers/manufacturers/duty holders and the regulators, approval/certification authorities and the commercial support entities. A smart solution needs to tackle the soft as well as the hard aspects of this deficit.

The scope of this research is largely the processes and methodologies employed in safety assessment of products, processes, systems, undertakings and services. However, the so called Smart Safety Assessment Framework developed is principally focused on the requirements of the complex and technology intensive case of High Speed Railways.

1.1 European Common Assessment Methods (CAMSS)

The European Commission is driving a new initiative for the assessment of standards and specifications largely for application in the ICT environment. This is referred to as the Common Assessment Method for Standards and Specifications (CAMSS).

The main objective of CAMSS in Europe (Europa, 2010) is to establish a neutral and unbiased method for the assessment of technical specifications and standards in the field of ICT. CAMSS is intended to be used by Member States and EU Institutions to avoid ICT vendor-pecific lock-in when defining ICT architectures and establishing European public services.

Whilst regarded as a largely commercial and procurement related issue, the need to an agreed, unbiased common assessment method for a large body of ICT based standards and specifications is aligned with the concepts in Smart Safety Assessment (SSA).

2 Assessment, a Rational Process

2.1 Generic Assessment

Assessment is a largely human function that is exercised to support and drive decisions in all walks of life and on a daily basis. A quick glance of the uses of assessment indicates that is largely seen as a domain related endeavor such as:

* Educational Assessment, the process of documenting knowledge, skills, attitudes, and beliefs;

- Health assessment, a plan of care that identifies the specific needs of the client and how those needs will be addressed by the healthcare system;

- Risk assessment, the determination of quantitative or qualitative value of risk related to a concrete situation and a recognized threat;

- Tax assessment, value calculated as the basis for determining the amounts to be paid or assessed for tax or insurance purposes

- Political assessment, assessment of officeholders for political donations.

There are a number of essential constituents in the cognitive assessment function that comprise:

1. An understanding or framing of a concept, physical or virtual entity and the scope;

2. A particular focus on the object or concept;

3. A set of criteria/features to support evaluation;

4. An evaluation (qualitative or quantitative) of the key features/criteria;

5. An appreciation of the criteria for acceptance/rejection;

6. A mental balancing act or value judgement between evaluated criteria and the benchmark for acceptance;

7. Coming to a decision by taking the local evaluation and larger relevant factors into account.

This however poses a constrained view of the nature and function of assessment. A more holistic perspective is depicted in the Weighted Factors (WeFA) schema in Figure 1. The WeFA methodology (Hessami, Grey 2003) represents a bipolar and holistic understanding of the dimensions, factors and constituents of a concept, physical or virtual entity or aim.

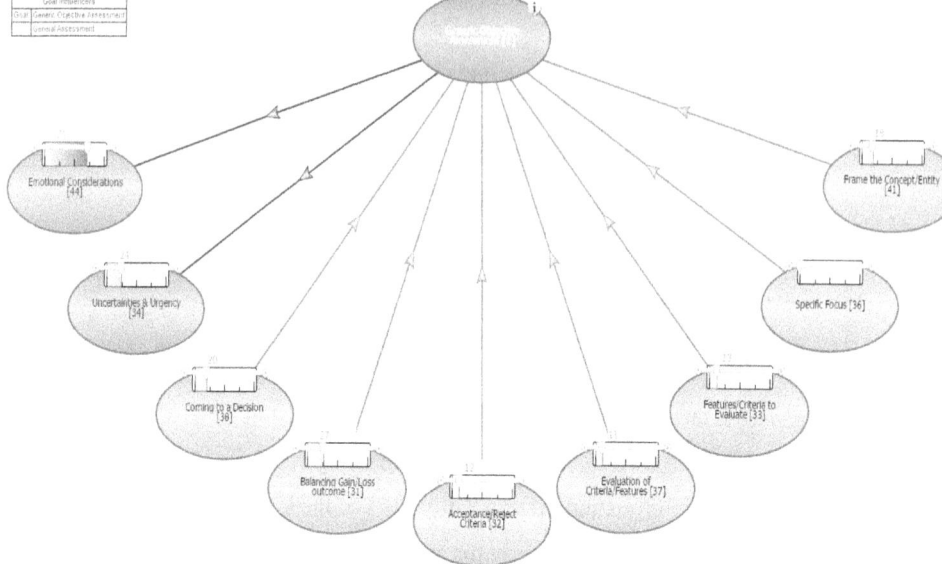

Fig 1. Schema for Generic Objective Assessment

It is evident from the schema of Figure 1 that apart from the seven *drivers* detailed above, there are a number of key *inhibitors* comprising:

1. prevalence of emotional attachment and personal issues influencing the decision;

2. degrees of uncertainty involved and the potential need for urgency that may over-ride or undermine other considerations.

This model-ased perspective paints a more comprehensive view of generic assessment that is influenced by rational, emotional and circumstantial factors. It is also worth considering the execution model for the generic assessment in that it may comprise a set of cognitive, sequential and concurrent processes and functions. Many contextual and emotional factors may impact on the objectivity and necessity for prior processes and outcomes before a decision is arrived at.

2.2 Safety Assessment

Bearing in mind the generic assessment theory as depicted in Figure 1, a specific instantiation of this ontology/Class is safety assessment. Employing the WeFA schema, a representation of Generic Safety Assessment function/process is depicted in Figure 2.

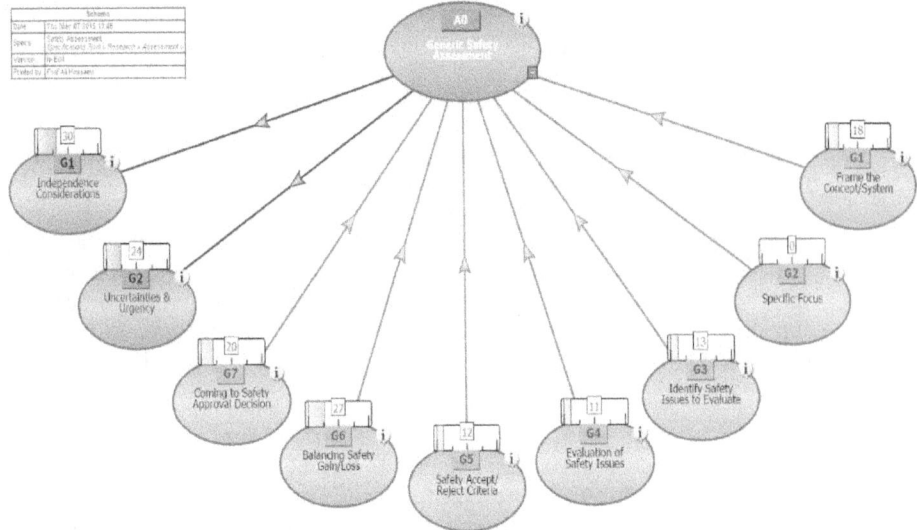

Fig 2. Schema for Generic Safety Assessment

The safety assessment schema (Object) derived from the generic objective assessment model (Class) has more specific safety focus and customisation on many of the *drivers*. One change in the context of the G1 *inhibitor* is replacing emotional influence with matters of independence (organisational, functional, matters of influence or self- interest). It is important to note that safety assessment is not confined to post-development evaluation and certification activities. The generic schema in Figure 2 equally applies to safety considerations during design, development and production activities for products, systems and services. In this spirit, it can be applied iteratively to all safety tasks and processes to arrive at a rational basis for taking positive actions for safety enhancements where appropriate.

The safety assessment schema highlights the essential constituents and activities inherent in evaluating a product, process, concept, system or service from safety performance perspective. These are detailed as follows:

G1- Frame Concept/System: this in principle is the same as a comprehensive system/concept definition, specification and scoping including architecture, design, implementation and integration as the available information may allow;

G2- Specific Focus: this implies the sub-system or type of safety (system versus occupational matters or type of assessment e.g. functional, technical, operational, environmental aspects);

G3- Identifying Safety Issues to Evaluate: this activity is equivalent to manual, semi-formal, formal or automatic forms of identifying the system's anomalies that may impact on safe performance. This is in essence the intended or un-intended yet undesirable hazardous modes/behaviours/states/failures of the system under analysis;

G4- Evaluation of Safety Issues: this amounts of frequency and consequence estimations from the identified safety issues with a view to estimate likely risks in expected value terms. The evaluation can take the qualitative, semi-quantitative and quantitative form and can the form of pure expert judgement to data driven complex models. Whatever the approach, the evaluation process for each safety issue shall enable some form of integration in order to support arriving at an overall safety risk profile for the product, process, system or service. This is a major shortfall of the current practice that tackles issues individually not systemically;

G5- Safety Acceptance Criteria: this amounts to having clear understanding of the measures of risk that can be accepted, tolerated or rejected as unacceptable at sub-system and entire system safety performance level. However, given safety is a regulated aspect of performance, these criteria should ideally align with the qualitative and quantitative legal requirements in safety critical products/systems/undertakings and services or changes imposed on a steady state safe system.

The acceptance criteria could alternatively be a complex set of rules, standards and regulations that should be employed as baseline for conformity evaluation and assessment. This complicates the assessment since a rule, standard or code of practice is not a single criterion case but may comprise many such requirements. In such cases, the requirements of the process and their conformity takes precedence over the evaluation of safety features/issues on the assumption that following the rules guarantees safety;

G6- Balancing Safety Gain/Loss: this amounts to providing a structured and preferably documented analysis and argument supported by evidence (Safety Case) that the product/concept/system or service under consideration is likely to create an acceptable level of safety harm/risk. In principle, this argument should additionally allow for the benefits and gains arising from the deployment of the entity under consideration.[1]

---

[1] Alas, consideration of benefits is completely ignored in the Western legal system dominated by single sided focus on harm and minimisation of risk and not a net social benefit. This concept has roots in the Eastern Taoist mind set of balance of Yin & Yang and is more likely to be a good candidate for a legal principle for tolerability of risk as opposed to the British concept of ALARP. The argument is whereas consideration of risk alone is a Western occupation, it also lacks a total systemic consideration and can lead to an unfair punitive regime on innovation whereas a balanced approach is effectively a whole system emergence approach and more inclusive

It is also customary to ask for an independent verification of the safety claims contained in the Safety Case by the supplier/service provider. The Independent Safety Assessment (ISA) is now a component of European safety approvals.

It is also worth noting that the assessment of risks and rewards may lead to the reconsideration of the system and alterations/improvements so that the level of risk is reduced to an acceptable or tolerable level. This iteration is the outcome of evaluation and may involve redesign or modification of the entity under consideration and contrasting against acceptance criteria until an overall and satisfactory level of safety performance is arrived at.

G7- Coming to a Safety (Approval) Decision: the principal outcome of assessment is a decision on what has been assessed in terms of the desired property being acceptable at a desired or acceptable level or not. This is often called conformity and requires criteria or standards as a reference to judge conformity against. In the context of safety and given legal implications, this is also equivalent to attaining endorsement for the safety claims and evidence in the safety Case and obtaining a Safety Approval from the Safety Regulator/Authority. The Independent Safety Assessor (ISA) is usually employed as an advisor and expert companion for the supplier or service provider to support their difficult technical decisions. ISA is also intended to provide a useful function an independent reviewer/auditor of the approach, methods, tests and outcomes to hopefully identify systematic errors that may have crept into the product/service during specification, design and development processes.

Finally, the quality and rationality of the decision is dependent on the skills, expertise and competence of the decision makers. However, competence is a context-ensitive quality or attribute of a person, team or even an enterprise. To ensure quality (fit for purpose) assessment, the roles and the competence of those fulfilling those roles need to be systematically addressed.

G1- Independence Considerations: there are two fundamental reasons for consideration of independence in safety assessment. The appearance of this factor as an *inhibitor* in the schema of Figure 2 indicates that if seeking independent input is not adopted and ensured, the quality of assessment suffers as a consequence. The principal reasons for consideration of independent input are:

a- Diversity principle and

b- Expert support to the client/duty holder organisation.

The diversity principle implies an entity that is not part of a project, design or implementation can, subject to having adequate competences, identify more errors of omission and commission. This principle is employed in systems safety standards as a normative practice to reduce systematic errors in complex technological systems. The second issue is largely due to the expertise required in systems safety processes, methodologies, tools and applications that may not adequately exist at a client/customer organisation. The concept of ISA, enshrined in the European approach to conformity is largely a response to this aspect of the schema in Figure 2.

G2- Uncertainty and Urgency: The second *inhibitor* in the safety assessment schema relates to the impact of degrees of uncertainty on the quality and objectivity of the assessment carried out. To this, matters of urgency, i.e. the need for quick answers adds a dimension not entirely unfamiliar in real life. Overall, major items of uncertainty driven by lack of significant novelty or complexity, insufficient analysis, poor data, inappropriate level of expertise employed in the process or function of assessment undermines the validity and quality of the decisions made as the outcome of this cognitive process.

3 Safety Assessment in Railways

3.1 A Critique of Current Safety Assessment Practice in Railways

In the absence of a formal and generally accepted theoretical framework, the assessment of safety is largely driven by codes of practice, standards and regulations. Where regulations exist, much of the emphasis is on the documentation and certification as opposed to the quality and objectivity of the assessment activities. The following observations (as a non-exhaustive list) relate to the current real and perceived deficits of safety assessment as practiced in the railways:

- No mention or justification of value generated, only costs are known or worked out but not the returns/benefits;
- Largely process based and document centric;
- Largely claim rather than evidence driven;
- Poor structuring of evidence and arguments;
- Poor or subjective/implicit assessment of evidence;
- Uncertainties not explicitly stated nor addressed systematically;
- Standards and methodology rather than systems framework driven;
- Largely ignores the influence of the requisite skills and competence of human agents;
- No process or data model in assessment;
- A linear and sequential perception of effort involved with certification as target;

- No views/measures of safety outcomes achieved at any moment in time and no benchmark/baseline requirements;
- No tailoring or risk-assed prioritisation of effort, process, resources;
- No scalable approach that can proportionately reduce the effort commensurate with the level of complexity and risks;
- No recognition of local and cultural influences;
- Ultimately judgement-driven so largely subjective and heavily dependent on quality of experts;
- Assessment is largely "ticks in the box" similar to quality and not based on scientific scales of measurement;
- No characterisation of required human skill and expertise in different aspects and activities;
- No/poor regulation of experts and organisations involved;
- Most assessment activities take place post-development;
- Adversarial culture between the regulators and duty holders that adds to bureaucracy, delays and costs.

The outcome of these current deficits is to give the perception that safety and quality activities are generally non-productive and hence wasteful. This poses a challenge to practitioners and duty holders alike: that unless safety engineering and assessment are seen as value generating activities, they may not attract sufficient support and resources hence not implemented adequately if at all.

Apart from changing perceptions, there's also a need to identify and reduce the scale and scope of activities that do not directly add value and benefit to safety. The non-exhaustive list of deficits above is a reminder that much can be done to improve on current "waste factories" using the terminology from Six Sigma. The nature of so called assessment is also a largely "ticks in the box" process similar to audits and quality checks and not based on scientific scales and measurement.

Finally, there's the soft issue of the dominant culture that pervades the relationship between the two major classes of stakeholders, the duty holders and the approvers (and the supporting commercial organisations). This dominant culture tends to reduce efficiency, timeliness of input and seeing the achievement of safety in products, processes and undertakings as a socially beneficial endeavor for all. The need for a more collaborative and supportive culture is largely felt in this arena and does not violate the legal liability or independence requirements. It complements any improvements made in the process and methodologies of assessment and it will be a folly to believe a polished and optimized approach focused on technicalities will have a major impact on the quality, efficiency and integrity of safety assessment.

3.2 The Dominant and Critical Role of Railway Safety Standards

The European CENELEC railway domain safety standards and their adopted variants (IEC 2002, 2003) are globally regarded as the most comprehensive and broadly accepted norms for assuring the safety of products and services in the railway context. In view of popularity and almost global adoption, these are also adopted for the global market by the IEC and re-issued as IEC standards. One common feature of these standards is the concept of life cycle and specifically, a safety life cycle for a product/system/service. The guidance and process in these standards revolves around safety activities at each phase of the development and deployment life cycle.

The generic product, system or process Safety Life Cycle comprises twelve phases as follows:

1- Concept Definition
2- Detailed Definition and Operational Context
3- Risk Analysis and Evaluation
4- Requirements (including Safety Requirements) Specification
5- Architecture & Apportionment
6- Design & Implementation
7- Manufacture/ Production
8- System Integration
9- Validation
10- Acceptance
11- Operation & Maintenance and Performance Monitoring
12- Decommissioning.

One key expectation and deliverable from the activities advocated in the standards is the documented evidence and argument for safety referred to as a Safety Case. Given products/systems/services exist at three levels of: generic product, generic application (of product) and specific application (of product), three types of safety cases are specified and expected to be produced depending on the relevant instance.

In principle, conformity with the standards and production of safety cases is a complex specialist activity that is often outsourced by duty holders/manufacturers to commercial safety services organisations. This market is generally not regulated so a wide range of commercial safety assurance organisations exist with varying degrees of process competence that trade in system safety services and certification.

The conformity is also largely a single product or system matter and does not easily extend to the complete extent of a large integrated system such as signaling. The required confidence for a large integrated system comprising a number of assured sub-systems is still a matter for safety assessment even though the existing standards barely address this issue adequately.

4 Smart Safety Assessment (SSA)

4.1 Aim and objectives

The general desirable qualities for a reliable, efficient and value driven safety assessment process and the long list of current deficits detailed above necessitate a serious review and rework of the current practice. This situation is more acute in the developing economies since the perception of cost without value return threatens the necessary and committed investment of time and expert resources in safety engineering of products and systems even those aimed at the global markets. Whilst the developed world employs a more stringent set of rules and a high cost burden for safety activities are often a source of disenchantment for many supply chain commercial enterprises other than major reputable multi-nationals. The cost burden is ultimately borne by the fee payers of transportation services and the substantial subsidies by public organisations. Even in the developed economies, there is a need to challenge the wasteful and costly practices conducted in the name of safety.

The principal goals in an endeavor to scrutinise safety assessment are:

1. To optimise the process and activities of safety assessment. This is not intended to do less, it is focused on doing the right thing at the right time to the right degree and by the right persons;
2. To question and justify all activities in terms of value brought to the cause of safety;
3. To prioritise and deliver the activities that address safety risks more effectively and responsively;
4. To provide the opportunity for timely application of corrective action to reduce delays, wasted effort and resources;
5. To emphasise value in "action & evidence" rather than "claim & documents";
6. To generate an instantaneous view of safety achievements, deficits and critical activities to arrive at an acceptable level of safety achievement;
7. To enhance transparency, accountability and collaboration between all stakeholders (the dominant largely adversarial culture) thus expediting the process and reducing delays, errors and wasteful repetition.

Given safety assessment is an iterative process and is applicable at all stages of design, development/realisation, deployment or approvals, any savings in time and effort is expected to have a large cumulative impact.

4.2 The SSA Architecture

This involves a number of meta-structures and supporting processes that attempt to deliver a total systems perspective on the safety assessment in the style of Lean Process (Lean Org). The fundamental building blocks of SSA are detailed below:

1- Selection of the Assessment Reference Model (ARM);
2- Transformation of ARM into Assessable Components;
3- Systematic Competence Assessment and Task Allocation (CARA);
4- Setting up Stakeholder Collaboration Rules and Work-space (CORS);
5- Initiation of Activity-Evidence-Assessment Cycle (AEAC);
6- Safety Assessment Claims & Outcomes (SACO);
7- Developing incremental Safety Value Case (SVC);
8- Capturing Process Learning in Enterprise Safety Repository (ESR).

The thrust of these activities implies that to have a high degree of confidence and trust in the safety of a product or system, we need to employ competent people who develop the product/system according to latest best practice and we assess the outcome efficiently, correctly and responsively. This is depicted in Figure 3.

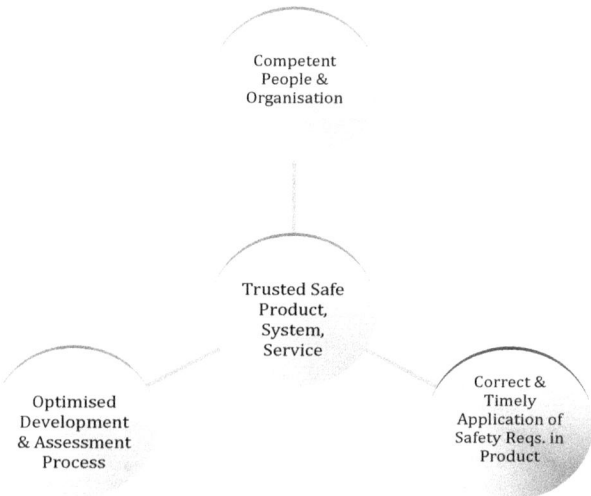

Fig. 3 Three constituents of high confidence in safety

The SSA constituents form a systems architecture. This by necessity has a structure as well as an execution model collectively referred to as SSA Architecture. The SSA architecture is depicted in the diagram in Figure 4.

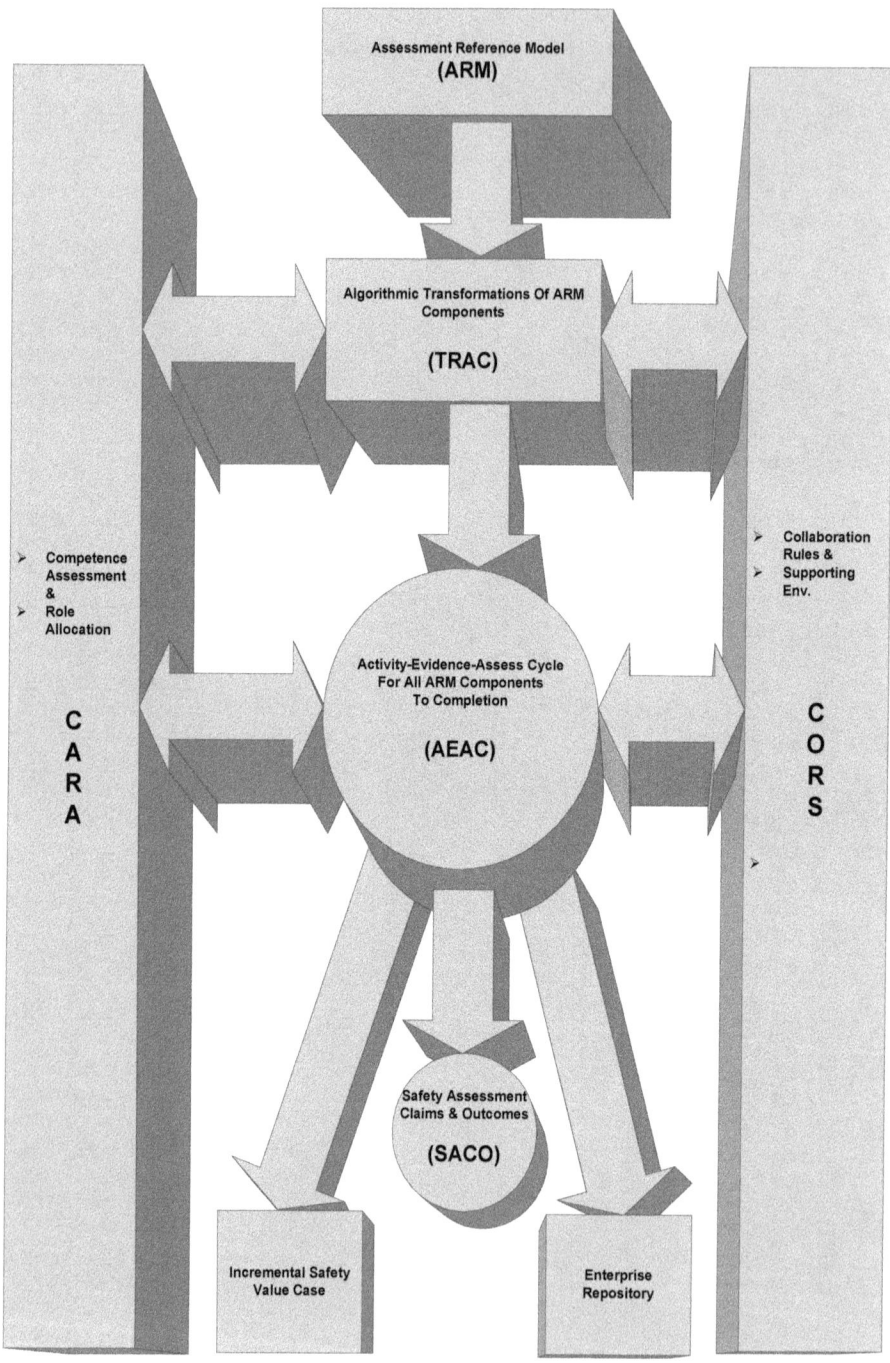

Fig. 4. The illustrative SSA Architecture

The execution model for the SSA architecture is driven by asynchronous/non-sequential concurrency and Near Real-Time (NERT) monitoring, awareness and control through a multi-stakeholder environment.

1.1   The SSA Processes and Supporting Tools

The SSA does not embody the body of knowledge required for the safe design, development, deployment and operation of products, systems and processes. It assumes these exist based on three classes of sources/approaches, i.e. standards, risk analysis or cross-acceptance that is referred to as the Assessment Reference Models (ARM). In this spirit, SSA is a meta-structure aimed at optimising the adoption and effective and efficient conformity with any set of chosen reference models, domain knowledge or expert driven process. It additionally strives to influence a culture change from adversarial to collaborative mode of inter-working between the key stakeholders in the design, development, deployment, approval/certification and operation of safety related and safety critical products, systems, processes and undertakings.

The simultaneous focus on the reduction of process inefficiencies and waste factories alongside a gradual change in the culture of safety are the cornerstone of SSA philosophy. It is recognized that a polished process without the supporting culture will amount to little again in the end and much of the shortfalls of the safety assessment detailed above will remain if the beliefs and behaviours are not tackled at the same time. This is illustrated in Figure 5.

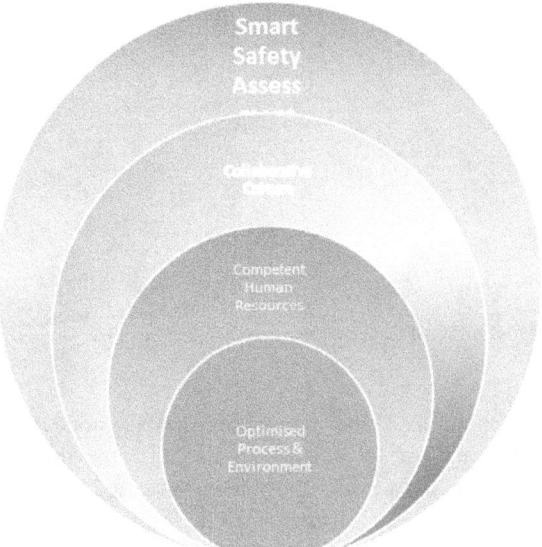

Fig. 5. The essential success factors in SSA philosophy

The SSA architecture components in terms of intrinsic process, exchanges and supporting tools are detailed below:

4.2.1 Selection of the Assessment Reference Model (ARM)

Three classes of approach to safety justification/assessment of products, systems and processes are depicted in Figure 6 but not detailed here for brevity. These constitute existing rules, codes of practice, procedures, national and international standards or historical evidence on safe performance. Each class of approach is a generic reference model for assessment and comprise:

1- Standards, codes of practice, national rules, expert knowledge;
2- Risk assessment due to novelty of the system;
3- Cross-Acceptance due to satisfactory record of safe performance or similarity with a Safe Reference System.

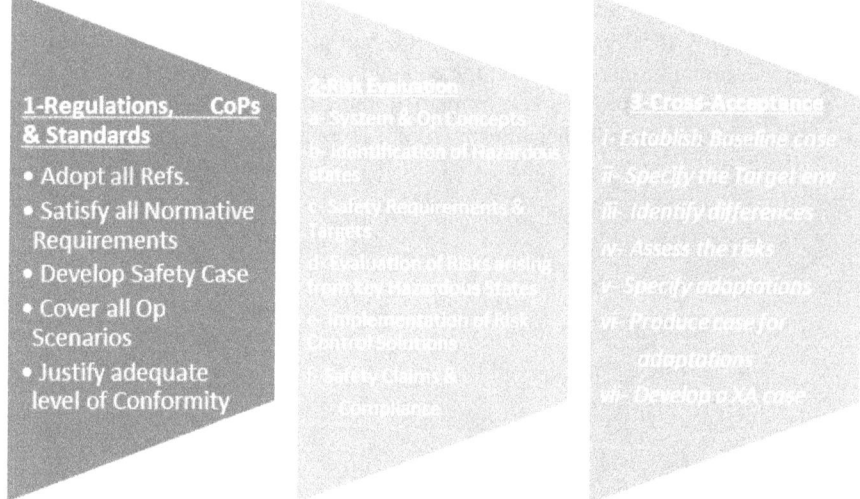

Fig. 6. Three Generic Reference Models for SSA

This is the first step in SSA and determines the backbone or the body of knowledge for the application of the smart assessment. In the simplest form, one class of Reference Model applies to the entity being assessed. However, complex cases may involve a combination of elements from all three classes. The SSA allows for such complexity.

This framework is more comprehensive than the guidance on Common Safety Methods developed by the European Railway Agency (ERA). The ERA guidance on acceptable methods for risk assessment refers to Code of Practice, Similar Reference Systems and so called Explicit Risk Estimation that also requires Independent Safety Assessment for verification.

Note that the ARM includes capture, verification and deployment of expert knowledge alongside standards, rules and regulations. In view of its tacit nature, the capture and formalisation of expert knowledge would be in the algorithmic and assessable form thus obviating the need for the transformation in phase 2 of the process.

Once an Assessment Reference Model is selected for the entity, all the relevant components i.e. rules, standards, regulations and processes must be clearly identified and recorded as the ARM Components. For practical purposes, the phases of the product/system or process relevant to assessment can be divided into three principal stages comprising:

1.   Development and Deployment;
2.   Approval & Certification;
3.   Operation, Maintenance and Upgrades.

The selection of the ARM should ideally be mirrored in the first two classes i.e. whatever is the ARM applied for development and deployment, it is the same ARM that needs to be declared and employed in the approval and certification activity.

The safety assessment applicable to operation, maintenance, upgrades is largely a matter for safety performance monitoring. The process for this and the required supervisory tools and environments are different to those in SSA and require a modification for the SSA process. This class is not currently covered by the SSA framework described in this research.

The final consideration in the process of selecting an Assessment Reference Model is also the operational states that have to be taken into account after the problem is framed (system is described and specified). As a minimum, the following should be included in the ARM and later assessment activities:

1-   Normal Operations;
2-   Degraded Operations (under constraints or maintenance);
3-   Failure states (when operations are severely hampered due to failures of rolling stock, infrastructure sub-systems or natural causes);
4-   Emergency states arising from train accidents, large scale damage to the infrastructure or natural disasters.

The above considerations apply to development and deployment as well as approval and certification classes of safety assessment.

4.2.2 Transformation of ARM into Assessable Components (TRAC)

Once the ARM components have been identified for the assessment, these need to be transformed into assessable representations. This is largely driven by the complex and wordy nature of rules and standards that are not generally developed with assessment and demonstration of conformity in mind.

The transformation process can be implemented in a multiplicity of forms and environments however, use of advanced computer tools capable of schematic capture, measurements of clauses, aggregation of results and reporting are recommended as illustrated in Figure 7.

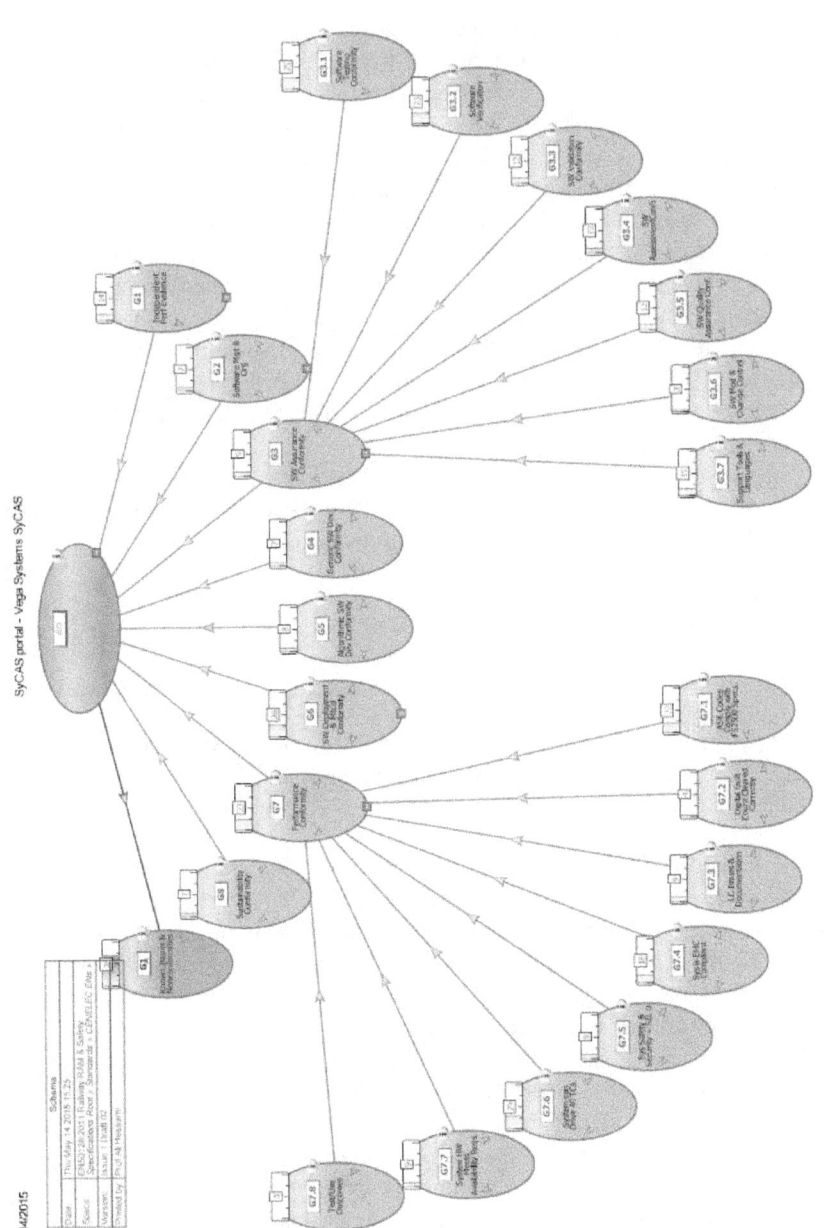

Fig. 7. Illustrative Schematic & Assessable Standard Representation

The transformation ideally involves taking a hierarchical view of the contents and structure and developing a taxonomic perspective on each element of ontology at the highest level of representation. The concept of product, system, process Life Cycle is an ideal candidate for this backbone and supporting structure. Many of the modern railway safety standards use the concept of Life Cycle to describe a progressive time sequence based series of activities from concept all the way to operation and even decommissioning. Taking a structure that represents the back bone for the rule, standard or procedure being transformed, all the sub-sections, activities and clause by clause requirements can be transformed into a hierarchical model. Ideally, the transformation environment provides the facility to delve into the requisite level of detail appropriate to the perspective and level of assessment. It should ideally be possible to assess a product, project, system or process at the highest level of the ontology or any combination of levels for each taxonomy to generate a truly flexible arrangement.

For every section, clause or requirement, an assessment object shall be created that should allow the following parameters to be declared:

1.   Identity of the clause/requirement
2.   Information to trace this to the exact location in the source
3.   Type of measurement required, units and scale
4.   The relative importance of the requirement in the local context of assessment
5.   Typical and benchmarks for acceptable score
6.   The nature or type of evidence required in satisfaction of the requirements
7.   Facility to rate/rank the quality of the evidence and the degree of conformity demonstrated

The transformation environment shall provide a facility to combine detailed conformity evaluations into a total for the whole product, project, system or process. Apart from providing a total perspective on the state of assessment, it would also be highly desirable to present a graphical perspective on the areas of poor conformity so that attention and energy can be applied in a timely manner on the deficits or shortfalls in safety performance.

The algorithmic transformation of the Assessment Reference Model components ideally generates a re-usable resource since this is a knowledge and effort intensive task. Much of the information required for the transformation comes from the components but the schematic assessable representation requires a fair degree of knowledge input from the domain experts in terms of how to measure each clause, the scale and units, acceptance benchmarks, requisite evidence and titles for the documents accepted as evidence. The transformation environment should also be capable of integrating multiple assessments.

By necessity, the complexity and assessment requirements in the transformation requires the use of advanced computer tools.

4.2.3 Systematic Competence Assessment and Task Allocation (CARA)

The influence of the knowledge, skill, motivation and innovation by the people responsible for safety related and safety critical tasks is largely established. This is largely based on the premise that with the exception of acts of nature, almost all other causes of incidents and accidents can be traced to human causes. In this spirit, competence is a general systemic attribute of the people charged with the specific tasks in safety assessment so that they perform the right work, with appropriate skill, efficiency, quality, drive and innovation and maintain this performance to avoid incidents and accidents.

The selection, development and appointment of competent resources to safety assessment tasks therefore constitutes a critical aspect of the SSA architecture. The implementation of this component of the architecture requires systematic identification of the critical roles, modelling, calibration and verification of a Competence Assurance framework for these role. Once the role specifications are completed, these are employed for the evaluation of the candidates and identification their further needs for development and allocation of roles and responsibilities. A generic Competence Assessment framework is depicted in Figure 8.

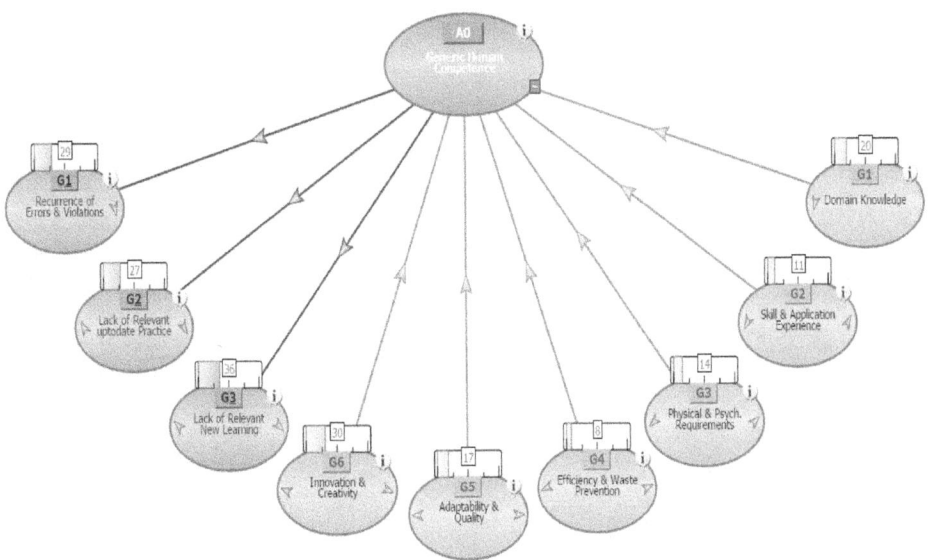

Fig. 8 Generic Competence Assessment Framework

The generic competence assessment framework depicts the key aspects of human competence that once adapted and adopted, require customization, analysis and calibration to provide a basis for specification of a given safety related or safety critical role.

A number of role specifications are recommended and necessary for the SSA environment as follows:

1. System Safety Analyst (SSAt);
2. System Safety Verifier (SSVf);
3. System Safety Validator (SSVd);
4. System Safety Assessor (SSAr);
5. Safety Technology Expert (STE);
6. Independent System Safety Assessor (ISA);
7. System Safety Manager (SSM).

The specification and calibration of role specific schema for each one of the above will pave the way for assessment of the appropriate candidates for each role, identification of the potential deficits and development of the individuals allocated the roles.

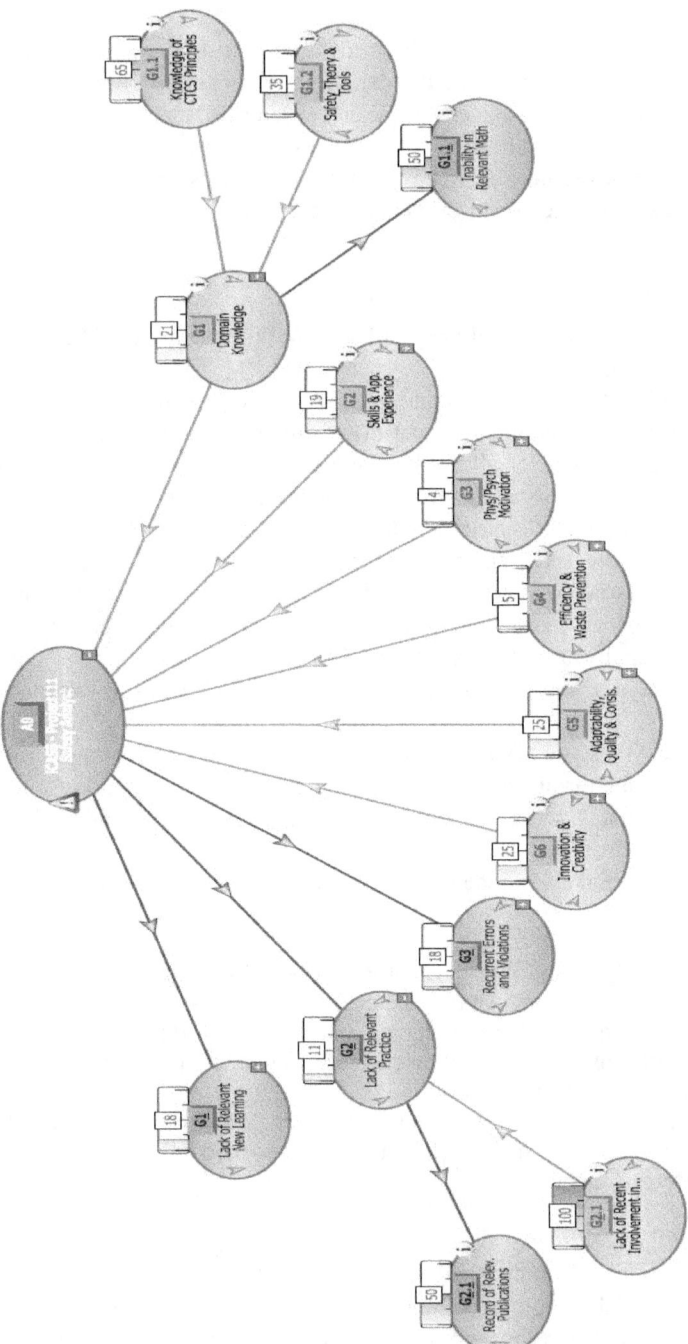

Fig. 9. Illustrative Role Specification for Safety Analyst

As illustration of a role specification for a safety analyst from a past project is given in Figure 9. This highlights that the detailed specification is based on the elaboration of the specific knowledge, skill, traits, efficiency, quality and innovation that is required and relevant to the role and its given context. The specification methodology employed here also allows for building measures of importance for each attribute based on the organizational culture and preferences.

4.2.4 Setting up Stakeholder Collaboration Rules & Work-space (CORS)

The success of a SSA effort is largely contingent on the development and deployment of an advanced IT based environment that facilitates efficient access, communication and collaboration for all the stakeholders in the assessment activities from the analysts and assessors to the process managers. Apart from engendering transparency, this environment is also intended to engage key stakeholders in safety at earlier phases of the development than currently achieved whilst also negating the largely adversarial culture and promoting a reinforcing and positive environment. This is intended to challenge the dominant processes and culture between the stakeholders towards a collaborative stance, treating attainment of safety in products, processes, systems and undertakings a social benefit of relevance and value to all.

Apart from the IT assisted collaboration environment, the orderly and error free access to the online resources requires certain degree of formalisation for the rules, access arrangements, work sharing, version control and configuration management for the work products.

A typical collaboration environment will comprise a set of essential supporting structures namely:

1. A repository of algorithmic Assessment Reference Model components in the form of standards, rules, regulations and codes of practice;
2. A controlled environment for creating and maintaining products and projects that permits the stakeholders to access the assessment environment and collaboratively work on the tasks;
3. A suite of rules, access arrangements/privileges such that various stakeholders can safety engage with the tasks and avoid access conflicts, loss of assessment effort or corruption of data.

Depending on the nature and scale of the assessment organisation and effort, all stakeholders should be briefed and trained in the functions and access arrangements for the collaboration environment. The next step is devising a competent and efficient organization for the SSA. A typical structure is depicted in Figure 9.

In the organisational diagram, NSA refers to the National Safety Authority and CSO to the Commercial Safety Organisation which has a largely an advisory role. The SAV represents the Safety Assessment Verification and Validation activity that needs a degree of independence from the safety assessment organisation.

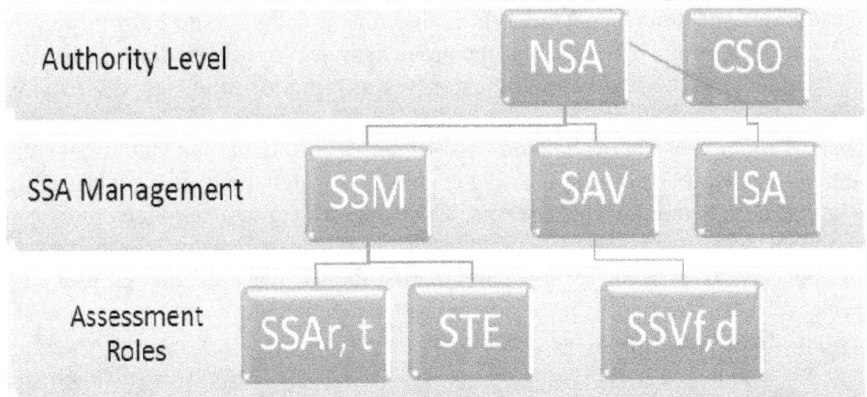

Fig. 10. Typical Organisational Structure for Safety Assessment

One key aspect to the SSA is the promotion of the culture of collaboration, work-sharing, transparency and efficiency. In this new paradigm, producers and duty holders communicate, engage with and collaborate with the regulator, independent assessment entities and the certification authority concurrently, a technology enabled culture change that is likely to change or transform the nature of the relationships that currently exist in the railway safety arena between stakeholders.

The Stakeholder Collaboration Workspace is the medium through which, it is anticipated that key stakeholders will have access to the approach, interim results and ongoing effort in developing, justifying and assessing safety of a product, process, system or undertaking. This workspace permits the justification against each clause/requirement to be backed up by any form of multi-media evidence that gets self-assessed by the provider and independently checked by other entities throughout the life cycle. The nature of evidence can be singular or multiple and will all be kept in a single accessible part of the SSA support environment.

This is a major culture change in the current context and should it happen, enabled by online secure access technology, is likely to change the focus from certification to getting the product, system or process safe as its development progresses. The implication will be that by the end of the development process and after commissioning phase, the product will be sufficiently scrutinised by key stakeholders to warrant approval and acceptance.

No specific aspects of the generic safety assessment criteria and framework are directly addressed by this phase since this is a largely productivity and collaboration environment.

4.2.5 Initiation of Activity-Evidence-Assessment Cycle (AEAC)

This phase involves an asynchronous and realistic application of the adopted standards, rules, regulations and codes of practice as the selected components of the Assessment Reference Model. In reality, this process should be followed by the duty holders and developers of the product, system or process through the implementation of the algorithmic components in terms of following the requirements (activity), capturing adequate and sufficient outcomes (evidence) and conducting evaluation and validation (assessment). This is carried out recursively until all the requirements of the ARM components (safety body of knowledge) are addressed, implemented and assessed. This is equivalent to a self-assessment and continual safety assurance carried out by the duty holders/developers. The task of Independent Assessment focused on checking the adequacy of the approach and evidence can subsequently be carried out by the assessment or certification authority potentially within the same collaboration environment.

This amounts to a non-sequential process and an incremental verification and continual validation that does not resemble the traditional life cycle model.

One advantage of a SSA architecture is the potential involvement of the assessment/certification authority in the course of this phase so that all activities, evidence and assessments are concurrently checked, corrected and endorsed leading to a highly efficient and collaborative approach unprecedented in the current setting.

Should the product, system or process be developed independently with claims about conformity with a given set of Assessment Reference Model or components, then the task of assessment is largely based on the evaluation of the evidence against the safety requirements in the algorithmically transformed components (ALC) since the activity aspect has been carried out by the duty holders. The supporting CORS environment allows association and attachment of evidence to the ALC components that further simplifies the task of independent assessment.

Given the hierarchical nature of the SSA phase two in algorithmic transformation of the ARM components, the evaluation, assessment, verification and validation can take place at any level and phase of the project based on the evidence, expert opinion and perceived/real priorities. This is likely to provide scalability, flexibility and reduce the so called waste factories so familiar in the safety domain.

The final consideration in the process of completing the evaluation of safety aspects of activity-evidence and assessment is also the major qualification and testing activities not limited to but including;

1-   Outcome of Factory Acceptance Test (FAT);
2-   Outcome of Site Acceptance Tests (SAT);
3-   Evidence that all safety requirements have been satisfactorily fulfilled.

These sources of often contractual compliance material contain evidence relevant to the safety of the product, system or process and should be included in the evidence considerations.

4.2.6 Safety Assessment Claims & Outcomes (SACO)

The cyclic process of activity-evidence-assessment (AEAC) in SSA will be performed asynchronously and in a Near Real-Time (NERT) basis until sufficient confidence is gained in the application of the ARM components. The algorithmic components essentially provide a road map for the conduct of the safety assessment while also indicating measures of significance for the attainment of sufficient confidence.

The selective and prioritized application of the algorithmic components (ALC) will generate an increasingly detailed view of the tasks in hand that should comprise a NERT perspective on:

1. what priority aspects have been achieved,
2. what are the remaining aspects,
3. what are the completed aspects that have proven unsatisfactory/unacceptable and need addressing,
4. what is the total level of confidence gained from what has been done at any cross-section in time.

This is in contrast with the current practice that is essentially sequential due to the nature of rules and standards and does not give an overall level of achievement until the whole process is completed. Also due to the risk/expert based prioritization of the assessment task using the ALC, this approach ensures work is primarily focused on the aspects that are logical and value added in the overall scheme of assessment.

The concurrent and non-sequential application of the ALC to the product, system, process or undertaking under development or assessment will generate the required deliverables, analysis, reports, tests and assurance components as required by the ALC or the original rules, standards, codes of practice and regulations. These will be arrived at much more efficiently and comprehensively than the traditional sequential process. The other key differentiator is the emphasis on activity and gathering of the evidence arising from the activity as opposed to documentation and often un-substantiated claims that are the currency of the current practice. The interim reports, strategic views of the degree of achievement and confidence can also provide a rich source of information and progression that are unprecedented in the culture of safety engineering and assessment.

In this spirit, SSA does not replace the body of knowledge required to engineer and assess a safety related/critical product, system or process. It strives to change the working environment into a collaborative setting whilst generating efficient NERT outcomes.

4.2.7 Develop incremental Safety Value Case

One critical shortcoming in the current approach to safety is lack of process and methodology to evaluate gains arising from the safety engineering and assurance activities. This creates a lopsided view of safety activities that are largely perceived as a cost issue/centre with no recognisable return other than receiving a deployment certificate.

This phase is intended to challenge and strive to remedy the current shortfall and develop an objective account of costs and value generated by the safety activities in the course of design, development, realisation and deployment of a product, system or process. This is referred to as Safety Value Case rather than the traditional Safety Case that does neither acknowledge the cost nor the benefits gained by following best practice, standards and codes of practice. The Safety Value Case is a separate deliverable of the SSA framework and is focused on the economic cost-benefit as possibly the social benefit dimensions that will complement a pure focus on technicalities of the traditional Safety Case.

The detailed requirements and process for the objective and multi-stakeholder driven evaluation of safety value and cost will be developed and presented in the next phase of this research project. All the requisite information for developing a Safety Value Case that complements the traditional Safety Case are generated by the safety analysis process so there will be no major need for a new cost accounting and value evaluation resource.

Whilst in view of the strong regulatory framework, compliance with the regulations and gaining safety approvals are common practice in the developed world, this is not generally the case elsewhere. The Safety Value Case is intended to objectively challenge the perception that safety is a cost to the duty holder, the customer and the society at large. It will go a fair way to provide a balanced perspective and encourage decision makers that engineering safe products, systems and processes makes good business sense as well as complying with the rules.

One final outcome of Safety Value Case is the development of arguments regarding the net balance of safety benefits and risks arising from the development or marketing of a product, system, process or undertaking. This is a largely missing component in the demonstration of compliance with the adversarial legal requirements that purely focus on risk minimization rather than balancing social benefit.

4.2.8 Capture Process Learning in Enterprise Repository

The SSA framework is principally founded on the Knowledge Management principles, recognising that the complexities of products, systems, services and undertakings pose a major challenge to the traditional model of assessment for simpler electro-mechanical systems. Two principal tenets of SSA are:

1. Timely application of relevant domain knowledge to the task of safety assessment;
2. A supportive collaborative culture between the principal stakeholders.

In this spirit, the process as well as the collaboration experience creates a new culture for doing the right thing at the right time by the right people to get safety right.

The capture, storage, reuse and enrichment of the new learning, algorithmic transformations of rules, standards, processes, assessment and project/product collaboration experience constitute a valuable outcome in this framework. The collaboration environment should provide a "corporate repository" or "process memory" for these valuable outcomes for re-use, enhancement and access by other stakeholders much in the spirit of Open Systems Development. This is justified on the grounds that safety is a social value/benefit and its experience and knowledge should be made openly available to the global stakeholders everywhere for sharing, enrichment and good social value.

Given the knowledge based nature of SSA, all experience, transformed reference models, assessments and process experience is captured in the form of reusable knowledge schema that lend themselves to re-use, assessment and customization for specific context and domains. The resultant knowledge schemas can ideally be stored in a suitably names and structure library in the Enterprise Repository.

The taxonomy of such libraries can be developed as a reference template for each product, system or process to ensure each project or undertaking focuses on the delivery of the knowledge components generated before the tasks are considered complete.

4.3 Expected Impact of SSA

The simultaneous attempt at developing a more credible and responsive process whilst also striving to influence a culture change for the inter-working between key stakeholders is expected to generate a number of benefits. These will be realized at different stages of development and degree of latency since culture change requires a shift in attitudes, behaviours and even the legal framework whilst the optimized process can deliver benefits more readily.

The expected impact of adopting the SSA philosophy is depicted below in no particular order of significance:

*   Continuous validation versus stepwise end of Life-Cycle (LC) sequence thus leading to an evolutionary level of confidence throughout the LC;
*   Evaluation and assessment on a continuous basis, initially based on expert opinion and increasingly updated based on evidence thus shifting from the document and claim centric current practice;

- Prioritised deployment of effort and resources based on perceptions of risk thus ensuring scarce resources are applied on aspects of highest value return to safety rather than the linear application of all the rules and requirements;
- Migration from "ticks in the box" towards a scientific scale of measurement thus enhanced resolution and credibility;
- Balanced view of cost and benefit/value to dispel the misconception about safety being only a cost and non-value generating activity that can be cut down or avoided to save costs and enhance profits;
- Near Real-Time (NERT) view of achievements and priority activities to enable timely action to be taken avoiding lengthy delays, waiting for audits and having a better understanding and control of the project;
- Determination of the shortfalls and deficits that can focus attention on the priority areas of shortfall in evidence or compliance with the safety requirements as the development progresses rather than stepwise audit periods;
- Collaboration between stakeholders that will engender a positive culture and re-enforcing of stakeholder knowledge towards a safer product, system or process/undertaking, saving on bureaucratic exchanges and protective attitudes and risk avoidance;
- Faster approvals due to work sharing since ideally approvals should take place when the product, system, process or undertaking is completed in physical works and is ready for deployment and operation. Unfortunately this is far from reality and often delays in tests, verification, validation, legal approval and certification end up frustrating the stakeholders' wish to deploy the completed works. The financial gains in such efficiency can be quite substantial even for small reduction in the time taken for acceptance and approval.
- Reduced risk of losing sight since lack of timely awareness of issues and shortfalls is the most wasteful aspect of development that wastes time and resources to undo the work done and correct the mistakes. It is generally agreed that the later the errors and mistakes are found, the higher the cost and impact on the product or project in terms of viability or even feasibility of fundamental change.
- 24/7 access and status that gives access to a view of the state of development/progress anytime and anywhere to all stakeholders. This could result in more timely identification and reaction to issues either technical, contractual or legal.
- Non adversarial focused on safety not certification which is probably a major aspect of culture shift in the current environment and international practice.

The realisation of benefits will be contingent on adoption, resourcing, training and development of competent staff, supporting tools and improving culture of collaboration. The success in this endeavor is contingent on the systemic attainment of improved process employing competent staff and operating within a supporting and re-enforcing environment where all stakeholders including the regulators focus on getting products safe, not documents polished for administrative or legal risk reasons.

5 Recommendations and Way Forward

The initial concepts and determination for the development of a smart approach to development of safe products, systems, services and undertakings arose out of observations about inefficiencies, high costs and concerns over paying lip-service to safety.

The approach and supporting processes highlighted in this document are a first attempt at identifying the wasteful aspects of the current practice that may give rise to misconceptions and even cutting corners in the development and certification of safety related products, systems and services. The so-alled Smart Safety Assessment architecture and supporting processes developed here come from a wide variety of sources, experiences in standardisation, safety regulation and safety engineering. The framework developed is also the outcome of systems view of the goal and how to reduce waste and enhance value in essence as well as in perception in the minds of key decisions makers and managers.

This is the first attempt at structuring and proposing a systemic perspective on developing, assuring, deploying, certifying and operating safety related complex systems and services. There is considerable scope in review, enhancement and further advancement of the principles and practices built into the SSA approach. Given safety is ultimately about avoiding unacceptable level of harm to people, it is a moral duty by all stakeholders to contribute to the betterment of safe performance.

The SSA is a small step in this context. It is expected that further attention on this demanding and challenging aspect of developing new and often complex products, systems and services that can be trusted for safe performance will encourage experts, researchers and practitioners to innovate and progress this cause further.

References

http://ec.europa.eu/idabc/en/document/7407.html
IEC 62278:2002 – GB/T 21562-2008. Railway Applications - Reliability, Availability, Maintainability and Safety
IEC 62279:2002 - GB/T 28808-2012. Railway Applications – Communications, Signalling & Processing Systems, Software for Railway Control & Protection
IEC 62425:2007 – GB/T 28809-2012. Railway Applications - Safety Related Electronic Systems for Signalling
IEC 62280-1:2002 – GB/T 24339.1-2009. Railway Applications – Communication, Signalling and Processing Systems-Safety Related Communications in Closed Transmission Systems
IEC 62280-2:2002 – GB/T 24339.2-2009. Railway Applications – Communication, Signalling and Processing Systems-Safety Related Communications in Open Transmission Systems
TR 50506-1:2007. Railway Applications-Communication, Signalling and Processing Systems-Application Guide for EN50129-Part 1: Cross-Acceptance
IEC 61882:2003. Hazard and Operability Studies (HAZOP studies) - Application Guide
IEEE/ISO/IEC 29148-2011. Systems and software engineering -- Life cycle processes -- Requirements engineering

A. G. Hessami and R. Gray (2003). Creativity, the Final Frontier, Proceedings of "The 3rd. European Conference on Knowledge Management", ECKM (2003)
http://www.lean.org/WhatsLean/Principles.cfm

Abbreviations

AEAC	Activity, Evidence, Assessment Cycle in SSA
ALC	Algorithmically transformed Components in SSA
ARM	Assessment Reference Model in SSA
CARA	Competence Assessment and Role Allocation in SSA
CCSC	Conditional Certificate of Safety Conformity
CMP	Configuration Management Plan
CORS	Collaboration Rules & Supporting environment in SSA
COTS	Commercial-Off-The-Shelf
CRS	Customer Requirements Specification
CSC	Certificate of Safety Conformity
CSNC	Certificate of Safety Non-Conformity
DRACAS	Data Reporting And Corrective Action System
FAT	Factory Acceptance Test
FMECA	Failure Mode Effects And Criticality Analysis
FRACAS	Failure Reporting And Corrective Actions System
FTI	Formal Technical Inspection
FTP	Field Trial Plan
FTR	Field Trial Report
FSR	Functional Safety Requirements
HAZAN	Hazard Analysis
HAZOP	Hazard And Operability Study
HRC	Human Resource Competence
HRCC	Human Resource Competence Certification
HSR	High Speed Railway
ICT	Information & Communications Technology
IHA	Interface Hazard Analysis
IPTL	Independent Product Test Laboratory
ISA	Independent Safety Assessor
IT	Information technology
LRU	Line Replaceable Unit
NCAA	National Certification & Accreditation Administration
NERT	Near Real-Time, a highly responsive system
OPSEC	Operational Scenarios
OSHA	Operation and System Hazard Analysis
PW	People Ware, function that have to be performed by humans in accordance with written procedures and rules
PHA	Preliminary Hazard Analysis
PLC	Programmable Logic Controller
PSP	Product, System or Process
QAP	Quality Assurance Plan
QMS	Quality Management System

RAM-P	RAM-Plan
RSAB	Railway Safety Assessment Board
RSCA	Railway Safety Certification Authority
SACO	Safety Assessment Claims & Outcomes in SSA
SAD	System Architecture Description
SADT	Structured Analysis and Design Techniques
SARP	Safety Assessment Review Panel (of experts)
SAP	Safety Plan
SAT	Site Acceptance Test
SHA	System Hazard Analysis
SIL	Safety Integrity level
SLSR	Railway System Level Safety Requirements
SMS	Safety Management System
SRS	System Requirements Specification
SSHA	Subsystem Hazard Analysis
SSRS	Subsystem Requirements Specification
THR	Tolerable Hazard Rate
TRAC	Algorithmic Transformation of ARM Components in SSA
VAP	Validation Plan
V&V	Verification & Validation
VTR	Validation Test Report
V&V	Verification & Validation

Accidents and Incidents:
Viewing the World through Data Eyes

Paul Hampton

CGI UK Ltd

London, UK

Mike Parsons

NATS

Fareham, UK

The role of data in influencing the safe operation of systems is just as important but has not attracted the same level of attention; there is no standardisation and little guidance on how the risks associated with data should be managed. There has been a marginalisation of data (inadvertent or otherwise) as a contributor to accidents and incidents, and it is clear there is an "elephant in the room" (Hampton and Parsons 2015). The problem is becoming more acute as many types of data are now used to specify, deploy, configure, operate, test and justify safety systems, moreover the volume of data in systems is also growing at an unprecedented rate. This paper is a retrospective reappraisal of selected historical accidents from the aviation and marine sectors, but viewed afresh from a data perspective. The paper shows that we do have a data problem; in fact we've always had a data problem.

1 Introduction

On the 9th May 2015 an Airbus A400M military transport aircraft took off from San Pablo Airport in Seville, Spain, on its first test flight but crashed shortly afterward. Surviving crew told investigators that "the aircraft had suffered multiple engine failures" (AINonline 2015). One would have been forgiven for thinking, prima facie, that this tragic accident was caused by catastrophic hardware failure or perhaps even a previously undetected software fault in, say, the engine management system. Investigations are still underway but early indications suggest (BBC 2015) that neither was the case: the engines were perfectly serviceable and the software was performing as designed. It is however, the accidental deletion of data files containing torque calibration parameters during a software installation process that is thought to be the root cause.

The parameter files are used by the Engine Control Units (ECUs). The ECUs have a critical part to play– they take the pilot's inputs and optimise the engine's performance under their control. To do this they need information about the torque generated by each engine and the parameter files are there to allow the ECU to interpret the sensor data. Without the files, the ECUs could not interpret this data and the crew could not command the desired power after takeoff.

Furthermore, the crew had tried switching what they considered to be malfunctioning engines into "flight idle" mode - their lowest power setting - in an attempt to tackle the problem. However, without the parameter files the engines were left stuck in this mode through design – intended as a safety feature to avoid out-of-control engines powering up.

This is a recent example where data has had a significant contribution to an accident. No doubt the software and hardware in this case would have met the relevant requirements for airworthiness certification, but what about the data files? If data's contribution to the accident is as suspected then clearly data should be subject to an equivalent level of assurance as the hardware and software.

In the marine sector, data as a contributory cause to an accident is rarely mentioned explicitly, but occasionally implied in regard to navigation errors. From an aviation perspective we often hear of pilot, maintenance or air traffic control errors leading to accidents; sometimes catastrophic hardware failures are also cited, less often "software glitches" or "computer problem", but data is rarely mentioned.

To really understand the contribution data has been making to accidents, we need to reassess historical accident reports to look afresh at them with "data eyes" – analyse sufficiently deeply to see whether data is a contributory factor.

For this retrospective analysis the aviation and marine sectors have been chosen as these have detailed and formal accident reports. The following sections will take these sectors in turn, provide insights into the common or recurring data related contributions to accidents, and will draw conclusions from the analysis.

An appendix gives details of the progress being currently made in introducing guidance into the management of data safety risks.

2 Aviation

For the aviation sector, accident and incident reports have been taken from the SKYbrary Accident and Incident Reports website (SKYbrary 2015). 762 reports were analysed covering a time period from 1969 to 2015.

2.1 Method

The approach to the analysis was to assess each report and categorise the factors that contributed to the resulting accident/incident in a significant way. There could of course be multiple factors; in these cases, each contributory factor has been assigned in equal proportions[1].

There are many ways in which accidents/incidents can be classified and the SKYbrary database has its own classification scheme. However, for the purposes of obtaining a meaningful data-centric classification, the following classifications were chosen[2]: Pilot Error, Procedural Error, Operations Error, Maintenance Error, Exceptional Events, Hardware, Software, Passenger/Cargo and Data (see Annex A for a full description of each type).

2.2 Analysis

The following figure shows the overall results of the analysis.

---

[1] . It is possible that this could be refined at a later stage by estimating a percentage contribution for each, but equal apportionment was considered sufficient for the purposes of the analysis.

[2] Note that the classifications of accidents and incidents are based on the authors' own opinions when reading the accident/incident reports.

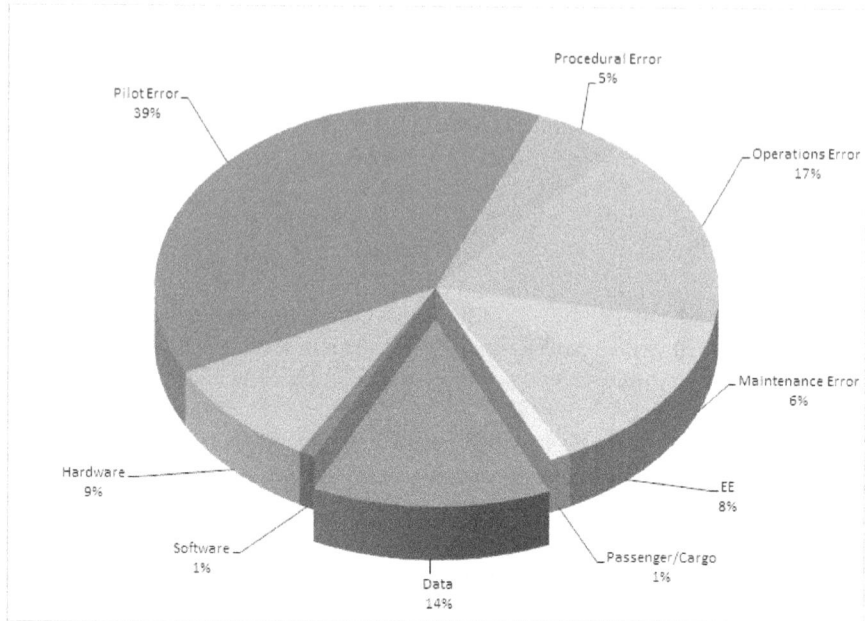

Fig. 1. Contributing Factors to Aviation Accidents/Incidents

While pilot error remains the greatest contributory factor, data has a proportional contribution greater than both software and hardware put together. This perhaps should not be surprising: much effort has been placed over the years in improving the reliability of software and hardware and data has not had the same level of attention. It should be noted that aviation data guidelines, (DO-200A 1998) have been in existence for some years but only cover some types of data.

So what is the nature of the historical data contributions? Each of the accident/incidents where data was a contributor were further analysed and classified. The following table shows the resulting classifications:

Table 1. Aviation Accident/Incident Data Type Classifications

Data Type	Description
Terrain Mapping	Data that records the height, profile, features and position of terrain and other obstacles.
Loading	Data recording the loading weights, locations and balances for cargo, crew, passengers and fuel.
Weather	Weather data relating to environmental conditions such as windspeeds, temperatures, convective weather positions and strengths, levels of runway contamination.

Data Type	Description
Pilot Nav Entry	Data used by pilots for the purposes of navigation such as barometric pressure values for altimeter adjustments, air traffic controller radio frequencies, compass heading values, navigational beacon names and locations.
Charts	Graphical data representing the physical world such as airport schematics, terrain maps, flight approach and takeoff profiles.
Tech Log	Data capturing issues noted by pilots for the attention of maintenance personnel.
Runway Configuration	Airport operational data relating to runway usage such as active runway selection schedules.
Maintenance Configuration	Aircraft configuration data set by maintenance staff.
Maintenance Documentation	Documentation used by maintenance staff such as service bulletins, wiring diagrams and maintenance procedures.
Pilot Performance Calculations	Data used in calculating the aircrafts takeoff and landing parameters such as flap & slat settings, thrust levels, roll speeds and required runway lengths.
Maintenance Technical Data	Aircraft data such as sensor telemetry data for use by maintenance staff.
Weather Radar Configuration	Configuration data that controls weather radar behavior.
Sensor Data	Data generated by sensors such as airspeed, altitude, cabin pressure, angle of attack, fuel gauges etc.
Notification Data	Advisory and warning data provided to pilots in form of NOTAMs.
Operator Documentation	Documentation relating to aircraft operations such as standard operating procedures.
ATC System Configuration	Data used to configure the systems used by air traffic controllers such as radar systems and warning systems.
Pilot Documentation	Documentation used by pilot such as Quick Reference Handbooks.
Airbridge Configuration	Data that controls the extension length of airbridges.
ATC System Data	The data used or presented by systems used by air traffic controllers.
Language	Controller/Pilot phraseology.

The following figure shows the proportion of accident/incidents assessed as being related to the data categories given in the previous table.

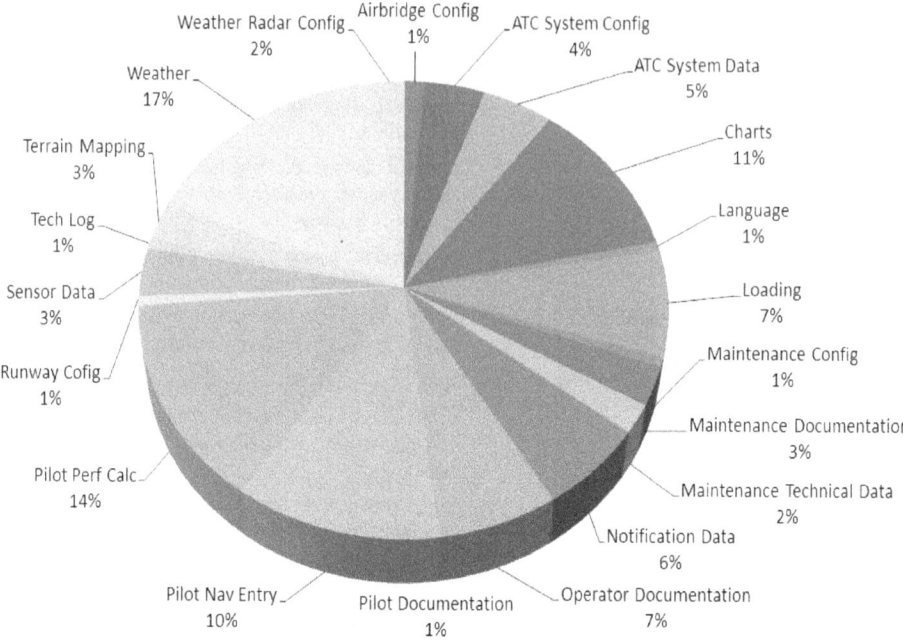

Fig. 2. Distribution of Data Contributing Factors to Aviation Accidents/Incidents

Hence, as can be seen in figure 2, some data types are recurrent contributors to aviation accidents/incidents. The following sections explore some of the more significant contributor types along with real examples to illustrate their significance.

2.2.1 Weather Data

For the crew, weather reports are significant for the safe operation of the aircraft as weather systems can bring significant risks.

On 9 June 2006, an Airbus 321-100, operated by Asiana Airlines, encountered a thunderstorm accompanied by hail at an altitude of 11,500 ft, while descending for an approach to Gimpo Airport, South Korea (ARAIB 2008). In the absence of up-to-date weather data, the crew flew into a convective weather system. The hail experienced was so severe it damaged the windscreen and caused the radome to detach.

Fig. 3. Hail damage from entering a convective weather system

Not only is there risk of physical damage to the aircraft, the noise and vibration caused by severe weather can render communication between pilots and controllers impossible and make it difficult for crew to read the aircraft instrumentation.

Information on the weather conditions at the destination airport is also critically important. In adverse weather situations, crew may need to make a decision quickly on whether it is safe to attempt a landing. For example, data values that record the level of windshear or tailwind are relied upon when making a decision to land or go-around (or indeed switch to the alternate destination entirely).

If this information is incorrect or absent, then this can have significant safety implications as can be seen in the following 4 examples.

Runway excursion on takeoff – wind information not disseminated to pilot (NTSB 2010)

Runway excursion on landing – inadequate report of runway braking conditions (NTSB 2007)

Runway excursion – runway braking action not as reported (BFU 2013)   Runway excursion – up-to-date tailwind information not used by pilots (TSB 2013).

Fig. 4. Implications of incorrect or absent weather data at destination airports

2.2.2 Pilot Performance Calculation Data

A detailed calculation needs to be made to determine how the aircraft should be "configured" for takeoff. This configuration includes thrust levels, rotation speed, the speed at which it is safe to abort the takeoff and whether the Centre of Gravity (CoG) for the aircraft is within safe limits. The CoG calculation factors in the weight of the aircraft when empty, the payload weight (passengers, crew, baggage and cargo) and its distribution, the ambient temperature, fuel weight, runway conditions, etc.

The calculations are now largely computerised either as fixed in-cabin or portable devices such as a PDA (see figure 5) or as an application on the crew's Electronic Flight Bag (EFB). Calculations can also be performed by ground based systems.

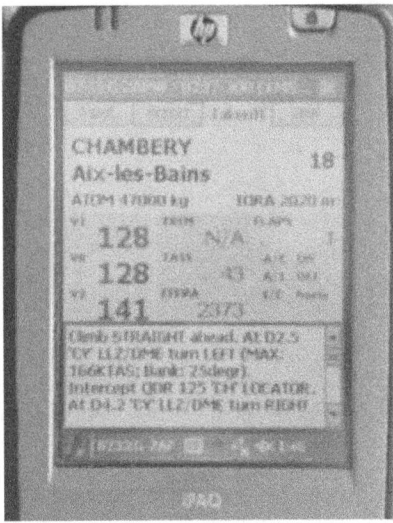

Fig. 5. Example of computerised support for performance calculations

While the correct operation of the software applications can be effectively tested, the actual data used as input in the calculations can significantly affect the outputs. If the input data is incorrect, then the calculated configuration could mean that the aircraft fails to become airborne, experiences a tail strike or overruns the runway.

Brussels, Belgium. Overrun on rejected takeoff due to birdstrike. However, aircraft lined up at the B1 intersection although the take-off parameters had been computed using the full length of the runway (AAIU 2009).

Cotonou, Benin. Aircraft had attempted take off at a higher weight than the crew believed was the case and that they had assumed a more aft centre of gravity for the elevator trim setting than was actually the case (BEA 2009).

Halifax, Nova Scotia. Crew calculated wrong V speeds and thrust settings using EFB (TSB 2006).

Chester, UK. Fuel starvation prior to landing EFB calculated fuel consumptions below actual. (AAIB 2014).

Fig. 6. Implications of incorrect data on performance calculations

2.2.3 Chart Data

Graphical data is intended to represent aspects of the physical world such as airport layouts, terrain maps, and flight approach/takeoff profiles. This information may be held as a physical paper chart or other document but could be held electronically in an onboard navigation system. These diverse forms and types of data are bundled together under the generic title of 'Chart Data'.

On 11 March 2005, an Airbus A321-200 operated by British Mediterranean Airways executed two unstable approaches below applicable minima in a dust storm to land in Khartoum Airport, Sudan (AAIB 2007). The pilots were attempting a Managed Non-Precision Approach (MNPA) to the runway; this requires the autopilot to follow an approach path defined by parameters stored in the aircraft's commercially supplied Flight Management and Guidance System (FMGC) navigation database.

The FMGC had recently been updated to reflect a change in the position of the final descent point to 4.4nm from the threshold. However, the pilot's approach chart (also commercially supplied) still retained the previous position at 5nm from the threshold. This led the pilots to believe they were high on approach and they disconnected the autopilot to increase the descent rate, which precipitated an unstable approach. This, in conjunction with other factors brought the aircraft in close proximity to the ground to the extent that an Enhanced Ground Proximity Warning System (EGPWS) audio warning was triggered.

The following figure shows the chart used by the pilots. The Hasan final approach waypoint is shown as 5nm from the threshold whereas the FMGC controlling the descent had the correct figure of 4.4 nm.

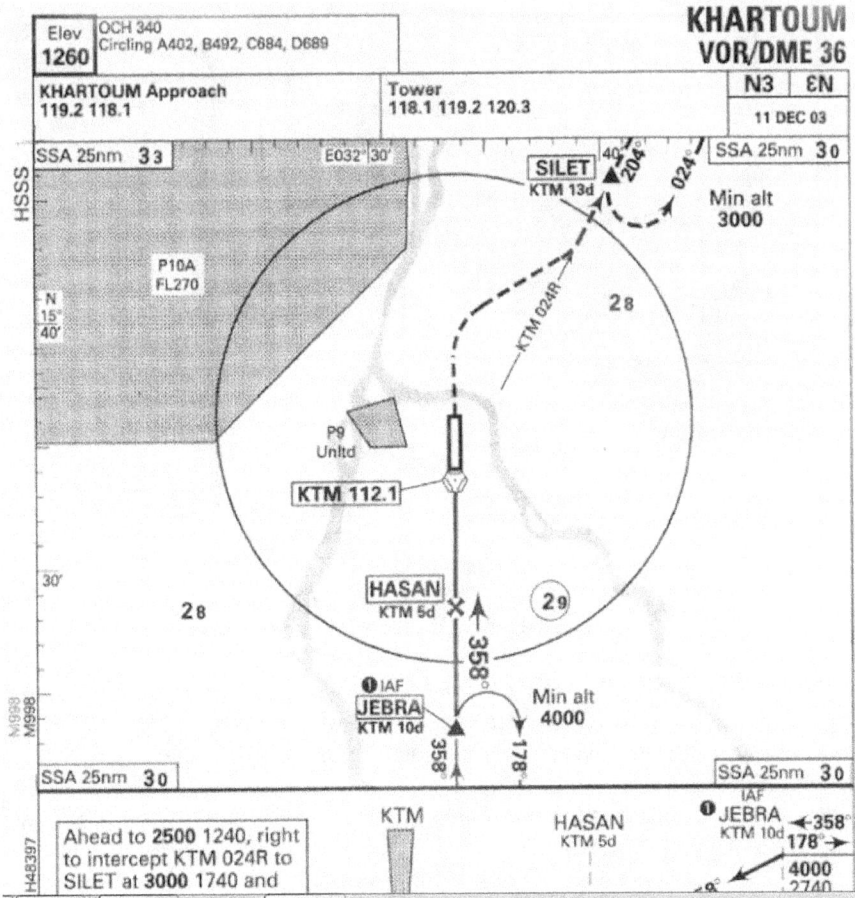

Fig. 7. Inaccurate chart used by pilots

In this example, confusion occurred as the data did not faithfully reflect the actual real world physical conditions. The final two examples in this section relate to the charts used to depict the layout of the airport showing runways, taxiways, buildings, obstacles and the visual cues such as lights, markings, stop bars etc. that the pilots can expect to see while taxiing.

The following figure shows an annotated aerodrome chart used by pilots at Nice airport in 2012.

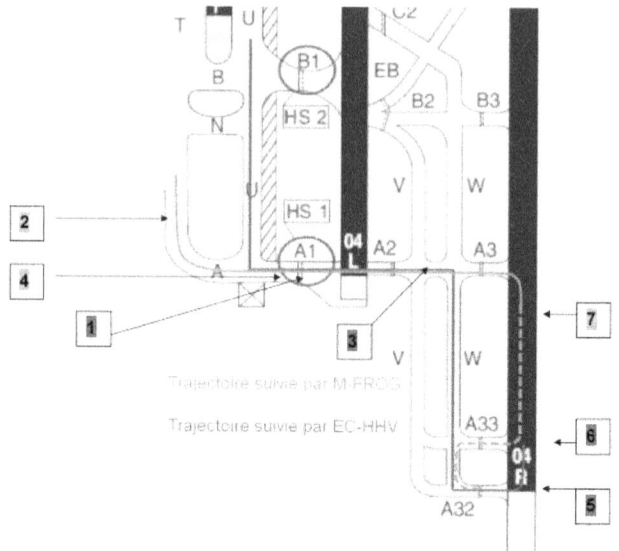

Fig. 8. Runway chart at Nice airport

A Bombardier CRJ200 had commenced its takeoff from runway 04R via taxiway A32 ahead of a Raytheon 390 Premier 1A, which having missed the correct taxi runway turn (into taxiway W) had found itself on active runway 04R from taxiway access A3.

The accident investigation concluded that both the AIP taxi chart (illustrated above) and the proprietary Jeppesen chart based on it, failed to correctly depict the detail of the movement area layout at the junction of taxiways and this was a key factor in contributing to the runway incursion.

In this case, the crew of the Bombardier were able to stop in time with no injuries. However, the crew and passengers in the next case were not so lucky.

On August 27 2006, a Bombardier CL-600 was taking off at night from Lexington, Kentucky when it inadvertently lined up for takeoff on the wrong runway, which was significantly shorter than the runway they were cleared to depart from (NTSB 2007). The crew noticed the error too late and could not get the aircraft airborne in time before colliding with the airports perimeter fence, trees and terrain. Out of 50 passengers and crew, only the first officer survived.

At the time the airport was undergoing a multiyear construction project, which included shifting a runway 325 feet to the southwest to accommodate a longer runway safety area. This involved relabeling of existing taxiway connectors, demolishing existing taxiways and creating new ones.

Although there were other contributory factors, it transpired that all the Comair pilots had been using an airport map with outdated information at the time of the crash. New maps with additional caveats were quickly released after the crash.

2.2.4 Pilot Navigation Data

As well as data being crucial in getting an aircraft airborne it is important in supporting the crew in safely navigating the aircraft to its destination. Also, incorrect navigation data may prolong journey times or cause increased workload for crew and air traffic controllers. An example is a Polish LOT flight from Heathrow on 7th June 2007 where longitude was entered as East instead of West into the two Inertial Reference Systems (AAIB 2008). This led to a meandering tour around the London airspace until the problem was detected.

With regards compass settings, less fortunate was the First Air Boeing 737-200 making an instrument approach to Resolute Bay, Canada on 20 August 2011 where an error of -8° in the compass systems, set during initial descent, contributed to an unstable approach and collision with terrain (TSB 2014).

By far the most frequent contributor to navigational accidents and incidents is the data used to calibrate the aircraft's altimeter. Barometric altimeters use the ambient air pressure to determine how far above sea level the aircraft is flying. As a plane ascends, the air pressure decreases and this can be used to determine the aircraft altitude. However, the barometric pressure is subject to the prevailing weather conditions and so the crew need to calibrate the altimeter with data provided to them. Around aerodromes, the calibration figures are provided by the air traffic controllers typically as the "QNH" code in hectopascals (hPa).

Errors in the QNH data can lead to the crew's instruments misleading the pilots as to their real altitude. On 31 December 2011 a USAF Beech 200 C12 on an instrument approach to Stornoway came into conflict with a Loganair Saab 340 on final approach. Traffic Collision Avoidance System (TCAS) warnings and the Saab pilot's initiative helped resolve the conflict, which would have otherwise brought the aircraft within 300 feet of each other (UK Airprox Board 2012).

The investigation found that the C12 crew had interpreted the QNH figure of 990 hPa given by air traffic control as 29.90 inches, the subscale setting units used in the USA.

On 11 January 1995 a Learjet 35 crashed into the sea while conducting an instrument approach to Masset, British Columbia, Canada (TSB 1995).

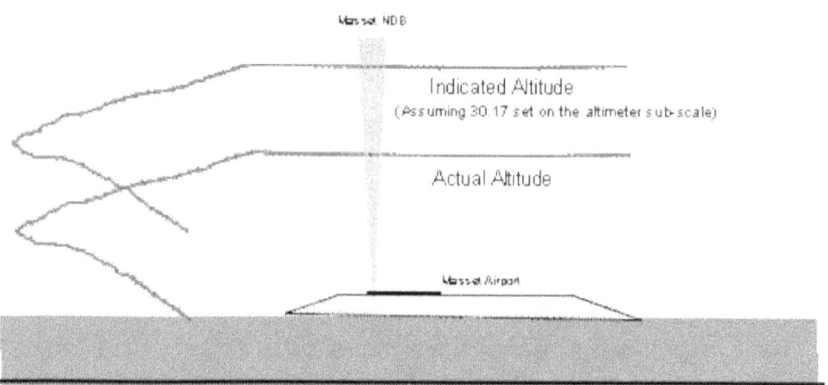

Fig. 10. Aircraft descent profile comparison

The figure above shows the actual (lower line) versus the indicated flight profile (top line) of the Learjet. The investigation concluded that the crew most likely conducted the instrument approach with reference to an unintentionally miss-set altimeter.

Even when heading and height data is correct there is still a large amount of data that pilots rely on for safe navigation such as controller radio frequencies, airspace class designators and navigational waypoint locations. Errors in each of these have had a part to play in some accidents and incidents.

The data used to support navigation can become critical when maneuvering through hazardous terrain in the absence of visual cues such as when flying at night or with poor meteorological visibility. On December 20 1995, a Boeing 757-200 flying at night from Miami to Cali in Colombia, was running behind schedule and the pilots accepted an alternative runway suggested by the controller on approach to Cali to make up time. The pilots however inadvertently cleared all the navigational way points from the aircraft's flight management system and the pilots had to reprogramme the waypoints by hand with the assistance of paper maps. Having discovered that they had just flown over the Tulua approach waypoint, they attempted to program the navigation computer for the next approach waypoint, Rozo, which was identified as R on their charts.

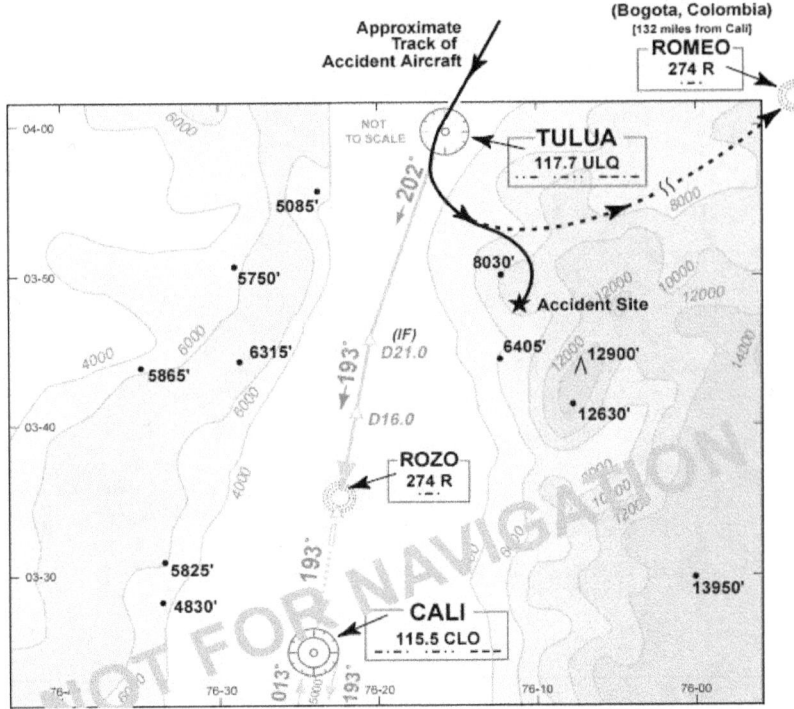

Fig. 11. Chart for approach to Cali, Colombia

However, Colombia had duplicated the identifier for the Romeo beacon near Bogotá, and the computer's list of stored waypoints did not include Rozo as "R", but only under its full name "ROZO". Where duplicate identifiers are allowed, these are often listed with the largest city first. By picking the first "R" from the list, the captain caused the autopilot to start flying a course to Bogotá, resulting in the airplane taking a wide turn east putting it on a collision course with a 3000m high mountain. Despite the Ground Proximity Warning System being activated, the crew were unable to avoid impact with terrain in time.

2.2.5 Loading

Information on factors such as the weight and distribution of the passengers, crew, payload and cargo are normally captured on a loading sheet as shown in the figure below:

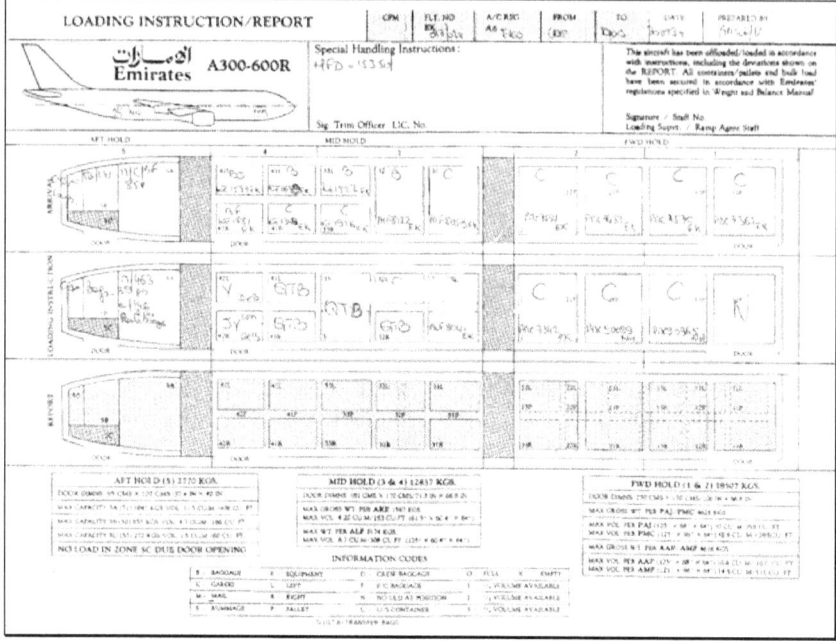

Fig. 12. Example of a loading sheet

On 30 July 1997, an Airbus A300-600 departing from Paris Charles de Gaulle, pitched up during the takeoff roll and its tail struck the ground violently.

The incident occurred due to the incorrect distribution of weight in the aircraft, with the centre of gravity a long way aft of the authorised limit. This distribution was the consequence of a ground agent keying error at the beginning of the process of calculating the loadsheet.

Another loading issue with a Qantas Boeing 737 occurred as recently as 2014 (Hampton P, Parsons M 2015). For this flight, there were 150 passengers, 87 of which were primary school children. These children were all seated together at the rear of the cabin. All had been mistakenly assigned an adult weight of 87 kg (ATSB 2014). During take-off the aircraft appeared nose heavy. Significant back pressure was required to rotate the aircraft and lift off from the runway. The aircraft exceeded the calculated take-off safety speed by about 25 kt.

There are also hazards associated with the loading of cargo planes. Consider the case in 2013 in Papua New Guinea (PNG AIC 2013) where a cargo plane loaded with 330 cartons of locally-manufactured cigarettes had been placed on the aircraft, each containing between 5000 and 10,000 cigarettes. The assumed weight of each carton was 12 kg but the shipper subsequently provided the investigation with average carton weights and it was found that the majority of cartons loaded had been 10% heavier than their assumed weight thereby contributing to an actual aircraft weight nearly 600 kg more than the crew's calculated maximum. The resulting overweight led the pilots to have difficulty getting airborne and the aircraft overran the runway during the subsequent attempt to abort the takeoff.

Fig. 13. Runway excursion – overweight during loading

3 Marine

3.1 Method

Publically available marine accident reports from the UK Marine Accident Investigation Branch website (MAIB 2015) were analysed to look for data aspects. 40 reports were examined from the last 3 years; these yielded 7 accidents where data was considered a factor by the authors of this paper, either in the actual accident event itself or in the subsequent investigation. Further analysis of historical accidents would undoubtedly provide more examples. Note that there is no claim being made that data was the primary factor in these accidents, but it was a certainly a contributory aspect.

3.2 Analysis

Within the 40 reports considered, the data involved in these accidents was found to be of 6 types:

1. Hydrographic survey data;
2. Passenger data;
3. Records of certification, maintenance and crew data;
4. Vessel identification data and chart display;
5. Navigational fix data (and falsified data);
6. Propulsion parameters.

3.2.1 Hydrographic Survey Data

Out of date or inaccurate hydrographic survey data is clearly a major safety issue for shipping as it is can lead to vessels encountering underwater hazards, leading to severe hull damage.

The ro-ro ferry Commodore Clipper grounded in the Little Russel channel approaching St Peter Port, Guernsey on 14 July 2014. The vessel raked over two granite pinnacles at full sea speed of about 18 knots; its hull was breached and seawater flooded into double-bottom void spaces (Commodore Clipper 2015).

Fig. 14. The Commodore Clipper

The vessel grounded on a rocky shoal charted at 5.2m. This depth was based on 1960's data, and the information on the Electronic Navigational Chart (ENC) defined it as CATZOC[1] B, which meant an error of +/- 1.2m should have been applied. While a more recent (March 2014) survey established a depth of 4.6m in the area of the grounding, this was within the tolerance denoted by the original CATZOC. Had the bridge team produced a berth to berth voyage plan, taking the chart accuracy into consideration, a worst case depth of 4.0m (5.2m – 1.2m) for the rocky shoals adjacent to Roustel and Boue de la Rade would have been applied; the area where the vessel grounded would have been identified as unsafe and avoided. This is shown in figure 15 below:

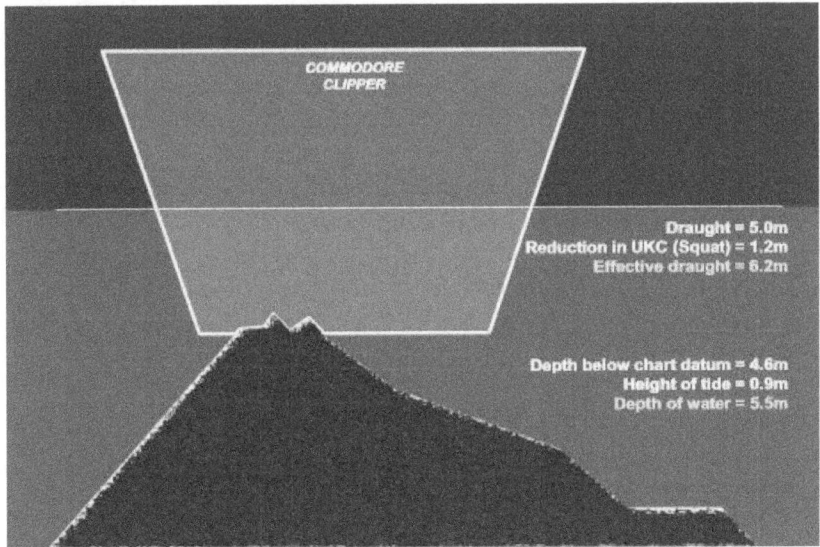

Fig. 15. Assessment of draught

In addition the route planning was considered deficient. The master had selected company 'Route 06' for Commodore Clipper's voyage to St Peter Port, but neither the track nor the ECDIS (Electronic Chart Display and Information System) safety

---

[1] CATZOC (Category of Zone of Confidence):
A1 - All significant seafloor features detected; very high accuracy survey
A2 - All significant seafloor features detected; high accuracy survey
B - Uncharted features dangerous to navigation are not expected but may exist; medium accuracy survey
C - Depth anomalies may be expected; low accuracy survey or passage soundings
D - Large depth anomalies may be expected; poor quality data
U - Quality of bathymetry yet to be assessed

features were checked as safe. The ECDIS safety contour[1], safety depth and XTD (Cross Track Distance) should have been adjusted for each passage taking the local conditions into account. The ECDIS safety depth setting of 7m was appropriate for the voyage, but the safety contour value of 5m was not. The safety contour could have been set at a minimum value of 6.3m and the ECDIS would have defaulted to the next deeper ENC contour of 10m. However, a 10m safety contour would have given the impression that the Little Russel was impassable. It would have been possible to adjust this for local conditions, but this was apparently never done.

SOLAS (Safety of Life at Sea convention) requires that governments make every effort to ensure that hydrographic surveying is carried out, adequate to the requirements of safe navigation. At the time of the grounding, data from the survey conducted in March 2014, which showed a depth of 4.6m in the grounding location, was still being processed by UKHO (UK Hydrographic Office). This would have produced more accurate results, but in fact had the old data been used with the larger error bounds the same safety margin would have resulted. Hence this is actually *not* an issue with the survey data itself, but with the *use* of the survey data in planning.

3.2.2 Passenger Data

Not knowing the number of passengers on board a vessel is clearly a safety issue, as in the event of an emergency it is unclear whether search activities (which may create additional risk) should be undertaken in looking for passengers who may or may not be on board.

A class V passenger vessel, Millennium Time, collided with the motor tug Redoubt and its tow on 17 July 2014 on the Kings Reach area of the River Thames. There were no serious injuries or significant pollution but both vessels suffered moderate damage and were out of service for a number of weeks (Millenium Time 2015).

---

[1] On ECDIS the user needs to select a safety contour. If a safety contour is entered that is not available on the particular ENC in use then the system will use the next deepest available contour. If within a specified time set by the user, the ship is about to cross the safety contour an alarm will sound. (NI 2015)

Fig. 16. The Millennium Time

The Thames Safety Inquiry (Thames 2000) highlighted the importance of knowing the number of passengers on board a passenger vessel in the event of an emergency, yet the number of passengers reported to be on board Millennium Time when this collision occurred was very inaccurate. The number of passengers counted by the MPU (Marine Policing Unit) on board Millennium Time following the collision with Redoubt was 362. This was 64 people fewer than the number transmitted on the vessel's AIS (Automatic Identification System), which indicated that 426 passengers were on board. It is evident that this discrepancy was due to the AIS having not been updated following the disembarkation and embarkation of passengers at Westminster and Waterloo piers. Consequently, the vessel did not meet the relevant regulation, which requires that an accurate record is held on board and ashore at all times. Even had the passenger numbers been updated on Millennium Time following each of its stops, it is unlikely that the passenger count would have been correct. In common with many operators on the Thames, City Cruises relied on its crew to use a hand-operated tally counter to count the passengers on board its vessels. Counting thousands of passengers on and off over a dozen stops each day using this method is extremely difficult and is prone to error.

3.2.3 Certification, Maintenance and Crew Data

It is important that records of important activities such as crew training and maintenance on-board ship reflect reality. In this case the records for these types of activities were found to be fictitious, contributing to an injury.

On 9 April 2014 a lifeboat on the refrigerated cargo vessel Nagato Reefer fell from its davit while being secured, following an abandon ship drill. A crewman was injured and the lifeboat was damaged as a result of the accident (Nagato Reefer 2015).

Fig. 17. The Nagato Reefer

During the inspection the PSCO (Port State Control Office) noted the following deficiencies:

- Certificates and flag state endorsements for some of the officers had expired;
- On board training – officers were not familiar with the operation of essential safety equipment;
- Records of hours of rest were found to have been falsified.

The lack of competence displayed by the crew when required to carry out emergency drills as part of the PSC (Port State Control) inspection was also evident when additional fire-fighting training was provided to them following the accident. Some of the crew had been on board Nagato Reefer for 19 months, and the documentation provided to the vessel regarding the operation of the LRRS (Lifeboat Release and Retrieval Systems) was sufficient to form the basis of a crew training programme in the safe operation of the lifeboats and their release gear. Despite this, no training had taken place and the investigation found that the training records on board had been falsified to indicate otherwise.

Comprehensive maintenance schedules for the lifeboats and the LRRS had been incorporated into the vessel's SMS (Safety Management System). The crew were required to inspect the release gear on a monthly basis; this included a requirement to check the "status of the reset" and "that there was no dirt or foreign matter on the moving part" [sic]. The vessel's maintenance records indicated that the release gear had been checked and was in "good" condition just a few weeks before the accident. In reality, the moving parts of the hook release mechanism on the port lifeboat had been painted and were dirty, and the reset indicator had been painted over. Additionally, the release gear cables were found seized and damaged when inspected after the accident. The release gear on the starboard lifeboat was in a similarly poor condition, requiring an hour of effort to free it during the drill held on 12 April, 3 days after the accident. It is apparent, despite records to the contrary, that no maintenance or inspections had been carried out since the annual inspection and service in October 2013.

3.2.4 Vessel Identification Data and Chart Display

If a system appears to work well, people get used to it and become over-reliant on it (Reason 1997). This applies to many systems and includes data. In this case an AIS which is an automatic tracking system used on ships and by vessel traffic services (VTS) for identifying and locating vessels by electronically exchanging data with other nearby ships, AIS base stations, and satellites. AIS data is supposed to supplement marine radar, which should be the primary method of collision avoidance for water transport (AIS 2015).

Two examples are given below:

Karen and Sapphire Stone

On 22 January 2014, the fishing vessels Sapphire Stone and Karen collided. At the time of the collision, Sapphire Stone was steering a north-westerly course towards Campbeltown, Scotland to land its catch, while Karen was towing its nets on a westerly course. Karen was struck on its port quarter and was severely damaged. Its hull was opened to the sea, which caused the aft crew accommodation and main engine room to flood rapidly, resulting in its foundering within 3 minutes of the collision occurring (Karen and Sapphire Stone 2014).

Fig. 18. Karen and Sapphire Stone

Since fitting an AIS unit and integrating it with the vessel's chart plotter Sapphire Stone's skipper had increasingly relied on AIS information for detecting and monitoring other vessels. Karen was not fitted (nor was required to be fitted) with AIS, which meant that Sapphire Stone's skipper was neither alerted to Karen's presence, nor able to assess whether risk of collision existed in the manner to which he had become accustomed. Due to the speed of the sinking no radio alert was issued.

The skipper had attempted to send a DSC (Digital Selective Calling) distress alert. This required him to press the activation button for a number of seconds (a feature to prevent false alerts); however, he became increasingly concerned about the vessel's stern trim and he left the wheelhouse before the alert was activated.

The absence of a DSC alert and/or distress message being transmitted on VHF radio channel 16 caused an unnecessary delay in the coastguard responding to the emergency which, under different circumstances, could have led to far more serious consequences.

(Note the successful activation, transmission, and receipt of the alert from Karen's EPIRB (Emergency Position Indicating Radio Beacon) demonstrates the value of fitting the equipment.)

Rickmers Dubai and Walcon Wizard

The multi-purpose cargo ship Rickmers Dubai collided with the un-manned crane barge Walcon Wizard which was being towed by the tug Kingston in the southwest traffic lane of the Dover Strait TSS (Traffic Separation Scheme) (Dover 2015). The accident occurred as Rickmers Dubai was overtaking the tug and its tow (Rickmers Dubai 2014).

Fig. 19. Rickmers Dubai

Fig. 20. Walcon Wizard and Kingston

Immediately following the collision, the towline caught on Rickmers Dubai and Kingston was towed stern-first through the water until the towline ran free from its tow winch. Walcon Wizard was badly damaged and Rickmers Dubai's hull was punctured above the waterline.

In view of Rickmers Dubai's OOW (Officer of the Watch)'s comments to Dover Coastguard following the collision, which highlighted that Kingston had not been transmitting on AIS, it is almost certain that his late detection of Kingston and his ignorance of the proximity of Walcon Wizard were due to his reliance on AIS information shown on the ECDIS.

ECDIS is capable of providing a wealth of information to the user, including charts, waypoints, safe water and overlaid AIS information. However, it should be used in conjunction with other aids to navigation and collision avoidance, particularly radar and visual lookout. In this case, the OOW had set up the X-band radar display at the start of his watch. However, VDR (Voyage Data Recorder) evidence indicates that the OOW did not see the radar targets of Kingston and Walcon Wizard, or that he saw them but did not acquire them using ARPA (Automatic Radar Plotting Aid). The absence of AIS information for the radar targets associated with Kingston and Walcon Wizard should have prompted the OOW to use ARPA to determine if a risk of collision existed. The detection of radar targets directly ahead and closing should also have prompted the OOW to look out of the window and attempt to correlate the targets with visual information. As the OOW took neither of these actions, and the targets were on the X-band display for almost one hour, it is likely that the OOW was not monitoring the radar display.

The second officer's reliance on the AIS information displayed on ECDIS for collision avoidance strongly indicates that he was not aware that many vessels, such as small fishing vessels, leisure craft, warships and vessels under 300gt might not be displayed.

The report also noted that the USB dongle used to save the VDR data following the accident had previously been used to store a movie. The abuse of dedicated VDR equipment in this way, whether by ship's crews or service technicians, risks corrupting valuable data.

3.2.6 Fix Data and Falsified Data

Not only did the second officer make some navigational errors in this case, but he then falsified records to try and cover his tracks. When this sort of activity happens it is apparent that data might be used to confuse investigators and shift the blame for an accident, thereby causing recommendations to be misplaced or missed altogether, potentially leading to further accidents.

On 3 January 2014 the Liberia registered liquefied gas carrier, Navigator Scorpio, ran aground on Haisborough Sand in the North Sea, England. The vessel was undamaged by the grounding; 2.5 hours later, it refloated on the rising tide. The investigation found that the vessel ran aground in restricted waters after the officer of the watch had become distracted and lost positional awareness (Navigator Scorpio 2014).

Fig. 21. Navigator Scorpio

When the 2/O (Second Officer) arrived on the bridge to take over the watch, he did not check to identify what sightings, particularly navigation marks, would be made or the potential dangers that lay ahead for the watch. He also made no assessment of the expected effects of tidal stream or wind. Both these actions are standard navigation practices and were a requirement of the vessel's safety management system. As the vessel approached the turn to 283°T[1], the 2/O was not expecting the radar waypoint alarm as he had no appreciation of the significant northerly set affecting the vessel. Thus, his decision to check the position by plotting a fix was understandable but unhelpful. When the decision was eventually taken to alter course, the vessel was already well to starboard of the new track. The 2/O appreciated this but did not take a fix immediately after completing the turn. This lost him the opportunity to build up a plot of the course made good or calculate an effective EP (Estimated Position); either of these actions would have been opportunities to determine an appropriate course-to-steer. Consequently, his choice of 270°T and the subsequent adjustment to 267°T were not effective in regaining track.

Although it was correctly recorded in the log, the fix taken at 1515 was incorrectly plotted 1 mile to the south of the vessel's actual position. When plotting the fix, the OOW's understanding of the situation was that the vessel was regaining track. As a result, it is highly likely that he plotted the fix where he perceived the vessel to be based on this incorrect assessment.

Having realised the vessel was aground and concerned that he had not followed the master's instruction to fix at 5 minute intervals, the 2/O fabricated two fixes on the chart, shown at 1505 and 1510. The 2/O's intent was to create a chart which

---

[1] True bearing

showed fixes at 5 minute intervals to give the impression that he had followed the master's instructions. The post-grounding actions of the 2/O are not uncommon; similar chart alterations post-casualty are often discovered during MAIB (Marine Accident Investigation Branch) accident investigations. The chart in use at the time of a marine casualty is critical evidence and, in the interests of safety and in accordance with accident reporting regulations, should not be altered after the grounding.

3.2.7 Propulsion Parameters

Many functions onboard a modern ship are computer controlled and these often rely on specific calibration parameters. In this case it appears that some crucial parameters were incorrectly set some years before the accident and never fixed.

On 20 December 2013, the 93m chemical tanker Key Bora made heavy contact with the western approach jetty at Alexandra Dock, Hull. The vessel's CPP (Controllable Pitch Propeller) system had a history of responding slowly to demands for astern pitch, and did not respond in time to the pilot's order of full astern to prevent the bow striking the quay. The bulbous bow was holed above the waterline (Key Bora 2013).

Fig. 22. Key Bora

Key Bora made heavy contact with the jetty because its CPP did not respond adequately to the full astern order from the pilot. The strong tidal stream and the delay in dropping the anchor caused by the communication difficulties between the bridge and the anchor station were contributory factors. Key Bora's recorded history of inadequate astern response dates back to November 2010. However, it is almost certain that the CPP anomalies were introduced when the system was commissioned. It was confirmed that the values of the CPP control parameters recorded in March 2014 were incorrect. These parameters were most likely set during commissioning as the vessel's crew were never provided with the password to access the system, and the only time a CPP service engineer had visited the vessel since commissioning was after the accident.

4 Conclusions

Transportation involves the construction of large and complex high energy physical systems. The safety of the passengers and crew and other individuals on the water and in the air can depend on the correct operation of these systems. Industries such as aviation have spent many decades in identifying and mitigated these risks and this had led to a body of standards and guidance to improve the reliability of systems. Reliability of the hardware has been a key focus and then software as it increasingly augmented and in some cases replaced the hardware functions.

The contribution of the people and procedures involved in the system have also been subject to scrutiny and the total corpus of work has led to significant changes in these industries.

The analysis has shown that, when we reassess these accidents and incidents from a data perspective, there is, and has always been, a significant contribution of data to the causal accident or incident chain. It is not that the reports have ignored these contributions; it is that the data issue tends to be masked by one of the other established factors such as human error.

The analysis has shown that the data contributions are considerable – as with the A400M example, the hardware and software assurance rigour can be completely undermined by the (mis)handling of data with catastrophic results. Furthermore, data contributions to accidents are not isolated occurrences: in the aviation reports studied, it is the view of the authors that data is implicated in accidents and incidents more often than both hardware and software put together.

We therefore conclude from an analysis of historic reference material that data safety is not a new issue; it has been contributor to accidents and incidents over many years but has not been subject to the same level of attention as other factors.

5 Further Work

This work has focused on two sectors where there have been good reference sources of accidents and incidents. There are other sectors, such as Healthcare, where data is known to be an issue and a similar analysis of accidents from a data perspective is a possible area for future consideration. A further study area is to allocate a percentage to each factor to reflect their contribution to the accident/incident.

The work has focussed on showing the extent to which data has contributed to accidents and incidents and implies that there is need for more guidance on how the risks associated with data should be managed. Annex B describes the work that an industry wide working group has been doing over the last two years to help address this and further work in developing standardisation is a key area.

Acknowledgments
Figure 13: reproduced from Flight Safety Foundation, Flight Safety Digest May-June 1998, Author: David A. Simmon.
All other figures are reproduced from the respective referenced accident reports.

References
AAIB (2007) AAIB Bulletin: Aircraft Incident Report No: 5/2007 (EW/C2005/03/02) http://www.skybrary.aero/bookshelf/books/878.pdf
AAIB (2008) AAIB Bulletin: Aircraft Incident Report No: AAIB Bulletin: 6/2008 (EW/C2007/06/02) http://www.skybrary.aero/bookshelf/books/626.pdf
AAIB (2014) AAIB Bulletin: 11/2014 G-BXUY EW/C2013/11/03. http://www.skybrary.aero/bookshelf/books/3033.pdf
AAIU (2009) Final Report on the accident occurred on 25 May 2008 At Brussels Airport. http://www.skybrary.aero/bookshelf/books/835.pdf
AINOnline (2015) http://www.ainonline.com/aviation-news/defense/2015-05-11/fatal-a400m-crash-disrupts-production-recovery-plan, Accessed 24th August 2015.
AIS (2015) Automatic Identification System, https://en.wikipedia.org/wiki/Automatic_Identification_System, Accessed 28 Sep 2015.
ARAIB (2008) Aircraft Accident Report, Hail Encounter During Approach, Asiana Airlines Flight 8942, A321-100, HL7594 20 miles South East of Anyang VOR, June 9, 2006. http://www.skybrary.aero/bookshelf/books/707.pdf
ATSB (2007) Aviation Occurrence Report 200501977: Collision with Terrain 11 km NW Lockhart River Aerodrome. http://www.skybrary.aero/bookshelf/books/1121.pdf
ATSB (2014), Loading issue involving a Boeing 737, VH-VZO at Canberra Airport, ACT. Australian Transport Safety Bureau. http://www.atsb.gov.au/publications/investigation_reports/2014/aair/ao-2014-088.aspx. Accessed 24 Oct 2014.
BBC (2015) http://www.bbc.co.uk/news/technology-33078767, Accessed 24th August 2015.
BEA (2001) Rapport Incident grave survenu le 29 mars 2010 sur l'aérodrome de Nice. http://www.skybrary.aero/bookshelf/books/2722.pdf
BEA (2009) Accident on 25 December 2003 at Cotonou Cadjèhoun aerodrome (Benin). http://www.bea.aero/docspa/2003/3x-o031225a/pdf/3x-o031225a.pdf
BFU (2013) Investigation Report. http://www.skybrary.aero/bookshelf/books/2742.pdf
CAD (2011), Report on the Serious Incident of an Attempted Take-off on Taxiway Flight FIN070 Airbus A340-300 OH-LQD at the Hong Kong International Airport. http://www.skybrary.aero/bookshelf/books/1832.pdf
Commodore Clipper (2015), Grounding and flooding of ro-ro ferry Commodore Clipper, https://www.gov.uk/maib-reports/grounding-and-flooding-of-ro-ro-ferry-commodore-clipper, accessed 28 Sep 2015.

DO-200A (1998) DO-200A Standards for Processing Aeronautical Data, RTCA

Dover (2015), Dover Strait crossings: channel navigation information service (CNIS), https://www.gov.uk/government/publications/dover-strait-crossings-channel-navigation-information-service/dover-strait-crossings-channel-navigation-information-service-cnis#tss, Accessed 25 Oct 2015

DSIWG (2014) Data Safety, SCSC. http://scsc.org.uk/paper_127/Data Safety (Version 1.0).pdf?pap=954 , Accessed 6 Sep 2015

DSIWG (2015) Data Safety, SCSC. http://scsc.org.uk/paper_128/Data Safety (Version 1.2).pdf?pap=958 , Accessed 6 Sep 2015

Hampton P, Parsons M (2015) The Data Elephant. In Parsons M, Anderson T (eds) Engineering Systems for Safety. SCSC

Karen and Sapphire Stone (2014) Collision between stern trawlers Karen and Sapphire Stone resulting in Karen sinking, https://www.gov.uk/maib-reports/collision-between-stern-trawlers-karen-and-sapphire-stone-off-cambeltown-scotland-resulting-in-the-loss-of-karen, accessed 28 Sep 2015.

Key Bora (2013), Failure of controllable pitch propeller of chemical tanker Key Bora resulting in contact with jetty https://www.gov.uk/maib-reports/failure-of-the-controllable-pitch-propeller-of-chemical-tanker-key-bora-resulting-in-heavy-contact-with-jetty-at-alexandra-dock-hull-england, Accessed 28 Sep 2015.

MAA (2014) Interim report from the service inquiry investigating the incident involving voyager ZZ333 on 9 Feb 14, http://www.skybrary.aero/bookshelf/books/2718.pdf NTSB (2007) Runway Overrun and Collision Southwest Airlines Flight 1248. http://www.skybrary.aero/bookshelf/books/2445.pdf

MAIB (2015) Marine Accident Investigation Branch reports, https://www.gov.uk/maib-reports, accessed 28 Sep 2015.

Millenium Time (2015), Collision between passenger vessel Millennium Time and motor tug Redoubt, https://www.gov.uk/maib-reports/collision-between-passenger-vessel-millennium-time-and-motor-tug-redoubt, accessed 28 Sep 2015.

Navigator Scorpio (2014), Grounding of liquefied gas carrier Navigator Scorpio https://www.gov.uk/maib-reports/grounding-of-liquefied-gas-carrier-navigator-scorpio-on-haisborough-sand-off-norfolk-england, Accessed 28 Sep 2015.

Negato Reefer (2015), Accidental release of lifeboat on Nagato Reefer with 1 person injured, https://www.gov.uk/maib-reports/accidental-release-of-lifeboat-on-nagato-reefer-with-1-person-injured, Accessed 28 Sep 2015

NI (2015), Safety contours, http://www.nautinst.org/en/forums/ecdis/ecdis-issues--enc.cfm/E9safetycontours , accessed 25th Oct 2015

NTSB (2007) Aviation Accident Report Attempted Takeoff From Wrong Runway Comair Flight 5191. http://www.skybrary.aero/bookshelf/books/1062.pdf

NTSB (2010) Aviation Accident Report Runway Side Excursion During Attempted Takeoff in Strong and Gusty Crosswind Conditions. http://www.skybrary.aero/bookshelf/books/1327.pdf

PNG AIC (2013) Papua New Guinea Accident Investigation Commission Preliminary Report AIC13-1007 http://www.skybrary.aero/bookshelf/books/3121.pdf

Reason (1997) Managing the Risks of Organizational Accidents, James Reason, Ashgate Publishing Limited, ISBN-13: 978-1840141054

Rickmers Dubai (2014) Collision between multipurpose cargo vessel Rickmers Dubai with a crane barge, Walcon Wizard, being towed by tug Kingston, https://www.gov.uk/maib-reports/collision-between-multipurpose-dry-cargo-vessel-rickmers-dubai-with-crane-barge-walcon-wizard-being-towed-by-tug-kingston-in-the-dover-strait-off-the-south-east-coast-of-england, Accessed 28 Sep 2015

SHT (2007) report on the aircraft accident at Bodø airport on 4 December 2003 involving Dornier DO 228-202 LN-HTA, operated by Kato Airline AS. http://www.skybrary.aero/bookshelf/books/329.pdf

SKYbrary (2015) http://www.skybrary.aero/index.php/Category:Accidents_and_Incidents, Accessed 24th August 2015.

Thames (2000), Thames Safety Inquiry: final report by Lord Justice Clarke, Cm. 4558, Lord Justice Anthony Clarke, ISBN 9780101455824, 2000

TSB (1995) Aviation investigation report A95P0004.
http://www.skybrary.aero/bookshelf/books/427.pdf

TSB (2006) Aviation investigation report A04H0004.
http://www.skybrary.aero/bookshelf/books/593.pdf

TSB (2013) Aviation Investigation report A12A0082.
http://www.skybrary.aero/bookshelf/books/3044.pdf

TSB (2014) Aviation investigation report A1HH0002.
http://www.skybrary.aero/bookshelf/books/2732.pdf

UK Airprox Board (2012) Airprox report No 2011167.
http://www.skybrary.aero/bookshelf/books/2227.pdf

Annex A: Aviation Accident/Incident Classification Types

The following classification types were used in categorizing the causal factors of aviation accidents and incidents.

Table 2. Aviation Accident/Incident Classifications

Title	Description
Data	The failure to preserve the properties of a critical piece of data.
Software	Software behavior or programming errors, algorithmic or logic errors.
Hardware	Physical failure of an aircraft component such as an engine, elevator, gear, air conditioning, hydraulics, sensors etc.
Pilot Error	Errors of judgment, failure to follow standard operating procedures.
Procedural Error	Processes were followed but the processes themselves were flawed.
Operational Error	Any error relating to aircraft operations such as air traffic control, aircraft loading, ground movements.
Maintenance/Manufacturer Error	Errors carried out during routine maintenance of aircraft or flaws in the original manufacturing process or materials.
Passenger/Cargo/3rd Party/Deliberate	Issues arising from passenger luggage, cargo, 3rd party interference (eg. Aircraft shot down by military personnel), and deliberate acts.

Title	Description
Exceptional events	Rare or unanticipated events such as bird strikes, sudden extreme weather conditions (lightning strikes, microbursts), crew incapacitation.

Annex B: The Data Safety Initiative Working Group

The SCSC Data Safety Initiative Working Group (DSIWG) was formed in January 2013 with the aim of making data a "first-class citizen" in safety systems development and providing concrete guidance on how to manage the risks associated with data. The group is a cross-industry initiative operating under the auspices of the SCSC but funded through the good will of the participating organisations.

The work of the group is being encapsulated in a guidance note that aims to form the foundation of more formal standardisation either as a standalone guidance document or to support the incorporation of data safety management in existing standards and guidance such as DEF-STAN 00-55.

An initial version of the Data Safety Guidance Note, (DSIWG 2014) was published in 2014 and made available at SSS'14, and was well received. A second version of the Guidance Note was then produced, (DSIWG 2015) addressing review comments and including new material at SSS'15.

The latest version of the guidance (presented at SSS'16) provides the following:

- A language, dictionary and ontology for articulating ideas and concepts across different sectors in an unambiguous way;
- Techniques for quickly assessing the organisational risk of data so that an organisation can understand its high level risk exposure and to inform and justify further activities;
- A framework for conducting a more detailed analysis exploring tangible hazards in the use of data in a structure that aids review and comparison between systems;
- A classification scheme for assessing and categorising the criticality of data and its properties;
- Models of different data lifecycles and guidance on when assurance techniques should typically be applied for a given data type;
- Recommendations for assurance techniques to be applied for a given data type and data integrity level;
- A set of cross sector data type definitions that allow different types of data to be categorised to allow more structured analysis and treatment of the safety risks they can give rise to;

- A set of cross sector principles that represent good practice in data safety assurance case development;
- An introduction to the concept of perspective when analysing risks for an organisation that is part of a wider data supply chain;
- Guidance on guidewords that could be used in Data HAZOP meetings to help elicit hazards that data could give rise to;
- An analysis of the relationship between safety and security from a data perspective;
- A proposed questionnaire to help understand the level of data safety awareness and data safety culture within a project or organisation.

Clearly, much progress has been made but there is still a lot of work to do. The group is continually looking for new members and contributions from as wide an industry cross-section as possible.

Safety Justifications for use of Smart Devices in Existing Nuclear Power Stations - "Getting the Balance Right"

John Delafield

EDF Energy Generation

Barnwood, UK

Abstract   *Modern smart devices such as pressure transmitters, controllers and valve actuators provide many key advantages but there are well-known difficulties in providing evidence to support the associated safety justifications. This paper reminds the reader of these difficulties but focuses on the need for an "As Low as Reasonably Practicable" approach and hence the requirement to use expert engineering judgement to weigh up the advantages/disadvantages of using a smart device against other possible options. The paper discusses the use of engineering judgement in safety justifications for installing smart devices and highlights that there are more than just software faults to consider. Issues covered include: allocation of 'best estimate' reliability data for use in Probabilistic Safety Assessments, 'proven in use' arguments and the importance of understanding the wider safety case picture.[1]*

1 Introduction

EDF Energy aims for an average of 8-year life extensions across all of its Advanced Gas-cooled Reactor (AGR) stations (Figure 1) and a 20-year extension for Sizewell B, the only Pressurised Water Reactor in the UK. Ensuring continued safe and reliable generation to support these life extension aims requires replacing ageing control, instrumentation and protection systems as well as installing addi-

---

[1] *This paper was first presented at the 8th IET International System Safety Conference in 2013 but has been revised to reflect the author's experience over the last two years as an independent assessor of safety cases and justifications.*

tional systems to provide enhanced control, monitoring and protection for the power station systems.

Fig. 1. Hunterston AGR Power Station

Modern commercial "off-the-shelf" control and instrumentation (C&I) devices provide many key advantages with improved performance and additional features, such as sophisticated self-diagnostics and plant condition monitoring. However, these devices are complex, as they include embedded software (firmware). Where these devices have a single function with the user limited to configuring the device by setting various parameters, these are known as smart devices. Examples include: valve actuators, motor variable speed drives, pressure transmitters, trip amplifiers, controllers and recorders, figure 2. (The nuclear industry definition of smart devices does not include smart meters, smart phones etc.)

Fig. 2. Examples of smart devices: valve actuator, pressure transmitter and trip amplifier

All modifications to nuclear power station plant require a safety justification to support the station's safety cases. A modification can be as simple as replacing an obsolete analogue pressure transmitter with a smart pressure transmitter, to the installation of a completely new system where a smart device is only a small part of the system. EDF Energy's safety justifications are based on the 'claims', 'arguments' and 'evidence' structure. The extent and required rigour of the justification essentially depends on the following:

i.   The impact on the nuclear risk should the modification fail to meet its safety duty;
ii. The claimed integrity of the modification;
iii. The complexity and novelty of the modification.

Taking the example of a smart pressure transmitter; although functionally it is not complex or unlikely to be novel, the actual device, within the nuclear industry is considered complex due to embedded software (firmware). Even relatively simple smart devices in terms of function, may contain significant software code. Because of the presence of software, it is difficult to provide evidence that the device will meet the claimed integrity requirement and in practice impossible to fully test.

There is a fundamental legal requirement for risks from operating the power stations to be ALARP (As Low as Reasonably Practicable). In this context, the term "reasonably practicable" is interpreted from UK case law to mean that measures should be taken to reduce risk unless the "money, time and trouble" of implementing those measures is grossly disproportionate to the risk averted. What does this mean in practice within the field of C&I modifications on our existing nuclear power stations? Faced with ageing C&I on the plant and the potential to reduce risks by installing additional C&I systems, at the highest level, doing nothing is not going to be ALARP. Likewise it is clearly not ALARP just to install a smart device with a safety justification based on claims that are only supported by weak arguments and no real evidence. On the other extreme, it is also not ALARP to spend excessive time, resources and money on a single safety justification where the nuclear risk from inadequate conception or execution is limited. Hence this paper will address the challenges of 'getting the balance right' for safety justifications for installing smart devices on our exciting nuclear power stations.

Getting the balance right involves using expert engineering judgement to weigh up the safety and commercial advantages of utilising a smart device against the risks that the device could fail and hence potentially impact safety. Prior to discussing this 'balance' issue the paper provides some basic background information. Note that although smart devices on the power stations are not connected to the internet, cyber security also needs to be considered.

2   What contributes to the integrity of a smart device and why is providing arguments and evidence difficult?

Figure 3 illustrates the contributing factors to the integrity of a smart device when installed on the plant. The 'operational environment' has an impact on the random failure rate but it also has to be considered at the design stage. For safety justifications, it is important to consider both the normal operational environment of the smart device and, where claimed, the environmental conditions which could arise from an internal or external hazard, examples being high temperature from steam leaks to events such as earthquakes.

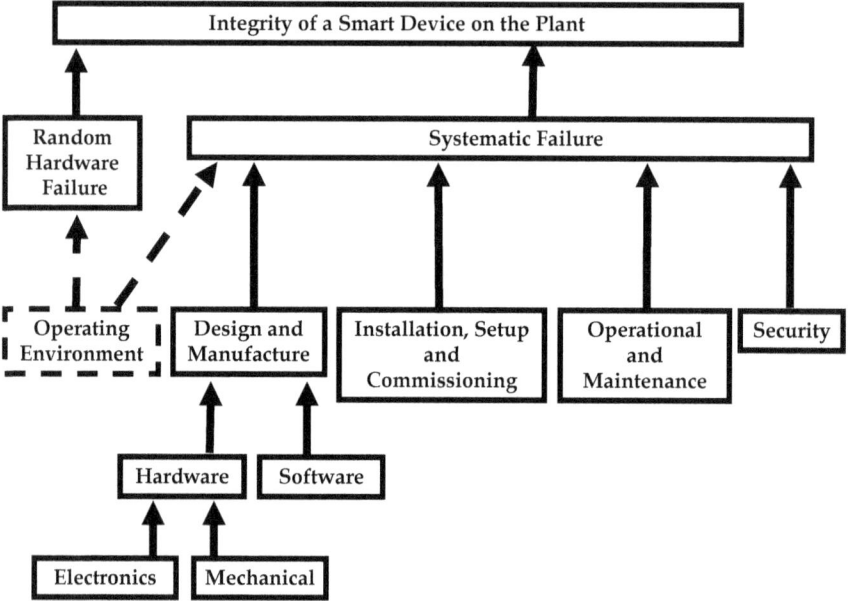

Fig. 3. Factors Contributing to the Safety Integrity of a Smart Device on the Plant

Taking the contributing factors from Figure 3 in turn:

2.1   *Random Hardware Failure*

This can be assessed from a conventional Failure Mode and Effects Analysis and/or field data. However, consideration has to be given to the operating environment as this can have a significant impact on the random failure rate of a device. Field data can provide the most accurate failure rate figure but this depends on the pedigree and scale of the source data as well as consideration of the operating environment.

2.2   Systematic failure - Design and Manufacture

This is applicable to any device, the key being the more complex and novel the design and/or manufacturing process, the greater the potential for a systematic failure. For simple hardware devices extensive testing, including type testing, can provide significant evidence that a systematic failure from design fault is unlikely. However, for any device using software it is, in practice, impossible to fully test it.

2.3   Systematic failure – Installation, Setup and Commissioning

Incorrect installation of a smart device on plant, which importantly includes justification and set-up of the various user configurable parameters and potentially inadequate commissioning, can have a significant impact on safety. From the author's perspective, this area has not always been adequately justified.

2.4   Systematic failure – Operational and Maintenance

This includes routine operator interaction, routine testing and breakdown maintenance. The potential for systematic failure by the user can be minimised by good design of the human machine interface (HMI) but also requires high quality supporting operational and maintenance (O&M) documentation from the manufacturer. When considering routine testing, a well designed smart device may significantly reduce the potential due to a human error compared to an older analogue device, however there can be a significant increased risk associated with configuration of the device (setting up of parameters). Consider a simple 3-term controller based on discrete electronic components compared to a modern smart controller.

Good HMI design and manufacturer's documentation will minimise the potential for incorrect parameter settings. However, the smart controller will inevitably have many more configurable settings compared to the discrete component device; potentially the impact of some of these parameters may not always be totally clear from the O&M documentation. Even if the documentation is clear, there is a further issue associated with training: will an experienced but conventionally trained C&I engineer fully understand the implications of all the parameter settings?

2.5   *Systematic failure – Security*

Although not discussed within this paper the general issue of security (including cyber security) also has an impact and needs appropriate consideration. Just because a smart device is not connected (wired or wireless) to other computer system or the internet, does not mean the security aspects can be dismissed.

3   Current Approach for Providing Safety Justification for Smart Devices

Within the UK civil nuclear industry a two-legged approach of 'production excellence' (PE) and 'independent confidence building measures' (ICBMs) for producing a safety justification for smart devices has been successfully used. The PE leg reviews how well the device's lifecycle, from concept and specification to the production stage, has been undertaken in accordance with the basic all-industry standard (IEC 61508:2010). The ICBMs are carried out independently from the manufacturer and include various testing techniques, and where higher integrity claims are required, an independent analysis of the software. The approach is graded requiring a more detailed PE analysis and additional ICBMs for the higher claimed integrity levels. Producing the PE leg requires considerable co-operation with the manufacturers once a claim of 10^{-2} failure per annum (fpa) or failures per demand (fpd) or lower is required. The approach has tended to concentrate on the software, as this can often be the area where a latent systematic design error occurs and where it is recognised that testing of the device cannot be totally relied upon to detect such an error. A powerful tool that can be used as an ICBM is statistical testing. However, it can be complex and expensive to apply and relies on the expertise of the test team to derive a representative set of tests that takes account of the application in which the smart device will be used. As the number of tests is large for high integrity, high confidence level claims, there is a potential for human error to undermine the results both from the difficulty in defining appropriate tests and then from analysing the results.

The general approach within EDF Energy is to carry out a generic justification of a smart device and hence permit the device to be used across the fleet up to the accessed integrity level providing a number of application specific requirements are met and these are addressed in the safety justification for each application. There are a number of challenges, including:

- The generic justification is carried out on a specific version of the smart device and often that version will be upgraded within a relatively short period.

- For the higher integrity justifications, the ICBMs require access to the source code and manufacturers are justifiably not always willing to release this.
- A smart device will be designed to fail to a 'safe state' by use of its internal self-diagnostics. However depending on the safety case the 'safe state' may not actually be safe. This aspect is discussed in more detail in section 6.1.

The approach currently taken to address the first point above involves carrying out an impact assessment by first understanding the change(s) and then carrying out any additional PE and ICBMs deemed necessary. This impact assessment can provide its own challenges, as it inevitably requires further co-operation from the manufacturer and an understanding of the significance of the change and also the potential to undermine previous arguments and evidence in the original assessment.

4   Proven in Use Arguments

Proven in use arguments and evidence for any mass produced device can potentially provide a strong leg to a safety justification; however there are problems when focusing on smart devices which include:

- Uncertainty that all faults are reported. From the software viewpoint, a device could fail and the user just powers off and on the device to reboot it! Even on the hardware side, as the devices are relatively cheap, failures may not be reported.
- The operational experience is likely to be across different versions of the device.
- The number of operating hours alone does not provide the required evidence for proving software; consideration has to be given to the extent of changes to the internal software states and inputs over time. (e.g. A smart trip amplifier could operate for 10 years and not be subject to any trip demands.)
- To provide high confidence the number of operating hours is significant when supporting all but the lowest fpa integrity claims.

5   Probabilistic Safety Assessment

One of the tools used to help demonstrate that the risks are ALARP is Probabilistic Safety Assessment (PSA). It is used, not only to help demonstrate the risks are considered ALARP and are within the probabilistic limits contained in the Office of Nuclear Regulation's, Safety Assessment Principles (www.onr.org.uk/saps - accessed 14 November 2015), but also more importantly to identify the main contributors to the overall risk, and hence allow resources to be focused at reducing this overall risk.   In order to get the most realistic assessment of risk the PSA should use data based on "best estimate". This applies to all data e.g. the frequency of events such as earthquakes, plant faults (such as control system failures) and the probability of failure of protection systems.

Depending on how a smart device is claimed in the PSA, the required data may be either per hour/annum or per demand.

6   Getting the Balance Right - Introduction

As mentioned in the introduction to this paper, there is a fundamental requirement for risks from operating our power stations to be ALARP. For modifications, this means there is a requirement to consider the current level of risk and weigh this up with the risks during and following the implementation of the modification.

The following sections of this paper address various aspects of 'getting the balance right'.

6.1   Getting the Balance Right - Probabilistic Safety Assessment and 'best estimate'

A PSA is an 'aid to judgement' and there will always be significant uncertainties in some of the data used, be it for a smart device or the probability of high integrity mechanical system failing, or indeed the return frequency of flooding from sea surges. Hence, sensitivity analysis within the PSA is used to provide an insight into the significance of uncertainties in data on the overall nuclear risk.

A generic assessment (not application specific) of a smart device provides a sensible approach to reducing the overall costs of safety justifications where the device will be used in multiple applications. However, for a generic assessment to conclude a smart device can be claimed in a PSA with a specific failure rate/probability can be misleading. This is best illustrated by an example such as a smart controller where a generic assessment may conclude a claim of 1E-2 fpd/fpa. The controller may well have been designed to ensure a high 'safe failure fraction' and include self-diagnostics to detect hardware, data or software faults and hence fail to a known 'safe' state, such as freeze its output and raise a fault alarm. However, if a 'safe' failure of the device causes an initiating event, for example by failing to control a plant parameter which then results in a demand on independent protection systems to prevent a radiological release, then the device's 'safe' failure is in fact not safe when viewed from the safety case perspective. The safety case requires a device with low 'safe' and 'unsafe' failure rates.

Conversely, especially for less onerous integrity claims, the 'best estimate' pfd value may well be significantly better than the figure justified from the approach summarised in section 3 as it is generally agreed this approach is conservative especially where there is redundancy and the assessed systematic failure dominates.

In conclusion, when allocating a 'best estimate' value for use in a PSA it is critical to understand the nuclear safety role that the smart device is performing and hence the failure modes of the device that are of concern. Consideration has to be given to both systematic and random hardware failures when allocating a 'best estimate' figure and expert engineering judgement should be used based on the available evidence of the integrity of the smart device including 'proven in use' as discussed in 6.4 below.

6.2   Getting the Balance Right -
Deterministic Safety Requirements

Deterministic considerations are vitally important to consider; however much PE, proven in use and ICBMs evidence there is, there will always be the potential for any smart device to fail due to a random or systematic fault and as such the integrity claims must be limited.   For a single smart device where there is no redundancy or diversity (as in the above example) random and systematic failures will need to be considered.   Where redundancy has been employed this will provide mitigation for random hardware failures but not for a systematic failure (design, manufacture or O&M activities).   Diversity provides a robust safety justification argument but dependencies between diverse sub-systems always need to be considered and supported by an appropriate level of common mode/ cause failure assessment.   Where for example one protection system is based on C&I and another on a mechanical device the diversity arguments provide strong defence against common cause failures.   However, where both protection systems are utilising smart devices, even if they are from different manufacturers, then dependency arguments are much weaker and need justifying.

6.3   Getting the Balance Right –
Self-diagnostics and Plant monitoring

There is a need to weigh up the real advantage of using a smart device against the risks associated with potential systematic faults.   Self-diagnostics and enhanced plant monitoring are two key benefits of smart devices.   A couple of simple examples best illustrates this.   A smart valve actuator will not only include self-diagnostics that will continually confirm the actuator is healthy and ready to respond to a demand to move the valve, it will also provide condition monitoring of the actual valve by identifying changes in valve stiction and hence allow a more proactive preventive maintenance regime to be implemented. A preventive maintenance regime driven by conditioning monitoring has both commercial and safety benefits, as any maintenance activity has potential for introducing maintenance induced faults. As a further example, a smart differential pressure transmitter measuring flow can detect if the sample lines are blocked or have been inadvertently closed!

6.4   Getting the Balance Right – Proven in Use

Standard (IEC 62671:2013) covers requirements for claiming proven in use under the section on 'operating experience'. However the criterion is challenging, as it includes *'All of the credited evidence of operating experience shall be auditable'*. For many smart devices it will not be practical to satisfy this type of criteria fully, however it does not mean 'proven in use' arguments and evidence should be ignored.

There are some important questions that, in the author's view, can be asked of the manufacturer that can provide some valuable evidence on the device's history and hence reliability. These questions are:

- How long has the device been on the market?
- How many units have been sold?
- How do you deal with faults reported by users?
- How many 'upgrades' have there been?
- Of these 'upgrades' how many have been to fix design shortfalls?
- How significant were any 'design shortfalls' and when were they implemented in the device's history?

In the author's view, if a manufacturer can provide evidence to answer the above questions that indicates the device is mature, has sold in significant numbers, has not been continually 'upgraded' to fix design shortfalls and fault reporting is appropriately investigated, then evidence to support the PE leg of a safety justification is less important. The author's reasoning is that the positive 'device history' provides evidence of the quality of the PE.

Furthermore, if the device is reasonably mature and has been sold in significant numbers it is likely to have been used in a variety of applications. Operating experience in a large number of applications is very valuable as it increases the likelihood of the software being exercised over a larger number of internal software states, as mentioned in section 4.

The last question above is considered critical, since a response indicating very few design fixes have been required provides a strong indication of the quality of the product. Even if the latest version of the device has not been available for a significant length of time, it may be reasonable to assume the manufacturer's modification process is of reasonable quality.

Although the issues mentioned in section 4 must be considered, a pragmatic approach is required as 'proven in use' arguments can provide a significant degree of confidence for a mature smart device that has been used in a large number of applications.

6.5   *Getting the Balance Right - Optioneering*

An important part of demonstrating compliance with the ALARP principle is to consider all options to solve a problem; possible solutions could include:

(a)   A smart device;

(b)   A smart device but with additional 'wrap around' protection;

(c)   If available, a non-smart commercial 'off-the-shelf' device;

(d)   Designing a bespoke solution;

(e)   Re-engineering an existing non-smart device;

(f)   Improve the operating environment / maintenance regime of the existing C&I system.

Clearly, costs and timescales have to be considered for the various options. For modifications to a nuclear power station, there are the conventional engineering costs associated with any modification on an industrial plant and then there are the costs involved in delivering the safety justification. The safety justification must be based on (i) to (iii) in section 1 and not just on the integrity claim.

The chosen option for a particular application will depend upon a number of factors. Sophisticated self-diagnostics and plant monitoring available from smart devices can provide many key safety justification benefits, particular for emergency standby systems, where they can warn of a fault that would otherwise remain unrevealed until the emergency system is routinely tested or required to operate! These benefits should be appropriately considered during the optioneering phase.

Where the PSA indicates a higher integrity claim is required option (b) may be the ALARP option, as although the engineering and lifetime costs will be higher, applying the 'single failure criterion' deterministic principle is likely to highlight that there is a real improvement in nuclear safety compared to the other options; furthermore the safety justification costs should also be less.

When considering option (c), such a device is unlikely to have some of the key benefits of a smart device; however, depending on the pedigree of the device and the availability of proven in use data, the safety justification may be (but not always) easier to make compared to option (a).

Both options (d) and (e) will have significant higher engineering design costs and have a potential for systematic design faults although there are well-established tools and techniques that can be applied to minimise the risk of a systematic design fault. To minimise the potential for systematic fault the design can be kept simple but compared to utilising a smart device these options are unlikely to have sophisticated self-diagnostics. Although such additional complexity could be engineered these will increase the potential for a systematic fault.

In some cases option (f) may be appropriate where it can be argued the existing C&I will continue to meet the safety justification requirements for the expected station lifetime.

Engineering judgement has to be used to choose the ALARP option, it being necessary to weigh up the real nuclear safety advantages and disadvantages of the different options as well as consider the safety justification costs. From the author's perspective this is an area where improvements could be made, as the smart option (a) may be erroneously considered too difficult and risky to make the safety justification and hence discounted when sound engineering judgement indicates this is the ALARP option.

6.6   Getting the Balance Right – Data Diversity

Where the same type of smart device is used in redundant systems, current guidance only permits a claim up to the systematic limit. This is clearly appropriate where the inputs to the device are identical. However, there are applications where the redundant smart devices are monitoring different input signals, hence it may be possible to argue they will not fail at the same time. To make any claims on data diversity requires a full understanding of the plant dynamics and the safety significance of a systematic failure. As an example, will the first failure of a smart device in a redundant system be revealed, and is there adequate time for safety to be maintained, prior to a further device failing due to the same or another failure mode? It must also be appreciated there is more than just the smart devices inputs to consider; other software internal states have to be considered. As an example, in theory, redundant smart devices could fail due to say an overflow of a warning message log.

6.7   Getting the Balance Right – Understanding the 'bigger safe case picture' and concentrating on the weak links

It is clearly important for a safety justification to consider the overall nuclear safety risks and consider where the 'weak links' are and address them. There is a potential risk that too much focus can be given to the smart device, that is only part of a total system that ensures safe and reliable operation, and a possible weak link may not receive the appropriate level of attention. As a simple example, consider an obsolete analogue controller controlling a pneumatic control valve on a closed loop cooling system consisting of a pump and heat exchanger to remove the heat from the plant. The ALARP option is likely to replace the analogue controller with a smart controller. A safety justification will need to be produced to cover the replacement and include an appropriate level of justification for the smart controller. In this case, what are the consequences of failing to control the temperature too high or low, and what is the integrity of the independent protection? When considering the overall integrity of the cooling system it may be appropriate to use the available engineering resource and funding in other areas, rather than say providing additional justification of the smart controller by carrying out, for example, extensive statistical testing or static analysis of the software. Overall safety may be improved by using the resources and funds towards replacing the pneumatic control valve with an electric smart actuator as this could be the 'weak link' in the system.

7   Concluding Thoughts

Ensuring safe and reliable generation to support the life extensions for EDF Energy's fleet of existing power stations requires replacing ageing C&I plant as well as installing additional systems to provide enhanced control, protection and monitoring of the power station systems. Getting the balance right initially means ensuring the appropriate ALARP solution is chosen and progressed; this requires the application of expert engineering judgement to weigh up the real advantages of a utilising a smart device against the risks that the device could fail and comparing this with other possible solutions.

Currently, it is the author's view that costs and timescales involved in making safety justifications for smart devices are such that during the optioneering phase using a smart device may be discounted, yet it could have real advantages in reducing the nuclear safety risk from improved plant monitoring, such as the example of the smart value actuator (section 6.3).

The approach outlined in section 3 for justifying the integrity claims on smart devices has been used for a number of years. It would be informative to review the operational experience with smart devices in safety-related applications and refine the approach to ensure the resources involved in justifications of smart devices are directed in the appropriate areas. It is the author's view that this is likely to show that resources for safety justification of smart devices need to concentrate more on the application of the device and less on the software design aspects.

Once a decision has been made to use a smart device, getting the balance right requires further application of expert engineering judgement to ensure an ALARP justification is provided. It is hoped this paper has illustrated that it is not as simple as saying, for example a smart device claimed at 1E-2 requires x, y and z to be carried out.

The paper has highlighted that the risks are not just from systematic failures due to the software, but also from systematic failures associated with installation, set-up, commissioning and from routine O&M aspects.

Overall the paper has illustrated that 'getting the balance' right is not easy and requires the application of expert engineering judgement, which requires an understanding of the following:

- the potential for software and hardware failures of smart devices;
- the value of the various tools and techniques available that provide confidence in the integrity of the smart device;
- the impact on the nuclear safety risk resulting from the different failure modes of the smart device;
- the bigger safety case picture, so there is an appreciation where possible weak links are, including the significance of other uncertainties in the case and their significance compared to those associated with smart devices.

International standards such as (IEC 62671:2013) contain very valuable information, however they cannot be expected to (and do not) cover ALARP issues and the wider safety justification picture, hence 100% compliance with all the requirements of such standards may not be achievable or appropriate to achieve an ALARP position. It is important that engineering resources are directed in the right areas to improve safe and reliable operation of the power stations. Hence, the resources allocated to a safety justification for a smart device must acknowledge the real risk to nuclear safety and consider it in relation to other uncertainties in overall safety case.

EDF Energy has recognised the requirement to improve the application of engineering judgement within safety justifications and has rolled out additional training on this and the application of ALARP.

The key to any safety justification where expert engineering judgement is used is to clearly record the arguments and evidence supporting the judgements and to share them at an early stage with the various stakeholders, including independent assessors and approvers of the safety justification.

Many modifications on the power stations involve all engineering disciplines from civil, mechanical, electrical and C&I as well as consideration of human factors aspects. For such modifications, producing a safety justification that demonstrates the risks are ALARP, requires a multi-disciplined team that understands the relative uncertainties in all areas of the justification and the nuclear safety significance of these uncertainties. Hence the team can ensure the extent and rigour of the various aspects of the safety justification are appropriate.

Finally, although this paper is advocating the appropriate use and justification of smart devices on the power stations, it must be appreciated that there will always be the potential for smart devices to fail. Hence, like many other engineering devices/systems, safety case claims need to be limited and, where required, diverse protection systems need to be employed to ensure the risks within the overall safety case remain ALARP.

Acknowledgements The author would like to thank Dr Stephen Johnson (EDF Energy's Chief C&I Engineer), for reviewing drafts of this paper.

References

IEC 61508:2010 Functional safety of Electrical/Electronic/Programmable Electronic (E/E/PE) safety-related systems

IEC 62671:2013 Nuclear power plants - Instrumentation and control important to safety – Selection and use of industrial digital devices of limited functionality

Improving European Aviation Safety Approvals

Stephen Bull[1]

Ebeni Limited

Corsham, UK

Abstract *The European aviation industry is experiencing wide-ranging change including introduction of new technologies and operational concepts, while also facing demands for higher levels of safety performance. Existing approaches to gaining approval are often perceived as a barrier to adopting innovation and change; they can also miss significant interactions between parts of the system. The EC-funded ASCOS Project has developed a method and supporting tools to address these challenges. The ASCOS Method uses modular safety arguments to provide a framework to integrate existing approval approaches while also providing the flexibility to adapt the approaches where necessary to enable the smooth approval of advances in aviation technology.*

1 Background

Fundamental changes in the institutional arrangements for aviation regulation in Europe, the introduction of new technologies and operations, and demands for higher levels of safety performance all call for the adaptation of existing approval processes.

The increasing amount of technological innovation within the industry challenges existing prescriptive regulations used in aircraft certification. These regulations are an established and effective way of capturing lessons from past experience to ensure that future implementations learn from these lessons, thus delivering safer operations. However, innovative solutions often cannot comply with such prescriptive regulations, making it difficult to demonstrate that solutions meet industry safety requirements. There is therefore a move towards performance based regulation where the applicant for approval must demonstrate compliance

---

[1] Stephen Bull is a senior safety engineer at Ebeni. He can be contacted at stephen.bull@ebeni.com

with a level of safety, but where there is freedom on the way in which that level of safety is achieved.

The shortcomings of existing processes are illustrated in the FAA Commercial Airplane Certification Process Study (FAA 2002), which concluded that there was no reliable process to ensure that assumptions made during solution development and assessment are valid with respect to operations and maintenance activities and, furthermore, to ensure that human operators are aware of these assumptions when developing their operations and maintenance procedures. It also became clear during the study that aircraft certification standards may not reflect the actual operating environment. Other studies have reached similar conclusions.

2 The ASCOS Project

In reaction to the drivers described above, the EU ASCOS[1] Project was established to develop a novel and innovative approach towards approval in order to ease the efficient and safe introduction of safety enhancement systems and operations. This novel approach was required:

- to be more flexible with regard to the introduction of new products and operations;
- to be more efficient, in terms of cost and time, than the current certification processes;
- to consider the impact on safety of all elements of the aviation system and the entire system life-cycle in a complete and integrated way.

Development of this novel approach was supported by safety-driven design methods and tools to ease the approval process. The project has followed a total system approach, dealing with all aviation system elements (including the human element) in an integrated way over the complete life-cycle.

The ASCOS programme was structured into six main work packages (WPs):

- WP1: Certification Process – Development of safety based certification process adaptations based on analysis of existing certification and evaluation of possible new approaches
- WP2: Continuous Safety Monitoring – Development of a methodology and supporting tools for continuous safety monitoring, using a baseline risk picture for all parts of the Total Aviation System (TAS)
- WP3: Safety Risk Management – Development of a total aviation system safety assessment methodology for handling of current, emerging and future risks

---

[1] Aviation Safety and Certification of new Operations and Systems

- WP4: Certification Case Studies – Application of the new certification approach to selected example case studies
- WP5: Validation – Validation of the new certification approach, supporting methods and tools
- WP6: Dissemination and Exploitation – Dissemination to ensure that results are correctly understood and exploited to the maximum extent.

There was also a seventh work package for project management. The relationships between these work packages are depicted in Figure 1.

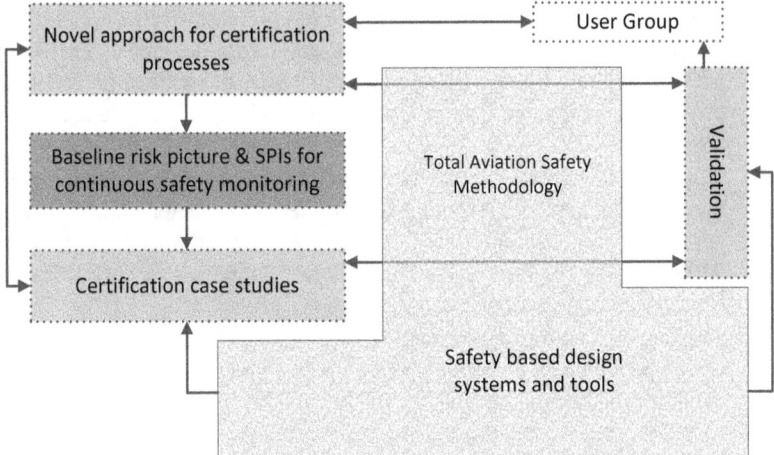

Fig. 1. Relationships between ASCOS work packages[1]

The final reports from the work packages are listed at the end of this paper and are available for download from the ASCOS Public Website[2].

The three year programme was an EC-funded partnership with participants[3] drawn from the aviation industry, including national Civil Aviation Authorities (CAAs), research organisations and consultancies. Ebeni led the effort to develop the novel approach for certification processes. A User Group was constituted to provide input from across the aviation industry.

This paper focuses on the development of adaptations to the existing certification and approval processes (i.e. WP1).

---

[1] Arrows indicate the flow of information between work packages.

[2] https://www.ascos-project.eu/

[3] A full list of participants can be found on the project website.

3 Key Concepts

The issue of concepts and terminology posed a real challenge, especially where the same term is interpreted differently across the aviation industry.

This section introduces some key concepts on which the ASCOS Method was built.

- The term **certification** is widely used in the aviation industry to describe the process of demonstrating that a physical item, or an organisation, meets a defined set of requirements and can therefore be issued a certificate to confirm this compliance. For example, certificates are granted for aircraft, air operators, providers of air traffic services. The ASCOS Method, although originally conceived as an adaptation of certification processes, is intended be applicable more widely, to any change requiring approval.
- **Approval** is a broader term than certification, covering the process where the relevant approver (usually a designated competent authority) gives approval for a change to the Total Aviation System (TAS). For example, approval may be granted to change the way in which an air traffic service is provided at a particular aerodrome.
- The **Total Aviation System (TAS)** refers to all elements of the system which provides an aviation service, including concepts, equipment, people and processes.
- The TAS is subdivided into **domains**; although domains are not formally defined in the method, they can be aligned to the structure of the applicable regulations under which approval will be granted.
- A **change** is any alteration to the TAS, beyond intended operational use or maintenance. Changes range in scope from upgrade of existing equipment items through to introduction of a new operational concept. Changes need approval before they are introduced into operational service; approval is usually given by the relevant authority (e.g. either the European Aviation Safety Agency (EASA) or the relevant national CAA).
- The concept of a modular **safety argument** will be familiar to most readers: in this context it is defined as a connected series of statements, with supporting evidence, used to persuade the reader of the correctness of an overall claim about the safety of a change. The argument is divided into **modules** (which may be aligned to the domains of the TAS) and the dependencies between modules are captured in **assurance contracts**.

4 Developing the Method

Initial research was undertaken to survey the state of certification processes in the aviation industry and to identify options for change (see section 4.1), published as

(ASCOS 2013a). As a result, an initial proposal for the ASCOS method was developed (ASCOS 2013b) (see section 4.2). This method was then applied using four case studies (see section 4.3) and evaluated through validation workshops (see section 4.4); the findings are summarised in section 4.5. The final version ASCOS Method (ASCOS 2015a) was then developed, based on these findings. Key features of the method are presented in section 5.

4.1 Options for Change

Initial research (ASCOS 2013a) into existing approaches and opportunities for improvement identified a number of principles to include in the ASCOS Method. These can be summarised as follows:

- taking a Total Aviation System (TAS) approach
    - provide a generic certification framework covering the TAS and the whole system lifecycle
    - standardise language across all domains
    - harmonise approaches between domains where appropriate
    - consider the balance between product and organization certification
    - promote flexibility within each domain to allow introduction of new technologies or procedures
- development of current processes
    - keep the existing approach where possible, minimising unnecessary change and recognising the good approaches already in place
    - simplify certification processes, where there are demonstrable benefits and no loss of confidence in the assurance of safety
    - learn lessons from other domains where this gives improvement
- fully considering interfaces between domains and organisations
    - provide rigorous management of interfaces, between domains and between the TAS and its environment, to ensure that all key safety issues are properly addressed and not lost at interfaces
    - establish a process for ensuring validity of assumptions, in particular with respect to operations and maintenance activities
- address known problems with existing approaches
    - reinforce existing techniques where they are appropriate but not consistently applied
    - provide a mechanism for identification and resolution of further bottlenecks and shortcomings
- specific issues
    - take more explicit account of electronic hardware in the proposed approach

-   consider that less experience is gained by flight crew when more au-
    tomation is used

The research also identified eight options for change. Evaluation of these options focused on identifying those with the greatest potential for safety and cost benefit, although it also considered other criteria. The four options chosen for further development were:

- Change between performance based and compliance based
- Use of Proof of Concept approach
- Enforce existing rules and improve existing processes
- Cross-domain fertilisation

The overriding consideration when designing the ASCOS Method was the need to develop a method which accommodates and integrates existing approaches while allowing for extension and adaptation of these approaches where necessary. The details of the method described in D1.5 (ASCOS 2015a) show how the principles and chosen options have been addressed.

4.2 An Eleven Step Process using Logical Argument

It was apparent early in the project that there is no single approval approach universally applied within the TAS, and that the role of ASCOS was to develop a framework which allows existing approaches to be integrated, and adapted where this is either necessary or beneficial.

The need to integrate multiple sources of information and evidence into what is effectively a single claim about safety led the team towards the concept of a safety argument. We also proposed to modularise the argument (see, for example, Fenn et al. 2007), aligning the modules to the boundaries of domains and organisational responsibility, with assurance contracts established between the modules to formally define and manage dependencies.

The use of logical argument was supported by an eleven step process as follows:

1. Define the change
2. Define the certification argument (architecture)
3. Develop and agree certification plan
4. Specification
5. Design
6. Refinement of argument
7. Implementation
8. Transfer into operation – transition safety assessment
9. Define arrangements for continuous safety monitoring

10. Obtain initial operational certification
11. Ongoing monitoring and maintenance of certification

Some written guidance was produced for each step, but the case studies (see below) applying the method were also supported by the team that had developed the method. Further written guidance was produced when the method was updated in light of this experience.

The report (ASCOS 2013b) describing this initial version of the method is available on the ASCOS public website. This initial description of the ASCOS Method went some way towards showing how concerns and options identified in the early work are addressed by the method.

4.3 Case Studies

When the eleven stage method had been developed, it was applied to the following four case studies intended to examine how it could be applied to realistic approval challenges:

1. Remotely Piloted Aircraft System (RPAS) failure management
2. Automatic Aircraft Recovery System (AARS)
3. Separate certification for on-ground de-icing
4. ATM / CNS[1] systems for improved surveillance

Each case study was developed over the period of a year (2014) by teams drawn from organisations participating in the ASCOS Programme. Each case study attempted to apply the ASCOS Method by developing a logical argument and approval plan, each with varying degrees of success.

As the case studies were desk top exercises, with limited time and effort available to them, they were only able to apply the early stages of the method. However, they produced valuable feedback which was used to improve the method.

A report on the case studies is published as (ASCOS 2015e).

4.4 Validation

The case studies were complemented by familiarization and validation workshops where the interim results of the ASCOS Programme were presented to the User Group: their feedback was elicited with the help of a questionnaire, completion of which was followed by a focus group meeting involving experts representing different aviation and certification domains. One of the validation exercises focussed

---

[1] Communication, Navigation and Surveillance

on the ASCOS Method itself (as initially presented as an eleven step method); the other exercises considered supporting tools.

A report on the validation activities is published as (ASCOS 2015f).

4.5 Recommendations

The case studies and validation exercises made the following recommendations for the improvement of the ASCOS Method:

- define the roles, responsibilities and team structures required to use the ASCOS Method
- use existing, and consistent, terminology where possible, and make this understandable to as wide a range of users as possible
- align the subdivision of the TAS into domains to the subdivisions contained within the EASA regulation structure
- refer to relevant assessment methods, in particular focussing on human factors and organisational assessment
- explain how to determine whether the ASCOS Method is applicable to a particular change
- explain how the ASCOS Method aligns to existing certification practice
- provide guidance on the definition of a change and its boundaries
- provide extensive guidance on the development of logical arguments
- provide guidance on how stakeholders should work together to apply the method, in particular where multiple domains with different approval regimes are involved
- provide guidance on how to develop a safety target which is unified across the TAS

Note: the recommendations have been simplified and combined for the purpose of presentation here. The full list of recommendations, including recommendations made to the EC and EASA is presented in a publicly available report (ASCOS 2015e).

5 The ASCOS Method

5.1 Introduction

The final ASCOS Method was developed (from the interim version) using the recommendations and experience from the case studies and validation workshops. The method is based on modular safety arguments and presents a framework of activities to be used to:

- understand and evaluate the change to the TAS
- identify the affected domains and stakeholders
- decide the path to be taken to gain approval (the approval path)
- agree the approval path with approvers via an approval plan containing an outline safety argument
- develop the modules of the safety argument in parallel with the development of the solution until the safety argument is ready to be presented for approval.

The method is illustrated in Figure 2, with the detail within the 'Develop solution' step illustrated in Figure 3. The detailed explanation of the method is published as ASCOS Deliverable D1.5 (ASCOS 2015a).

The ASCOS Method recognises that changes vary from simple equipment replacement to introduction of new complex concepts involving significant development, both of concepts and products. It also promotes the establishment of a TAS Engineering and Safety Group (TESG) for complex changes, responsible for co-ordinating the engineering and safety activities of all organisations involved in such changes and taking the role of argument architect.

The following sections present the key features of the ASCOS Method.

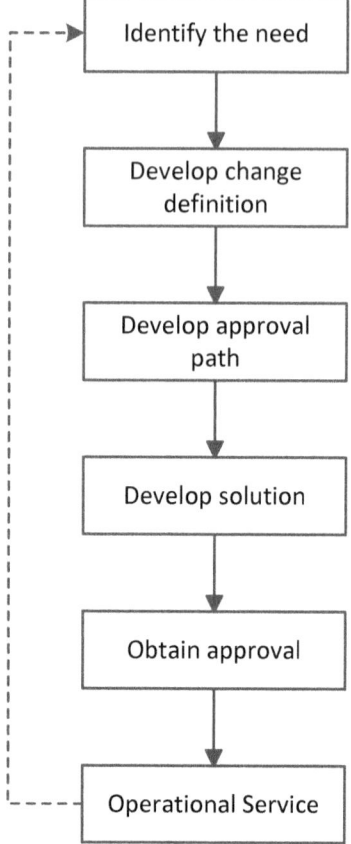

Fig. 2. Overall View of the ASCOS Method[1]

---

[1] The dashed line shows that experience from operational service may lead to identification of the need for further changes.

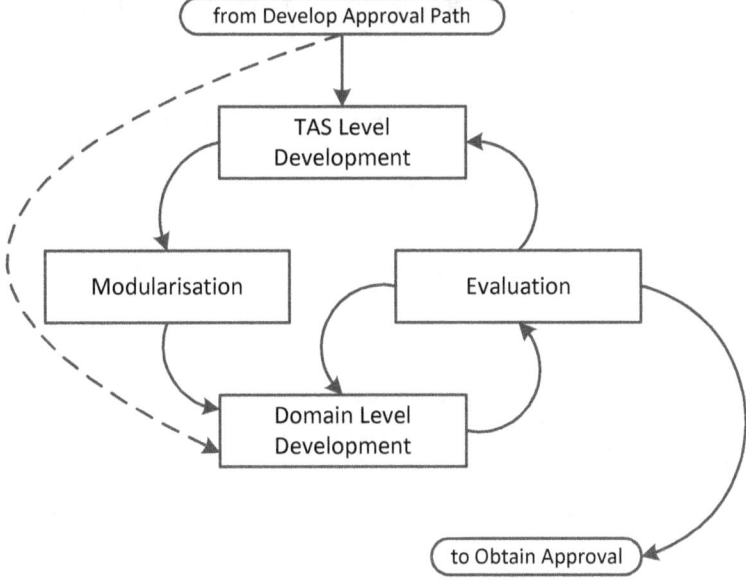

Fig. 3. Cyclic development of solution and safety argument[1]

5.2 Modular Safety Arguments

The concept of Modular Safety Arguments will be familiar to many readers, although such arguments are not used in all areas of the civil aviation industry. It is worth noting that even where the safety argument is not explicitly documented, any application for approval will rely on an underlying implicit argument, which may be embedded in the regulations against which the application is made. See (Holloway 2015) for a further discussion of implicit arguments underpinning standards. We also chose to express these arguments using Goal Structuring Notation (GSN) (GSN 2011), although this is not mandated and there are alternative ways to express safety arguments.

We proposed a top level argument (see Figure 4), based on a commonly used approach within Air Traffic Management (ATM) (EUROCONTROL 2010), with a top level claim that the change being made to the TAS is acceptably safe. The strategy chosen is to align the argument to stages of the lifecycle, including both

---

[1] Development may follow the dashed line if no further development of the solution at TAS level is required.

the transition of introducing the change and the continued monitoring of safety performance in service.

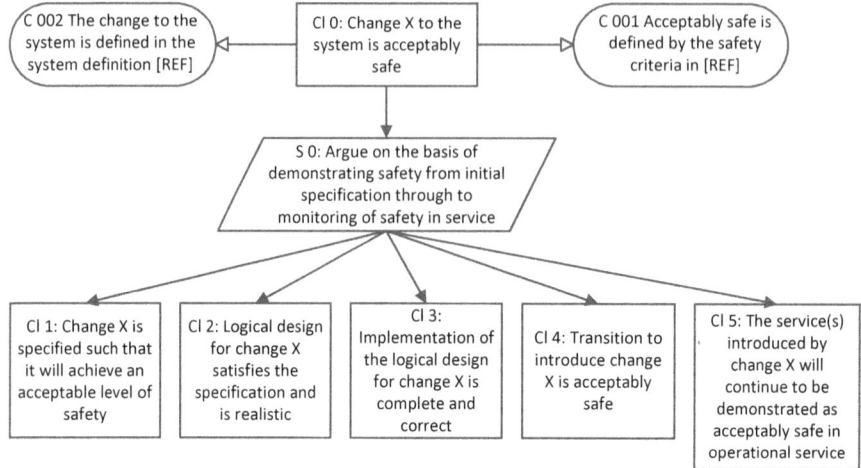

Fig. 4. Generic Top Level Argument

We also proposed to modularise the argument (see, for example, Fenn et al. 2007), aligning the modules to the boundaries of domains and organisational responsibility, with assurance contracts established between the modules to formally define and manage dependencies.

This argument is then decomposed in the usual way, with emphasis on introducing evidence from existing approaches and specifications as soon as possible. This minimises the amount of argument development needed, especially where the existing approaches are adequate. The main areas of argument development envisaged are:

- where the proposed change introduces innovation beyond the scope of existing standards
- where it is necessary to ensure that interfaces between parts of the system are fully addressed.

Modular safety arguments are not a universal panacea; however the concept provides a framework within which the applicant(s) can demonstrate to relevant approvers that any proposed change with the TAS achieves the defined acceptable level of safety. The framework provides a structure within which existing approaches can be evaluated and applied (and augmented where necessary) while also integrating them across the whole TAS.

5.3 Acceptable Level of Safety

The ASCOS Method focuses on demonstrating that the change delivers an acceptable level of safety across the TAS. In other words, the level of safety after the change must be acceptable to all competent authorities who will need to approve the change. *Note: this does not necessarily mean that an improvement in safety must be demonstrated.*

It is therefore necessary to determine appropriate safety targets in each domain affected by the change and demonstrate that each of these is met. Such criteria may be either absolute (specific safety objectives and integrity requirements based on apportionment of a safety target) or relative (comparison of the risk prior to the change against the predicted risk following the change). In the civil aircraft domain, the existence of the target for a catastrophic failure of 10^{-9} per flight hour makes it much easier to apportion absolute targets, whereas the absence of (and difficulty of defining and agreeing) similar absolute targets in other domains means that relative targets are often used.

As a result, each module of the safety argument will need to demonstrate that the change achieves the acceptable level of safety applicable in the domain for which the module is making the safety argument.

It should be noted that a change which increases safety risk in one domain is usually difficult or impractical to justify, even when it significantly decreases safety risk overall. Such a justification would need to be based on robust quantification across all domains which demonstrates a significant overall decrease of safety risk. Production of such a robust quantification is made more difficult by the fact that different domains use different types of targets (often with different units), making it difficult to create valid comparisons between domains. A similar assessment would also be needed in the event of a change with differing impacts on different sovereign states.

5.4 Developing and Agreeing an Approval Path

The overall intention of the ASCOS Method is to gain approval for a change to the Total Aviation System (TAS). Approval is granted by the approver on the basis of a safety argument (supported by evidence) justifying that the change will be acceptably safe.

The ASCOS Method can be viewed as establishing an approval path which, where possible, is based on existing approaches. For some changes, the approval path can be based entirely on existing approaches and appeal to the existing (possibly implicit) safety argument. An example of such a change might be the introduction of an upgraded equipment item on board an aircraft, where the new item has the same fit, form and function as the existing item. This could be visualised as a straight, already-established path, as shown in Figure 5.

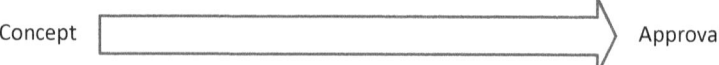

Fig. 5. Approval path using existing approaches

For other changes, established approaches will provide the majority of the evidence needed, but with some gaps. In this case, the approval path may be established by developing approaches which cover the novel solution. These approaches must be developed in a way which takes account of the interface between the novel parts of the solution and the rest of the solution, to make sure that these are fully considered and integrated. The development of these additional approaches provides the missing part (see Figure 6) of the path to solution. This must then be supported by a safety argument which demonstrates that the combination of existing and new approaches fully addresses the change and that the resultant solution is acceptably safe.

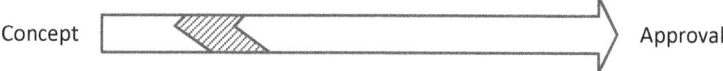

Fig. 6. New approaches developed to complete the approval path

The analogy can be extended to approval paths which improve the efficiency of existing paths, or which may need to be developed from scratch (Figures 7 and 8).

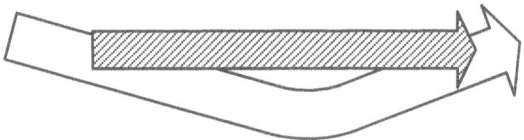

Fig. 7. New approaches developed to provide more efficient approval path

Fig. 8. Development of entirely new approval path

Complex or large changes may involve a combination of the above, as illustrated in Figure 9. Note that in these cases it is important to review the approaches against each other to ensure that the overall approach remains consistent in achieving the overall objective of a safe change to the TAS.

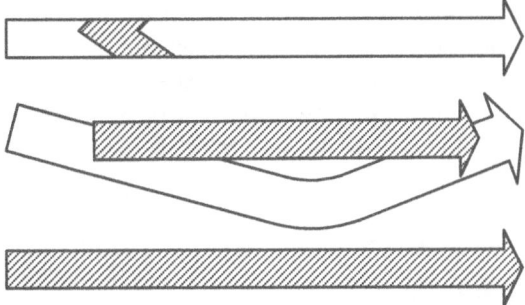

Fig. 9. Different approval paths for different parts of the system

In each case (with the possible exception of where the path exactly follows existing approaches), a safety argument is needed to demonstrate that the change achieves the defined acceptable level of safety. However, the scope of the argument required depends on the degree of novelty involved and on the degree to which the change spans multiple domains of the TAS.

The approval path should be documented in an approval plan. It is very strongly recommended that the safety argument and the proposed supporting evidence should be agreed between applicant and approver. If this agreement is not achieved early in the development, there is a significant risk that the safety argument and evidence produced by the applicant will not be acceptable to the approver. The applicant may then need to incur significant extra effort (and significant delay) in order to produce the evidence required. At worst, the approver may be completely unable to accept the proposed change.

5.4.1 Evaluating Existing Approaches

Great care is needed when evaluating existing approaches to determine whether they need adaptation or enhancement for the specific change. This is especially true where approaches or standards are being applied outside their usual context, or to novel solutions. It is necessary to ensure that the underlying assumptions remain valid; however, this can also be a very difficult task where assumptions are not clearly identified.

This can be illustrated with an example from the adoption of composite materials in airframes. Certification Specifications (CSs) specify the requirements which must be met in order to obtain a Type Certificate for an aircraft; however they are generally not specific to a particular type of material. As metallic structures have been the norm in airframes for many years, the CSs often (implicitly) assume a metal airframe. Some of these assumptions are evident, e.g. references to corrosion; others are less obvious. For example, the mechanisms for growth of cracks in metal structures mean that some cracks can be tolerated, as long as they are detected and monitored. However, damage growth in composites can be rapid, unpredictable and not readily detectable, meaning that a very different approach is needed for composites compared with metal structures.

This is an example of where review of context and assumptions has led to revision of the specification on which certification (and subsequently approval) is based and is addressed in EASA Guidance Material (EASA 2010). Further details can be found in (JAM 2014).

5.5 Modularisation

The ASCOS Method addresses the issue of interfaces within the TAS by introducing the concept of dividing the argument into modules aligned to domains of the TAS and organisational responsibilities. Assurance contracts are established between modules to define and manage dependencies between modules.

This approach has the advantages of:

- making the overall safety argument easier to visualise and understand
- allowing modules to be developed separately from one another in confidence that the final result will be consistent and correct
- partitioning the safety argument such that each approver needs only:
  - to consider specified modules of the safety argument
  - to be assured that the assurance contracts at the boundary of those modules are correctly implemented

Modularisation also allows assumptions and dependencies which might otherwise be lost at the interface between domains to be formally agreed and documented – although this is not sufficient on its own: affected parties must also fully understand their responsibilities and commit to meeting them. This is particularly important given that such interface issues are a key concern within the aviation industry: see for example (AAIB 2015). Further general discussion of modularisation techniques can be found in (Fenn et al 2007).

Modularisation in the ASCOS Method principally involves:

- decomposing the TAS level claims into a complete set of supporting claims spread across each of the domains of the TAS
- determining the assurance contracts between the domains which allow the claims to be made in each of the domains
- specifying the context within which each claim must be demonstrated is completely captured to ensure that the evidence delivered is valid – this includes ensuring that the relevant context is specified within the assurance contracts

Effective management of modularised arguments depends on the argument architect role described below.

5.6 The Argument Architect

As described above, modularisation allows subdivision of the argument into modules, with assurance contracts between these modules allowing them to be developed separately from one another in confidence that the final result will be a consistent and correct overall argument. However, this also introduces a significant risk of divergence between the modules in ways which were not envisaged when the modules were created. It is therefore necessary to ensure that the argument is properly maintained and integrated throughout the development of the change.

When engineering complex systems, the role of system architect is responsible for the design of the overall system; this includes ensuring the integration of the resultant modules. The ASCOS Method gives the argument architect the similar role of designing and maintaining the safety argument, which includes ensuring that the argument modules are correctly bounded and interfaced to other modules throughout the development.

When considering the number of organisations involved in the TAS and their disparate roles, it is often not easy to identify who should be the argument architect. This in part explains why a key concern within the industry is the inadequacy of the management of interfaces between domains; sometimes integration is supervised by the approver or even ignored altogether.

ASCOS proposes (ASCOS 2015c) that any complex development should be co-ordinated by a TAS Engineering and Safety Group (TESG); the TESG would be responsible for co-ordinating all the engineering and safety activities involved in the development of the change. The TESG would therefore play the role of argument architect for changes involving multiple organisations.

5.7 Developing the Safety Argument

The top level safety argument (see Figure 4) needs to be decomposed into subclaims until a level is reached where the claims can be directly supported by evidence.

Within the ASCOS Method, the safety argument is mainly being used to make the link between:

- the high level safety targets embodied in the regulations
- the evidence (to be) produced

The aim is for this evidence to be produced by following existing processes wherever possible and a part of the process of developing the safety argument is a search for links between the safety targets and the evidence produced by existing processes.

The development of the argument should be specific about the evidence required and why it is required; this supports the exercise of reviewing the argument to determine whether the argument claims are in turn satisfied by that evidence.

Where development of the argument leads to requirements for additional evidence which would not be produced by following the usual processes within the domain, this highlights the fact that additional approaches need to be defined and followed.

Where the evidence produced by existing approaches (e.g. standards, AMCs) is sufficient to support the claim, the safety argument should only be developed down to the execution of that approach, along with justification that the context assumed by the approach matches the context required by the safety argument.

Where new or adapted approaches are developed, it may be necessary to develop the safety argument in more detail in order to develop and justify new approaches to generate the required evidence. Especially where new approaches are developed, the safety argument should be specific about the evidence required and why it is required; this supports the exercise of reviewing the safety argument to determine whether its claims are in turn satisfied by that evidence.

5.8 Performance Based or Compliance Based?

Approaches to approval are often characterised as either performance based or compliance based. This terminology can be used to distinguish between:

- requirements or targets which are relatively high level and solution independent (performance based) and
- requirements which are expressed as a detailed set of constraints often assuming a particular solution

The terminology can also be used to distinguish between:

- the goal based approach often used in ATM and
- the certification based approach often used in the aircraft domain

Although there is overlap between these two different ways of viewing the approaches, it is useful to bear both views in mind.

One concern driving the ASCOS Project is that parts of the aviation industry have historically taken a compliance based approach to approval and that this approach stifles innovation because specifications based on historical solutions can be difficult to apply to novel solutions. The performance based approach has been suggested as a way of allowing developers the freedom to innovate and therefore develop optimal solutions. In practice most approvals use a mixture of approaches – for example, the main requirement in the Certification Specification (CS) for

Large Aircraft (EASA 2015) relating to failure analysis (CS25.1309) is goal based, whereas the rest of the CS largely contains prescriptive requirements.

The ASCOS Method allows a goal based approach using high level, solution independent targets to support the development and assessment of innovative solutions, while also allowing more detailed requirements to be used to ensure consistent application of established solutions. Prescriptive requirements (a compliance based approach) are also useful to constrain interfaces or express well established rules, especially where these relate to interfaces with parts of the TAS unaffected by a change.

5.9 Who and When

Table 1 illustrates the roles involved in applying the steps of the ASCOS Method.

This introduces the concept of a change leader, which is the organisation with the primary motivation to make the change to the TAS happen. This organisation will lead the application of the ASCOS Method with support from other organisations. The change leader is responsible for developing the overall plan for approval of the change: through the TESG the change leader will work with the other stakeholders to ensure that the change is developed in a way which is coherent across the whole TAS. The change leader is likely to be the organisation introducing the change into service and therefore likely to also be (one of) the applicant(s).

The approver is the organisation responsible for approving the change. A change may involve multiple approvers, or multiple disciplines within a single approver organisation. Often the approver will be an authority such as EASA or the relevant national Civil Aviation Authority (CAA).

Table 1. Involvement within the ASCOS Method

	Change Leader (supported by TESG)	Applicant	Approver	Argument architect	Manufacturer	Other affected organisations
Identify the need	The need for a change may be identified by one or more parties across industry: the type of need will then drive which organisation(s) become change leader.					
Develop change definition	Lead definition of change at TAS level	Support development of change definition	Support change definition	None	Provide product capability information; support concept development	Provide information about impact of change
Develop approval path	Lead definition of approval path, in collaboration with individual applicants where appropriate	Agree approval plan with approver	Review and accept approval plan	Develop safety argument modules as required to support approval path	Provide information about compliance with requirements	Provide information about impact of change
Develop solution	Lead development of solution at TAS level	Develop safety argument module and assurance contracts; generate evidence	None	Monitor argument and compliance with assurance contracts	Supply evidence to support relevant safety argument modules	Monitor impact of solution on organisation
Obtain approval	Co-ordinate applications for approval	Make application for approval	Review application and grant approval	None	Provide supplementary evidence as required	Provide supplementary evidence as required
Operational service	Monitor occurrences of precursor events or other incidents	Responsible for operation under terms of approval	Monitor operator's compliance with their SMS	Maintain argument based on monitoring of performance	Investigate occurrences of precursor events or other incidents	Monitor impact of operation on organisation

6 Conclusions

In response to the call to improve approval processes in the European aviation industry, the ASCOS Project has shown how modular safety arguments (as widely popularised in safety critical industries, assisted by the development of GSN) can be used to provide a flexible framework for developing safety arguments for changes to the Total Aviation System.

The ASCOS Method is a development of previous work by EUROCONTROL (EUROCONTROL 2010) and Single European Sky ATM Research (SESAR) (SESAR 2012), which introduced a high level safety argument for ATM safety cases. The developments here expand that work to ease its application across the TAS.

The ASCOS Method introduces a modular safety argument structure with modules aligned to domains and organisations within the TAS. Assurance contracts are used to capture and manage the dependencies between modules, addressing one of the critical issues in the development of safety arguments, where stakeholders make assumptions about parts of the TAS which are outside their control. Incorrect management of assumptions, especially those between domains, was a key concern which emerged from the research.

The modular safety argument framework allows approaches and techniques currently used to gain approval within European aviation to be retained where they remain applicable and provides guidance on how these approaches can be augmented where necessary to meet the challenges presented by the complex changes which are now being introduced.

This framework also recognises the importance of early co-ordination between all stakeholders in a change, including the applicant and the approver, to ensure that the safety argument and supporting evidence will be acceptable.

The ASCOS Method described here is the culmination of three years collaborative effort between the ASCOS participants. The method either addresses directly, or provides a framework for addressing, the principles and recommendations identified earlier within the project; further details are provided within the full description of the method (ASCOS 2015a). The report also makes a number of recommendations for development of supporting material and other activities which would support the implementation of the method, including:

- documentation of the implicit safety arguments currently used within aviation
- further definition of the domain structure of the TAS
- development of example safety arguments
- research into open sharing of safety risk information across the industry
- refinement of the TESG concept
- research into the trade-off of safety between domains

However, the most important next step is to apply the ASCOS Method to real life projects and use the experience from those projects to refine and improve the method.

Acknowledgments This technical publication has been realized partly with funding from the European Commission under the ASCOS (Aviation Safety and Certification of new Operations and Systems) Project (Grant Agreement No 314299). The support of the ASCOS consortium partners (see http://www.ascos-project.eu) and Dr Michael Kyriakopoulos, EC scientific officer for project ASCOS, is greatly appreciated.

Abbreviations and Acronyms

Acronym	Description
AAIB	Air Accidents Investigation Branch
AARS	Automated Aircraft Recovery System
AMC	Acceptable Means of Compliance
ANSP	Air Navigation Service Provider
ASCOS	Aviation Safety and Certification of New Operations and Systems
ATM	Air Traffic Management
CAA	Civil Aviation Authority
CNS	Communication, Navigation and Surveillance
CS	Certification Specification
EASA	European Aviation Safety Agency
EC	European Commission
FAA	Federal Aviation Administration
GSN	Goal Structuring Notation
JAM	Journal of Aviation Management
RPAS	Remotely Piloted Aircraft System
SESAR	Single European Sky ATM Research
SCSC	Safety Critical Systems Club
TAS	Total Aviation System
TESG	TAS Engineering and Safety Group
WP	Work Package

References

AAIB (2015) Report on the accident to Airbus A319-131, G-EUOE, London Heathrow Airport, 24th May 2013. https://www.gov.uk/aaib-reports/aircraft-accident-report-1-2015-airbus-a319-131-g-euoe-24-may-2013 Accessed 27th August 2015.

ASCOS (2013a) D1.2: Definition and evaluations of innovative certification approaches. https://www.ascos-project.eu/downloads/ascos_wp1_nlr_d1.2_version-1.4.pdf Accessed 17th September 2015.

ASCOS (2013b) D1.3: Outline Proposed Certification Approach. https://www.ascos-project.eu/downloads/ascos_wp1_ebe_d1.3_version-1.2.pdf Accessed 27th August 2015.

ASCOS (2014) D2.5: WP2 Final Report – Continuous Safety Monitoring. https://www.ascos-project.eu/downloads/ascos_wp2_ava_d2.5_version-1.3.pdf Accessed 11th September 2015

ASCOS (2015a) D1.5: Consolidated New Approval Method. https://www.ascos-project.eu/downloads/ascos_wp1_ebe_d1.5_version-1.1.pdf Accessed 20th October 2015.

ASCOS (2015b) D1.6: WP1 Final Report. https://www.ascos-project.eu/downloads/ascos_wp1_tr6_d1.6-version-1.3.pdf Accessed 20th October 2015.

ASCOS (2015c) D3.5a: Total Aviation System Safety Standards Improvements. https://www.ascos-project.eu/downloads/ascos_wp3_aps_d3.5a_version_1.4.pdf Accessed 17th September 2015.

ASCOS (2015d) D3.6: WP3 Final Report – Safety Risk Management. https://www.ascos-project.eu/downloads/ascos_wp3_aps_d3.6_version-1.2.pdf Accessed 27th August 2015.

ASCOS (2015e) D4.6: WP4 Final Report – Certification Case Studies. https://www.ascos-project.eu/downloads/ascos_wp4_nlr_d4.6_version-1.1.pdf Accessed 11th September 2015.

ASCOS (2015f) D5.5: WP5 Final Report - Validation. https://www.ascos-project.eu/downloads/ascos_wp5_dbl_d5.5_version-1.1.pdf Accessed 20th October 2015.

EASA (2010) Composite Aircraft Structure (AMC20-29 - Annex II to ED Decision 2010/003/R)

EASA (2015) Certification Specifications and Acceptable Means of Compliance for Large Aeroplanes (CS25 Amendment 17 - Annex to ED Decision 2015/019/R), 2015

EUROCONTROL (2010) Safety Assessment Made Easier – Part 1 Safety Principles and an Introduction to Safety Assessment, Edition 1.0

FAA (2002) Commercial Airplane Certification Process (CPS) Study: an evaluation of selected aircraft certification, operations, and maintenance processes

Fenn J., Hawkins R., et al. (2007) Safety Case Composition Using Contracts – Refinements Based on Feedback from an Industrial Case Study. In: Redmill F. and Anderson T. (Eds) The Safety of Systems. Springer

Fowler D. (2015) Functional Safety by Design – Magic or Logic. In: Parsons M. and Anderson T. (Eds) Engineering Systems for Safety. SCSC

GSN Committee (2011) GSN Community Standard Version 1

Holloway (2015) Explicate '78: Uncovering the Implicit Assurance Case in DO-178C. In: Parsons M. and Anderson T. (Eds) Engineering Systems for Safety. SCSC

JAM (2014) Safety Management, Certification and the Extended Use of Composite Materials in Large Passenger Aircraft Structures. In the Journal of Aviation Management 2014.

SESAR (2012) Safety Reference Material, Edition 00.02.01, Project ID 16.06.01, 30th Jan 2012

Industrial experience with Agile in high-integrity software development

Roderick Chapman[1]

Protean Code Limited

UK

Neil White[2]

Altran UK

Bath, UK

Abstract   *This paper reflects on the issues and opportunities raised by the use of Agile practices in the development of high-integrity software, based on the scientific literature, projects, and our own understanding of the relevant regulatory regimes, standards and markets. In particular, this paper considers the assumptions that underpin Agile practices and where these seem to conflict with the disciplines of high-integrity development. Conversely, we'll consider some opportunities where an Agile approach could be significantly improved by the adoption of high-integrity practices.*

[1] rod@proteancode.com

[2] neil.white@altran.com

1 Introduction

In the last decade, there has been much interest in Agile approaches to engineering and, in particular, software development. Proponents of the approach promote its flexibility, lean-ness and ability to manage changing project requirements, and deride the "Plan driven" or "Waterfall" approach. Detractors criticize the apparent free-for-all nature of Agile.

At Altran UK, we are used to a disciplined and planned approach to engineering, particularly when it comes to high integrity systems which involve safety, security or other mission-critical properties. A shallow analysis might conclude that Agile is anathema to the development of high integrity systems, but this is a naïve reaction.

Following a workshop on this topic (SCSC 2015), this paper reports our analysis of Agile, as seen through the lens of an organisation with a track-record in high-integrity software development. We also hope to address some of the following issues:

- Which Agile practices are suitable for and compatible with a high-integrity software development approach?
- Which Agile practices are inappropriate?
- Which Agile practices are underpinned by assumptions that don't seem to hold for high-integrity and/or embedded systems?
- Where does Agile have blind spots where the common good-practice from the high-integrity community could improve matters and potentially offer a "best of both worlds" scenario?
- What would happen if we add best practice from the Lean Engineering and Formal Methods communities to the mix?

We don't imagine having all the answers (yet) to these questions, but we hope this article continues to provoke debate on this important topic.

2 Background and Sources

We started with a literature survey. While many consider that "Agile" really began with the publication of Kent Beck's XP book (Beck 1999), the roots of the approach are much older. Beck points out that many of XP's core practices were well-established long ago—it was their combination and rigorous practice that was novel. A 2003 survey (Larman and Basili 2003) notes instances of both "Incremental" and "Iterative" styles of engineering being used in projects as far back as the 1950s. Redmill's work in the 1980s on Incremental (Redmill 1989) and Evolutionary Delivery (Redmill 1997) predicted many of the problems faced by Agile projects today. Barry Boehm's book (Boehm and Turner 2003) provides some useful insight, while our own work on the development of the MULTOS CA in 1999 (Hall and Chapman 2002) led us to compare our (rather formal) Correctness by Construction approach with XP (Chapman and Amey 2003) with the somewhat surprising result that the two weren't such strange bed-fellows after all. We have also taken inspiration from the Personal and Team Software Processes (Humphrey 2005).

We also looked for reports from industry of Agile methods being used in high-integrity development. These are less well reported, although there appears to have been some substantial take-up of Agile approaches in medical devices (Leffingwell 2011)—this may reflect the nature of FDA's regulatory regime rather than any great breakthrough or strength of the Agile community. In the late 90's, Lockheed Martin developed the Mission Computers for the C130J, combining a semi-formal specification, strong static verification, iterative development and a strongly Lean mind-set (Middleton and Sutton 2005). More recently, some use of Agile has been reported by Thales Avionics (Chenu 2009) in the development of civil aerospace software, while SINTEF have reported success with their SafeScrum approach (SINTEF 2015).

The most recent and plain-speaking evaluation of Agile comes from Bertrand Meyer. While the book (Meyer 2014) does not specifically deal with high-integrity issues per-se, his advice rings true. The book is short, mercifully free of waffle, and required reading.

3 Agile Assumptions and Issues

Agile practices make assumptions about the engineering style and context of an idealised project. How do these assumptions stack up against our norms for a high-integrity project? This section presents a partial list of the most obvious clashes.

3.1 Single Customer

Many Agile texts talk of the importance of involving "the customer" directly in a team, or at least in the management of the product and sprint backlog. The use of the definite article rings some alarms: "the" customer—just one? What about the procurer, end users, industry regulator, independent auditor and so on? In particular, regulators can wield substantial power and work with their own expectations, traditions and standards, which may or may not be compatible with an Agile approach.

Paige (Paige et al. 2011) suggest a "stakeholder consortium" to deal with this, while (Redmill 1997) points out the importance of "the customer" role having substantial decision making power. Agile depends on decisions (such as sprint backlog management) being made quickly—how can a project react if a large committee has to be convened every time a bug or a change has to be triaged? There may have to be distinct "rules of engagement" for different types of defect—from simple bugs that can corrected immediately, to those that involve substantial rework, impact safety, or even require a sprint to be cancelled completely.

3.2 Dependence on "Test"

Agile calls for Continuous Integration, use of a continuously-growing Regression Test suite, and a "Test First" development style, where each piece of functionality is associated with specific tests. Meyer calls these practices "Brilliant" in his summary analysis (Meyer 2014), but Agile seems to assume that dynamic test is the principal (possibly only) verification activity, and lets you know when refactoring is "done", or when the product is "good enough" to ship. Is that it?

The safety-critical community hit the limits of testing long ago. Ultra-reliability cannot be claimed from doing "lots of testing", while every week or so we see the latest security vulnerability, such as HeartBleed (CERT 2014), that has defied an arbitrarily large amount of testing and use. In high-integrity development, we are used to many more forms of verification, including strong (checklist-driven) personal and peer reviews, the use of sound and constructive static verification tools before testing is attempted, traceability analysis, structural coverage analysis and so on.

We don't see any barrier to the use of all of these verification styles in an Agile development, with as much as possible being automated as part of a continuous integration pipeline. The iFACTS project at Altran, for instance, precedes testing activities with a pre-commit static analysis of the software using the SPARK toolset (Chapman and Schanda 2014). The tools are now so fast that this can be done all the time by all the developers. Some would argue that XP's "Pair Programming" replaces personal and peer review, but the data remains inconclusive—our work on the MULTOS CA (Chapman and Amey 2003) led us to postulate that the use of strong verification and focussed review (i.e. don't review where the tools have already verified the absence of defect classes) is better in terms of both resulting defect density and productivity.

3.3 Up-front Activities and Architecture

The proverbial elephant in the room. Agile advocates a "minimal design" approach, with a focus on building what's needed now, and deferring decisions with the use of re-factoring to accommodate later changes. This assumes that the refactoring is cheap, fast and doesn't invoke significant rework other than changes to "source code". Agile also seems to assume that test cases and results don't change in a refactoring because they are all "Black Box" functional tests.

Imagine you're building an aeroplane: you start with a single-engine design for simplicity. Later, the customer requests that the plane can carry some fare-paying passengers. You go to your national aviation regulator, who informs you of the need for at least two engines to carry passengers, so you decide to "refactor in" another engine for added redundancy and safety. We think not.

In the design of safety- and security-critical systems, architecture (both software and hardware) remains our principal weapon in meeting non-functional requirements. Tools at our disposal include physical matters such as redundancy, replication, separation (of critical from non-critical), which remain rather expensive to get wrong and refactor late in the day. The trick, then, is to do "just enough" up-front activities to come up with a specification and architecture that can support a satisfaction argument for non-functional properties. Our experience also suggests that sound up-front work saves money overall by avoiding later (expensive) rework. NASA's study on flight software complexity (NASA 2011) uses a COCOMO-II model to yield a "sweet spot" for up-front investment that minimizes later re-work, depending on the size of the final system. At 1 Million lines-of-code, the model suggests investment in up-front activities of 29% of total budget.

3.4 User Stories and Non-Functional Requirements

For security and safety, a key goal is to make sure that our specification of system behaviour covers all possible cases of input and system state. For example: it seems reasonable that, for security, we would want a system to respond in a well-defined way to all possible input data, including (possibly formal) definition of data validity and all error-handling cases.

Our main concern with Agile in this context is the use of "user stories" as a substitute for requirements engineering. A set of stories provides a sampling of the cases of a system's behaviour, but provides no obvious verification of their completeness. What lies in the "gaps" between stories? Vulnerabilities, bugs, unexpected termination, and (at worse) undefined behaviour for starters. Meyer files user stories under "Bad and Ugly". We agree.

3.5 Sprint Pipeline, Iteration Rate and Acceptance

Most descriptions of Agile show a model of a single active "Sprint" which is immediately delivered to the customer. There are therefore a maximum of two builds of the system of interest at any one time:

- Build N, which the customer currently has "in operation" and is using and possibly reported defects against.
- Build N+1, which is the subject of the current development sprint.

If only it were that easy. This model assumes that the customer is always willing and able to accept a delivery of the product and put it into operation immediately. This might be true of web sites (where the users don't even get any choice of which build of the server they're using), but we know this is not realistic for real-world high-integrity or embedded projects. Our projects have successfully fielded an approach, though, that uses a deeper pipeline and *multiple* iteration rates. One major project, for example, has at least four builds in the pipeline, thus:

- Build N is in "live" operation with the customer.
- Build N+1 is undergoing acceptance test in the customer's test lab. This process is subject to significant regulatory requirements, and so can take months.
- Build N+2 is being developed and tested, based on a previously developed formal specification.
- Build N+3 is undergoing requirements engineering and formal specification.

All four pipeline stages are "running" concurrently. The project also successfully runs multiple iteration rates and delivery standards. For example, a rapid "nightly" build can be delivered to the customer but comes with substantial limitations on its "assurance package" and allowed use (i.e. it is expressly *not* for operational use, but may be appropriate for seeking feedback from the customer on a new feature, or validation of a defect fix). At the other end of the spectrum, a *full* build comes with a complete assurance package, including a safety case, and is designed to go into acceptance test and eventual operation. The trick is to make the iteration periods *harmonic* with the customer's ability to actually accept them.

3.6 Embedded Systems Issues

Embedded systems pose several further challenges. In particular, an Agile approach presumes the plentiful availability of testing resources and that these machines are fast enough to drive an Agile development pipeline. For embedded systems, this is likely to be a problem: many projects have just one or two "target rigs" on which testing can be performed, even when the hardware has been built at all. Such target machines are typically much slower than today's servers, and are difficult to "get at" owing to lack of traditional I/O, filesystem and so on.

Two approaches can mitigate this problem. First, we can reduce "on target" testing with more static verification and testing "on host" if we can justify it. Secondly, the advent of virtualized simulation of target hardware offers the possibility of each developer having a "virtual target" on their desktop, or possibly a cloud of such machines hosted on fast servers as part of an Agile deployment pipeline. While emulation and/or virtualization of popular embedded microprocessors are common-place, high-fidelity simulation of a target's I/O and operating environment remains a significant challenge.

4 Opportunities

We now turn to opportunities, where common high-integrity practices might improve or complement an Agile approach.

4.1 Static Verification and Pair Programming

Development of the MULTOS CA system led to two observations on this topic:

- Strong static verification tools tend to complement (not replace) human-driven review. The tools are very good at some problems (e.g. global data flow analysis, theorem proving) where humans are hopeless, and vice versa. If we do the static verification first, then we can adjust manual review processes and checklists to take advantage of this.
- The use of sound static verification (i.e. a tool which justifiably argues that it catches *all* the bugs, not just some of them) can support the complete elimination of subsequent verification activities. This point has been picked up standards such as (DO-178C 2011) and (DO-333 2011).

We do not propose to eliminate peer review, but to make it *more effective* through the use of tools allowing humans to concentrate on what they're good at such as the complex trade-offs between security, safety, performance, testability, maintainability and so on. Another way of thinking about it is that a programmer with a static verification tool *is* a form of "Pair Programming" so perhaps there's no clash here after all.

4.2 Automation of Assurance Evidence

A simple Agile deployment pipeline (Humble and Farley 2011) automates the generation of deliverable "packages" (mostly binary code, data, configuration files and so on), test results and possibly some documentation. We can certainly do better. For recent high-integrity projects, we start with the requirements for assurance evidence from whatever regulator and/or standard is relevant, then plan the development process and pipeline "right to left" to automate (as far as possible) the generation of these artefacts.

5 High-Integrity Deployment Pipeline—Turning the Dials Up

At Altran, various projects have fielded the ideas in this article, but we have yet to try all of them at once. For new projects, an Agile development and deployment pipeline would exhibit the following properties:

- Requirements engineering in the style of (Jackson 2000), concentrating initially on non-functional requirements and development of architecture, specification and an associated satisfaction argument.
- A "rolling" formal specification, with just enough formality and completeness to inform estimation of the remaining work and initiation of the opening development iterations. This is especially important in a commercial environment: if we are asked to prepare a fixed-price bid for the implementation phase of a project, then we need enough confidence in the completeness of the specification so we can win the work, but still make a profit.
- A continuous integration and deployment pipeline, combining static verification with the dials "turned up to 11" (Reiner 1984), continuous regression test, automated generation of documents and assurance evidence, plus a cloud of virtualized target platforms for integration and deployment testing. In this way, the standard Agile "Deployment Pipeline" (Humble and Farley 2011) becomes an "Evidence Engine", shown in Figure 1.
- A planned iterative development style, starting with a partial-order over system infrastructure and "features". A *partial* order reveals the "X must be built before Y" relationship between features, informing an iteration plan, but also exposing potential for parallel development of independent features. Early iterations will be planned in some detail, while later iterations will left open to avoid waste and accommodate change.

6 Conclusions and Further Work

The $64 million question, it seems, is how much up-front work is appropriate for a particular project. We doubt there's a one-size-fits-all approach, but surely the answer should be informed by disciplined requirements engineering of key properties (e.g. safety, security, primary functional behaviour and others) that can inform the design of a suitable architecture and its accompanying satisfaction argument.

As for further work, we are deploying these ideas on current projects, and look forward to being able to report the results. We hope others will do the same.

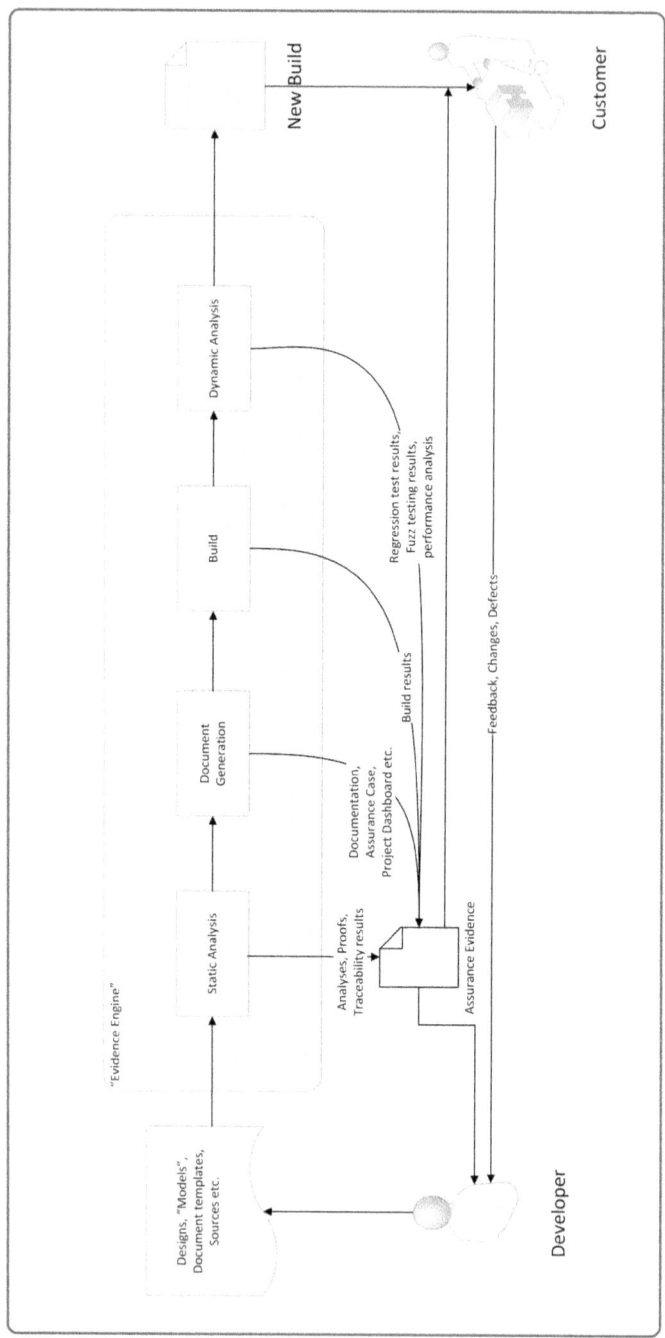

Fig. 1: High-Integrity Agile Evidence Engine

Acknowledgments   Many thanks to Felix Redmill, Jon Davies, and Mike Parsons for their comments on an earlier draft of this paper.

References

Beck K (1999) Extreme Programming Explained: Embrace Change. Addison Wesley, 1999

Boehm B, Turner R (2003) Balancing Agility and Discipline: A Guide for the Perplexed. Addison Wesley, 2003.

CERT (2014) "HeartBleed" OpenSSL Vulnerability. US Computer Emergency Readiness Team. https://www.us-cert.gov/security-publications/Heartbleed-OpenSSL-Vulnerability. Accessed 19th October 2015.

Chapman R, Amey P (2003) Static Verification and Extreme Programming. Proceedings of ACM SIGAda Conference, 2003.

Chapman R, Schanda F (2014) Are we there yet? 20 years of industrial theorem proving with SPARK. Proceedings of Interactive Theorem Proving (ITP) 2014. Springer-Verlag LNCS Vol. 8558, pp. 17-26. DOI: 10.1007/978-3-319-08970-6_2.

Chenu E (2009) Agility and Lean for Avionics. Thales Avionics, Valence. http://www.opendo.org/2009/05/07/avionics-agility-and-lean/ Accessed 6th October 2015.

DO-178C (2011) Software Considerations in Airborne Systems and Equipment Certification. RTCA Inc., 2011.

DO-333 (2011) Formal Methods Supplement to DO-178C and DO-278A. RTCA Inc., 2011.

Hall A, Chapman R (2002) Correctness by construction: building a commercial secure system. IEEE Software, vol. 19, no. 1, Jan/Feb 2002, pp. 18–25, 10.1109/52.976937.

Humble J, Farley D (2011) Continuous Delivery. Addison Wesley, 2011. ISBN 978-0-321-60191-9.

Humphrey W (2005) PSP: A Self-Improvement Process for Software Engineers. Addison Wesley, 2005.

Jackson M (2000) Problem Frames. Pearson Education, 2000.

Larman C, Basili V (2003) Iterative and Incremental Development: A Brief History. IEEE Computer, June 2003.

Leffingwell D (2011). "Agile Software Development with Verification and validation in High Assurance and Regulated Environments – A Rally Software Development Corporation Whitepaper." From scalingsoftwareagility.wordpress.com Accessed 6th October 2015.

Meyer B (2014) Agile! The Good, the Hype and the Ugly. Springer, 2014.

NASA (2011) NASA Study on Flight Software Complexity Final Report. From http://www.nasa.gov/offices/oce/documents/FSWC_study.html. Accessed 6th October 2015.

Middleton P, Sutton J (2005) Lean Software Strategies. Productivity Press, 2005.

Paige R et al. (2011) High-integrity agile processes for the development of safety critical software. International Journal of Critical Computer-Based Systems. Vol. 2, No. 2. 2011.

Redmill F (1989) Computer Systems Development: Problems Experienced in the Use of Incremental Delivery. Proceedings of SAFECOMP 1989. pp. 25-32.

Redmill F (1997) Software Projects: Evolutionary vs Big-Bang Delivery. Wiley 1997. http://www.safetycritical.info/library/NFR/ Accessed 6th October 2015.

Reiner R (1984) This is Spinal Tap. MGM Pictures 1984 (the scene with the amplifier.)

SCSC (2015) Agile Development for Safety Systems Workshop. Safety Critical Systems Club, London, June 2015. http://www.scsc.org.uk/e346. Accessed 19th October 2015.

SINTEF (2015) SafeScrum website. http://www.sintef.no/safescrum Accessed 6th October 2015.

Practical Statistical Evaluation of Critical Software

Peter Bernard Ladkin

University of Bielefeld and Causalis Limited

Bielefeld, Germany

Bev Littlewood

City University,

London, UK

Abstract   *In 2010, Rolf Spiker approached one of us with a query from a client concerning advisory material in IEC 61508 on the statistical evaluation of software. We realised that there is a dearth of practical guidance for those who wish to evaluate critical software statistically. We believe statistical evaluation of software is an increasingly important assurance technique. We commence with a brief introduction to some of the simpler statistics and then consider discursively the issues which arise during evaluation.*

1 Introduction

It is sometimes said that "software failures are 'systematic' and therefore it does not make sense to talk of software reliability in probabilistic terms". It is true that software fails systematically, in that, if a program fails in certain circumstances, it will always fail when those circumstances are exactly repeated. Where then, it is asked, lies the uncertainty that requires the use of probabilistic models and measures of reliability?

There are two main sources of uncertainty. First, there is uncertainty about which inputs, of the many possible inputs the software could receive, will result in failure (of the software to fulfil its intended purpose) when executed. Second, there is uncertainty about which inputs the software will in fact receive in the future as it executes: these inputs will depend upon the external operating environment, about which there will be uncertainty.

It follows from these two sources that there is uncertainty about when a program will receive an input that will cause it to fail. Failures thus form a stochastic process (a random process) as time progresses during execution of the software. There are some simple probabilistic models for such failure processes (as well as some complicated ones). We describe these briefly and show how they can be used to obtain quantitative probabilistic measures of software reliability.

Because software failures occur randomly, it follows that many of the classic measures of reliability that have been used for decades in hardware reliability are also appropriate for software: examples include failure rate (for continuously operating systems, such as nuclear reactor control systems); probability of failure on demand (pfd) (for demand-based systems, such as nuclear reactor protection systems); mean time to failure; and so on. This commonality of measures of reliability between software and hardware is important, since practical interest will centre upon the reliability of systems comprising both. However, the mechanism of failure of software differs from that of hardware, and we need to understand this in order to carry out reliability evaluation.

2 Simple probability models of the software failure process

In this section we outline two simple probability models that describe two common types of failure processes: a discrete-time (counting) model for on-demand systems, and a continuous-time model for continuously operating systems, respectively. Many software-based systems fall into one of these two classes, although there are, of course, exceptions: see discussion in Section 3.

2.1 On-demand software based systems

Consider a nuclear reactor protection system (or "safety system", as it is called; NRPS). An idealized view of the NRPS is that its role is to act only when the reactor enters a hazardous state (the "demand"), whereupon its function is to shut down the reactivity and keep the reactor in a safe state. Such demands upon the NRPS might arise because of the failure of a wider system – e.g. the continuously operating control system – and in a well-designed reactor they could be expected to be quite rare, say about once a year. A dangerous failure of the NRPS would be the system not responding to a legitimate demand. Part of the wider safety case for the reactor would contain a requirement that the probability of such a failure on demand (pfd) of the protection system be adequately small[1].

How could such a figure be claimed with high confidence? First of all we need a simple model for testing and operational use of this kind of on-demand software-based system.

We observe that there is a (probably very large) set of possible demands. Label these 1, 2, 3, ... Selection of successive demands by the operational environment (i.e. the wider reactor and its environment in our example) occurs randomly and, we claim, independently, with $P_i = Pr$ (demand i is selected) forming a probability distribution over all demands. Note that selection is not generally equi-probable (indeed this is usually unlikely). Each demand either results in failure, or does not. Define the variable:

$X_i = 1$ if demand i results in failure

$X_i = 0$ otherwise

It is easy to see that the probability of failure of a randomly selected demand is:

$$pfd = \sum_i P_i \times X_i \qquad (1)$$

In practice, we would not know the distribution $\{P_i\}$ completely; nor would we know which demands cause failure, so that the $\{X_i\}$ will also be unknown. It follows that (1) cannot be used to calculate pfd.

Instead, pfd can be estimated statistically from the results of operational testing, i.e. testing that selects the demands in exactly the same way they would be selected in operational use. Such testing is often based on simulation that uses an understanding of the physical world in which the computer-based system operates. In the example of a reactor protection system, this would require knowledge of the physics and engineering of the reactor, and of the reactor's operational environment.

In such testing we observe a sequence of trials, each of which will result in either success or failure. If the trials are statistically independent, and the probability of failure has the constant value pfd for each trial, they are called Bernoulli trials. A sequence of such trials forms a particularly simple stochastic process, called a Bernoulli process.

There are two random variables of interest in such a process. Firstly, the number of failures in a given number, n, of successive demands. This has a Binomial distribution:

$$\Pr\left(r \text{ failures occur in } n \text{ demands} \right) = \binom{r}{n} pfd^r \left(1 - pfd \right)^{n-r} \qquad (2)$$

---

[1] For example, in the case of the UK Sizewell B reactor, this figure was 10^{-7} (of which 10^{-3} was allocated to the software-based Primary Protection System (PPS), and 10^{-4} to the hardware-only Secondary Protection System (SPS), in this 1-out-of-2 configuration).

Secondly, the number of demands until the next failure has a geometric distribution:

Pr(number of demands up to and including next failure $= r$)

$$= (1 - pfd)^{r-1} pfd \tag{3}$$

Notice that this is true regardless of whether we count starting from a failed demand, or not: the Bernoulli process is said to be *memory-less*.

If we observe r failures in n trials it is straightforward to compute estimates of *pfd* as a function of r and n: details can be found in any introductory text-book on stochastic processes, e.g. (Siegrist 2014). In particular, if we see *no* failures[1] (i.e. r=0), confidence bounds for *pfd* can be obtained as in the Table 1. Table 1 includes numbers related the IEC 61508 SIL levels, taken "one-sided". They arise from the mathematics of the Binomial distribution in Equation (3).

Table 1: Numbers of failure-free demands required to obtain confidence in different pfd levels

SIL level	Acceptable probability of failure	Number of failure-free demands for 95% confidence	Number of failure-free demands for 99% confidence
SIL 1 or greater	$<10^{-1}$	3×10^1	4.6×10^1
SIL 2 or greater	$<10^{-2}$	3×10^2	4.6×10^2
SIL 3 or greater	$<10^{-3}$	3×10^3	4.6×10^3
SIL 4	$<10^{-4}$	3×10^4	4.6×10^4

Table 1 is simply for illustration. Generally, 95% confidence can be placed in a claim that the *pfd* is smaller than 10^{-x} if 3×10^x failure-free demands have been observed, and so on.

These results are based on two important assumptions, and a user needs to be confident that these are satisfied for hisher particular application.

First, the statistical properties of the test case selection need to be exactly the same as those of demand selection in operation. If the distribution of selection probabilities of the test cases was $\{P_i^*\}$, different from $\{P_i\}$, then the probability of failure on demand *in test* will be

$$pfd = \sum_i P_i \times X_i \tag{4}$$

---

[1] In some safety-critical industries, regulators will accept *only* evidence of failure-free working in support of *pfd* claims.

which will not be the same as *pfd*, (1), the probability of failure on demand in operation. In such a case, estimates of the former will not be accurate estimates of the latter.

Second, successive demands must be independent, with constant probability of failure. In our illustrative example of a reactor protection system this may be a plausible assumption since the demands will be far separated in calendar time. It seems reasonable to assume that today's demand is not affected by the nature of a demand that occurred last year: i.e. knowing whether or not last year's demand failed will not affect the probability that *this* demand will fail, which is just the constant *pfd*.

2.2 Continuously operating software-based systems

Many software-based systems operate in continuous time. Common examples include those that control complex hardware: e.g. automobile engine control systems, fly-by-wire airplane flight control systems, nuclear reactor control systems. In such examples, the state of the system under control will be determined by the elements of a many-dimensional vector of inputs – for example, in the case of a reactor control system: temperatures, pressures, coolant flow rates, etc.

For continuously-operating software-based systems, the vector of inputs forms an evolving *trajectory*, or path, in the multi-dimensional input space as (continuous) time passes. With this way of looking at things, software failures can be identified with regions of the input space. Call these *fault regions*. When the execution trajectory enters a fault region, a software failure occurs.

There are two sources of uncertainty, as in Section 2.1. First, there will be uncertainty about the nature ("shape") and location of the fault regions in the input space. Secondly, there will be uncertainty about the future direction an execution trajectory will take. Thus as time passes the occurrence of failures – points on the time axis – is random: it forms a *continuous-time* stochastic point process.

The simplest such process is called a Poisson process, and this will often be an accurate model of the failure process of continuously operating software-based systems. A Poisson process is characterised by a single parameter, λ, its *failure rate*, measured for example in failures per hour. As in the case of on-demand systems discussed in Section 2.1, there are two random variables of interest. First, the number of failures in a given interval, $(0,t)$, of elapsed time has a Poisson distribution:

$$\Pr(r \text{ failures occur in } (0,t)) = \frac{(\lambda t)^r e^{-\lambda t}}{r!} \tag{5}$$

Second, the time to the next failure has an exponential distribution with probability density function

$$\lambda e^{-\lambda t} \tag{6}$$

Notice that, as in the Bernoulli process, this is true regardless of whether we measure the time from a failure or not: the Poisson process is memory-less.

We can use test data to estimate λ. If r failures have been observed in elapsed time t it is a simple matter to estimate λ, calculate confidence bounds, etc. Many textbooks give the simple details, e.g. (Siegrist 2014). As before for on-demand systems, a particularly interesting case for safety-critical applications is where $r=0$. In Table 2 are some examples of confidence bounds based on IEC 61508 SIL levels.

Table 2: Numbers of failure-free hours required to obtain confidence in different failure-rate levels

SIL level	Acceptable probability of failure per hour	Number of failure-free hours for 95% confidence	Number of failure-free hours for 99% confidence
SIL 1 or greater	$<10^{-5}$	3×10^5	4.6×10^5
SIL 2 or greater	$<10^{-6}$	3×10^6	4.6×10^6
SIL 3 or greater	$<10^{-7}$	3×10^7	4.6×10^7
SIL 4	$<10^{-8}$	3×10^8	4.6×10^8

Again, these numbers are just illustrative. Generally, if it is required to claim a failure rate better than 10^{-x}, with 95% confidence, then 3×10^x hours or more of failure-free working need to be observed; and so on.

3 Some Observations on Applicability

The advantages of using these stochastic processes for interpreting software behaviour in situ are threefold. First, the pertinent mathematics of these stochastic processes are simple, clear, and well-understood - for Bernoulli processes for some 300 years! (Bernoulli 1713). Line engineers tasked with assessing software could be routinely expected to develop the pertinent mathematical skills. Second, the key parameters are few and clear, so that it is often a straightforward matter to identify these key parameters in system operation and be reasonably assured one has them right. Third, interpretations as Bernoulli resp. Poisson processes are indeed often feasible in software operation. However, whilst many systems fit into one of these classes – on-demand systems operating in discrete time, or continuously operating systems in continuous time – there are also many exceptions.

Indeed, the choice of which of the two interpretations to use can sometimes be a matter of convenience. Consider, for example, a safety-critical flight control system in a civil airplane: this is obviously a continuous time system. But it may be convenient sometimes to treat it as a discrete time system where the measure of interest is probability of failure per flight. Here, a "demand" is a "flight". A count of the number of demands – i.e. take-offs and landings – may better reflect the exposure of the system to possible failure than calendar time, which includes hours spent in straight and level flight.

In the example of a protection system used above, only failures to respond to a (genuine) demand were considered, and these naturally form a discrete time stochastic process in terms of the sequence of successive demands. Such failures are sometimes called "Type 1", in contrast to "Type 2" failures in which the protection system incorrectly shuts down the reactor when the latter is not in fact in a hazardous state. Type 2 failures, in contrast to Type 1 failures, form a continuous-time stochastic process of events in real time (i.e. clock, or calendar time). Type 2 failures are generally less serious than Type 1 failures, and may not impinge on system safety, but they certainly affect system *reliability*. It would be reasonable to have probabilistic requirements for both types, necessitating the use of both of the probability models described in Section 2.

We have endeavoured to make clear in the examples above that the discreteness or continuity of time concerns the world outside the system, and not the system itself. Whilst it is true that computer systems themselves can be thought to operate in discrete time – clock cycle time – this discreteness is entirely distinct from the worldly discreteness in a Bernoulli process which concerns successive demands upon a computer-based system. There is to our knowledge no simple way (indeed at time of writing we do not know of *any* reasonable way) to relate processor clock cycles to the demands in a legitimate Bernoulli process model.

Of course, not all computer system failures can be modelled by a Bernoulli or a Poisson process. For example, the assumption of constancy of *pfd* (or failure rate) will be violated if fault fixes are made (or attempted) when failures occur, because the code has changed. One would not be measuring the selfsame object after such a change. In such cases, it might be expected that there will be reliability growth, at least in the long run1. More complex reliability growth models (RGMs) are needed to represent such situations and there is now a large scientific literature on problems of this kind.

However, it is questionable whether such models are appropriate for safety-critical systems. They require assumptions about the efficacy of fault-fixing that are difficult to justify, and thus may not produce conservative results. The simpler models described here, in contrast, require that no changes are made to the system as failures occur and are thus guaranteed to be conservative in this respect. In fact, as has been remarked earlier, in many safety-critical applications there will be a requirement that *no* failures are observed.

---

1 Some fix attempts may not succeed. Some may introduce novel faults. But in the long run it might be expected that reliability will increase in spite of such reversals of fortune.

4 Determining Success and Failure

Talking about failure relies on having some notion of successful execution and non-successful (that is, failed) execution of software. There are generally two notions in common use when speaking about software execution.

First is when something happens which does not conform with the expectations of a user of the software. What is meant here by "user" is also a fluid notion. I can use software without having any defined stake in its evaluation or ability to report on its operation, for example if I use third-party web-application software to perform a transaction. The term "stakeholder" might be more appropriate. When using WWW software to perform a transaction, I certainly have a stake in its (to me) correct operation, but it may still do things I don't wish – and the other party to the transaction may wish it so. There is nothing prima facie to say who is right about whether the software is operating correctly. And some software may be designed to force a third party to use it in certain ways uncomfortable for them. It follows that this notion of correctness is a social construct.

Second is when there is a rigorous specification of software behaviour. A failure can be defined as a behaviour (or the outcome of a behaviour) which does not conform with the specification.

There are notions which transgress the boundaries of these conceptions. Say I have a specification, but the specification is in retrospect not quite right (which is often the case). There may be behaviours which may be what I want, but which do not conform with the inaccurate specification.

There is not space here to investigate the notion of failure of software in any detail. However, the statistical evaluation of software does depend on a coherent, deterministic notion of failure of software. One must be able to say in any given circumstance whether the software has failed or has not failed. In the absence of such a clear notion in a specific case, software cannot effectively be evaluated statistically using the methods we have indicated above.

5 Some Tricky Issues

None of the above says that interpreting software operation in situ as a Bernoulli or Poisson process is a straightforward matter. Indeed, there is a case to be made that the key skill for an engineer wishing statistically to evaluate critical software is interpretive rather than mathematical. The hard question is: are you sure your process really is legitimately Bernoulli, respectively Poisson?

5.1 "Easter Egg"-Type Behaviour

A major issue is that there is no useful constraining relation between the behaviour of the software on one set of inputs and its behaviour on another, closely related set of inputs.

In certain consumer software of the past (and present), programmers would occasionally include so-called "Easter Eggs" (Wikipedia 2015). When a specific combination of inputs was given, the software would cease functioning as intended and it would play a tune, or display a cute picture or greeting, or some such. The chosen trigger combinations were such as to be deemed to be extremely unlikely in normal operation, so only people who "knew" could usually evoke the Easter Egg behaviour.

Some software used in critical applications has a "debug" or "maintenance" mode (DMM) which allows a user access to internal data structures in the software. Giving the software input while in DMM results in output of interest to the maintainer, which will rarely be values appropriate for the critical function of the software. Thus this critical function will routinely fail when the software is in DMM. The software is switched into DMM by a specific combination of input values known to the developers/maintainers ("maintainer"), but not necessarily by the engineer wishing to use the software in a critical application and evaluating its use statistically ("client"). The maintainer knows about the quasi-Easter-Egg, the client not.

Suppose the software has been statistically evaluated on typical in-service inputs. Suppose future inputs are identical to those past inputs, with the sole exception that occasionally the DMM trigger input is seen. The software will fail (to fulfil its intended function) each time this trigger input occurs. The software failure behaviour in the future application will be decisively different (worse) than has been seen in the evaluation. But the difference in inputs from evaluation inputs to future inputs is just a single one of the input values! This shows clearly that the condition that future inputs must be the same, and occur with the same relative frequency, as in the evaluation, must be taken rigorously for predictions from the evaluation to be realised in the future use.

5.2 Masked Dependencies

Sometimes the behaviour of software is dependent upon input parameters which have not been explicitly recognised. If the behaviour of these parameters is different in the future use from that in the evaluation, then the software behaviour might well be different, even when the behaviour of the explicitly-recognised parameters stays the same.

A colleague tells of assessing a system for dependence on GPS. The developer assured the assessors that the software was not at all dependent on GPS signals: it had no function that would require location information; no such dependency had been deliberately implemented; indeed, an attempt had been made explicitly to avoid it. The software did not use library or other external functions that were known to rely on GPS.

The assessors brought in a GPS jammer and activated it. The software soon ceased to operate as intended because of the jamming. This, apparently, is not an uncommon occurrence (Thomas 2011, RAEng 2011).

5.3 Version Deviations

It is commonplace that minor changes to software may result in major changes in behaviour. Apple's "*goto fail*" bug in its TLS/SSL verification software, noted in February 2014 (Ducklin 2014), ensured that all WWW-site certificates were validated, no matter what their actual status as genuine or spoofed. That is a radical failure of intended function. However, the source code responsible was one line containing 11 ASCII characters that seems to have been spurious (an exact duplicate of the preceding line).

Since there is no estimable correlation between behaviour of software and source-code changes, there is no way of reliably estimating the failure behaviour of new versions of software based on the failure behaviour of previous versions and the nature of the changes. If, say, Version 1.2 has been evaluated, and a minor change has been made resulting in Version 1.3, then the failure behaviour of Version 1.3 cannot in general be reliably estimated from the evaluation of Version 1.2.

It may be possible to evaluate the behaviour of Version 1.3 if an impact analysis can demonstrate reliably that the changes made to Version 1.2 cannot affect the pertinent behaviour of the software. Such analyses move outside the realm of statistical evaluation and, to be performed reliably, likely involve the use of rigorous formal methods.

5.4 Failure Masking

Failure masking is a phenomenon often desired in fault-tolerant systems. Large parts of computer science have been devoted towards devising algorithms and techniques to tolerate failures, often but not always involving component redundancy. Failures so tolerated may not be apparent to the user; that is, may be "masked".

This is not the phenomenon usually meant when the term "failure masking" is used in statistical evaluation of software. Failure masking relevant to software evaluation occurs when a software component S fails or is imminently about to fail, but this failure is not registered because a larger or a different component C fails: the failure is registered as a failure of this second component C, and the state of software S is not registered. If software S is about to fail, or has failed unnoticed, then this would count for statistical evaluation as a failure of software S on the existing input. However, it is not registered. But is a precondition for successful statistical evaluation of S that all failures are registered.

One of the most well-known examples of failure masking concerns the meltdown of a Babcock and Wilcox 900 Pressurised Water Reactor (PWR) at the Three Mile Island Generating Station in Pennsylvania, USA, in March, 1979 (IAEA 2002). The primary coolant surrounding the reactors is itself cooled by a separate secondary cooling system, also water-based, and a heat-transfer mechanism. The secondary cooling system had stopped circulation, so the primary coolant was heating up. A relief valve (called an "electromatic relief valve" by the manufacturers, Dresser Industries (Perrow 1984) but "pressurizer relief valve" in (IAEA 2002), allows overheated primary coolant to overflow into a sink, relieving pressure in the primary containment (the pressure vessel holding the reactor core) due to the overheating. Enough primary coolant should remain after venting to continue to function, so the relief valve must close when pressure has reduced appropriately. The valve, however, failed to reseat and coolant continued to drain out; ultimately a third of it escaped through the valve. The valve position indicator itself had a fault and indicated to plant controllers that the relief valve was closed, when it wasn't. The failure of the indicator masked the failure of the relief valve. An animated image of the sequence of events is included in the U.S. Nuclear Regulatory Commission "backgrounder" (USNRC 2014).

Software failure masking occurred in the incident to Malaysian Airlines Boeing 777-200 9M-MRG in August 2005 (ATSB 2007). The software was fault tolerant, and, before and during the accident flight, operation of the software masked a previous failure of an air-data unit, whose erroneous values were treated as veridical by the primary flight control computer system, which then commanded significant and untoward deviations in pitch. The failure masking is considered in detail in (Johnson and Holloway 2007), which illustrates the difficulties that may arise in registering failures (of software or of hardware) accurately.

It is beyond scope here to consider failure masking in detail. Suffice it to say that considerable attention must be paid to its possibility where software is to be statistically evaluated.

5.5 Deviations from the Model

The conditions of memorylessness mentioned above are strong conditions on evaluation. An example is given in (Ladkin 2015) of non-memoryless behaviour in software with one failure condition. The particular interest of that example is that some people think that complex software such as real-time operating systems (RTOS) have "proved their worth" over sometimes millions of hours of "successful" operation and on this basis are appropriately dependable in new critical applications. There are many problems with such assertions, in particular the possibility of failure masking and version control issues mentioned above. A particular problem arises, though, if attempting to construe RTOS operation as a Poisson process for the purposes of statistical evaluation.

Suppose the RTOS has at least one failure mode. Then there is some short period of time, microseconds or milliseconds, just before such a failure when that failure becomes inevitable (for example, at the last instruction before a HALT). Let such a short time period be ε. Then in that period ε the probability of failure of the RTOS is effectively 1. However, consider a time period of ε from start of boot-up. The probability of failure in this time for any reasonably well-used RTOS is effectively 0. So, for some time intervals of length ε in operation the probability of failure is effectively 1 and for other time intervals of the same length it is effectively 0. But the memoryless property of a Poisson process requires the probability of failure in any time interval of length ε to be exactly the same, no matter where the interval occurs during an execution. It follows that the operation of an RTOS with at least one failure mode cannot be considered simpliciter as a Poisson process[1]. This argument and associated considerations is presented in more detail in (Ladkin 2015).

6 Inappropriate Evaluation Attempts

The international standard for functional safety concerning systems which include electrical, electronic or programmable electronic (E/E/PE) components, IEC 61508, includes a short guide to statistical evaluation in Part 7, Annex D. The second sentence of this Annex suggests that the methods, the construal of software operation as a Bernoulli or Poisson process as above, can be used to evaluate software libraries, compilers, even operating systems.

We have heard anecdotes from industrial assessors of people trying to do just that. For example, a client C comes to an assessor. C proposes to use a real-time version of an operating system to run critical software with OS+Software having a safety requirement of SIL 3. C claims that the operating system has more than enough hours without failure, for a particular safety function, to satisfy the reliability conditions for SIL 3 for that function. In particular, the function is continuous (rather than on-demand) and C has detailed logs of the order of 10^8 failure-free (for this function) operating hours on the software, way more than required (see Table 2 above).

From the discussion above, besides the logs, C will have to show accurate registration of all failures in previous operation (and lack of such), with particular consideration given to possible failure masking in operation of such complex software. C will have to address the issue of versions: are all instances of the OS,

---

[1]   We note, though, that there are more complex ways to consider such software, some of which relax some of the assumptions of a Poisson process.

whose behaviour has been registered, exactly the same version? Or are there "slight" divergences amongst them? And, finally, C must address the issue of whether the software operation is indeed memoryless in the required sense. These are all tricky issues, but only when they have been satisfactorily addressed can C "plug in the numbers" from Table 2 and draw the conclusion that OS operation fulfils the safety requirement. Then it is incumbent upon C to argue, and to ensure, that inputs to the future system have the same statistical properties as the recorded inputs from the past.

It follows that "plugging in the numbers" is not at all easy.

7 Extending Evaluation Techniques

The question arises whether there are more subtle, and/or more widely applicable methods of evaluating software behaviour statistically than the straight conceptions as Bernoulli or Poisson processes. The answer is yes. However, their application is currently a matter on which expert statistical advice is needed.

In many of the more sophisticated methods, the software architecture plays a key role. Individual behaviours of individual components of the architecture are statistically assessed, and the results are combined into an assessment of the whole architecture. In many cases, the criteria for statistical assessment of the individual components may be relaxed, but the combined assessment still enables the key Bernoulli or Poisson mathematics to be used.

The second author has devised and used such techniques, see for example (Bedford 2001, Chapter 12). Colleagues have recently communicated the successful use of such techniques in evaluating critical software for rail applications (Schäbe 2015). There is a recent method for assessment of two heterogeneous channels, of which one may be "possibly perfect" (Littlewood 2012).

8 Conclusions: Why Use Statistical Evaluation?

In light of the discussion above, the reader may well wonder why anyone would bother with statistical assessment of software proposed for use in critical systems. There are many reasons. We give some.

Suppose a particular software-based system component has an adequate record of past use. Suppose, indeed, it appears informally to be the "best kit for the [new] task". The safety requirements for each critical system or component are individual, special to the specific system. The kit has been used before successfully to execute a specific function, and this function may be required for the proposed new use. However, it may well be that the safety requirements for previous use differ considerably from the safety requirement for the proposed new use. It may indeed be that documentation of the kit does not exist sufficient to justify its inclusion "as new" in the proposed new use. This may well be the case if the kit was not originally intended for safety-critical use, but has established its dependability through experience. It may also be the case that the assessment requirements in previous uses were less stringent than those for the proposed new use. This could occur for two reasons: one, that safety standards have changed; two, as mentioned above, the safety requirement may be different.

If the kit is indeed the "best kit for the task", then there is good reason to use it. And there is good reason to be able to use statistical evaluation of previous use to make the case for its new use, if the statistics are available and adequate. A recent example of this industrial need has been communicated to the first author, but specific details are not available at time of writing (Kindermann 2015). We may speculate that such cases will arise more frequently, as more and more examples of relatively simple and reliable E/E/PE system components for specific critical functions come onto the market with time.

Acknowledgments We particularly thank Peter Bishop, Jens Braband, Wolfgang Ehrenberger, Rainer Faller, Andreas Hildebrandt, Bertrand Ricque and Hendrik Schäbe for detailed discussion of some of the issues arising here, and Bernd Sieker for valued help with the formatting.

References

ATSB (2007) In-flight upset, Boeing 777-200, 9M-MRG, 240 km NW Perth, WA. Investigation Number 200503722, 2007. Australian Transport Safety Board. https://www.atsb.gov.au/publications/investigation_reports/2005/aair/aair200503722.aspx , accessed 2015-02-03.
Bedford T, Cooke R (2001), Probabilistic Risk Analysis: Foundations and Methods, Cambridge University Press, 2001.
Bernoulli J (1713) Ars Conjectandi, Basel, 1713.
Ducklin P (2014) Anatomy of a "goto fail" - Apple's SSL bug explained, plus and unofficial patch for OS X! nakedsecurity blog, 2014-02-24. https://nakedsecurity.sophos.com/2014/02/24/anatomy-of-a-goto-fail-apples-ssl-bug-explained-plus-an-unofficial-patch/ , accessed 2015-11-03.
IAEA (2002) Tutorial on the Accident at Three Mile Island, 2002. International Atomic Energy Authority. https://www.iaea.org/ns/tutorials/regcontrol/assess/assess3233.htm , accessed 2015-11-03.
IEC (2010) IEC 61508, Functional safety of electrical/electronic/programmable electronic safety-related systems, Parts 1-7. International Electrotechnical Commission.
Johnson C W, Holloway C M (2007) The Dangers of Failure Masking in Fault-Tolerant Software: Aspects of a Recent In-Flight Upset Event, in Proceedings of the 2nd IET International Conference on System Safety, IET 2007. Availablemfrom http://www.dcs.gla.ac.uk/~johnson/papers/IET_2007/Chris_Michael_Upset.pdf , accessed 2015-11-03.
Kindermann M (2015) personal communication.
Ladkin P B (2015) Some Practical Issues in Statistically Evaluating Critical Software, in System Safety and Cyber Security 2015, ISBN 978-1-78561-092-9 e-ISBN 978-1-78561-093-6, ISSN 0537-9989 Reference PEP....U, IET, 2015. http://www.rvs.uni-bielefeld.de/publications/

Littlewood B, Rushby J (2012) Reasoning about the Reliability of Diverse Two-Channel Systems in Which One Channel Is "Possibly Perfect", IEEE Trans. Software Engineering 38(5):1178-1194, 2012

Perrow C (1984) Normal Accidents: Living with High-Risk Technologies, Basic Books, 1984.

RAEng (2011) Global Navigation Space Systems: reliance and vulnerabilities, 2011. Royal Academy of Engineering, http://www.raeng.org.uk/publications/reports/global-navigation-space-systems, accessed 2015-06-25.

Schäbe H, Braband J (2015) . Basic requirements for proven-in-use arguments, preprint 2015. Available from http://arxiv.org/pdf/1511.01839v1.pdf , accessed 2015-11-06.

Siegrist K (2014) Virtual Laboratories in Probability and Statistics, University of Alabama at Huntsville, 1997-2014. http://www.math.uah.edu/stat/, accessed 2015-06-25.

Thomas M (2011) personal communication.

U.S. Nuclear Regulatory Commission (2014) Backgrounder on the Three Mile Island Accident, 2014. http://www.nrc.gov/reading-rm/doc-collections/fact-sheets/3mile-isle.html   accessed 2015-11-17.

Wikipedia (2015), Easter Eggs https://en.wikipedia.org/wiki/Easter_egg_(media)   accessed 2015-11-17.

Why We Cannot (Yet) Ensure the Cyber-Security of Safety-Critical Systems

Chris Johnson[1]

University of Glasgow

Glasgow, UK

Abstract   *There is a growing threat to the cyber-security of safety-critical systems.   The introduction of Commercial Off The Shelf (COTS) software, including Linux, specialist VOIP applications and Satellite Based Augmentation Systems across the aviation, maritime, rail and power-generation infrastructures has created common, vulnerabilities.   In consequence, more people now possess the technical skills required to identify and exploit vulnerabilities in safety-critical systems.   Arguably for the first time there is the potential for cross-modal attacks leading to future 'cyber storms'.   This situation is compounded by the failure of public-private partnerships to establish the cyber-security of safety critical applications.   The fiscal crisis has prevented governments from attracting and retaining competent regulators at the intersection of safety and cyber-security.   In particular, we argue that superficial similarities between safety and security have led to security policies that cannot be implemented in safety-critical systems.   Existing office-based security standards, such as the ISO27k series, cannot easily be integrated with standards such as IEC61508 or ISO26262.   Hybrid standards such as IEC 62443 lack credible validation.   There is an urgent need to move beyond high-level policies and address the more detailed engineering challenges that threaten the cyber-security of safety-critical systems.   In particular, we consider the ways in which cyber-security concerns undermine traditional forms of safety engineering, for example by invalidating conventional forms of risk assessment.   We also summarise the ways in which safety concerns frustrate the deployment of conventional mechanisms for cyber-security, including intrusion detection systems.*

---

[1] Contact: School of Computing Science, University of Glasgow, Scotland, G12 8RZ.
http://www.dcs.gla.ac.uk/~johnson, Johnson@dcs.gla.ac.uk

1 Introduction

There is a growing threat to the cyber-security of safety-critical systems. This is, in part, due to the integration of a small number of Commercial off the Shelf (COTS) products across the supply chain of national critical infrastructures. In previous generations, critical infrastructures tended to rely on bespoke systems that were not reused across different industries (Johnson, 2015). Now COTS components in safety-related applications include, but are not limited to, Linux, VOIP and Satellite/Ground Based Augmentation Systems (SBAS) such as WAAS in North America and EGNOS in Europe. A growing number of potential attackers have the technical knowledge to undermine safety-related applications across the transportation, energy distribution, food and water industries, etc.

At the same time, we have seen the rise of a new generation of terrorist threats, based around semi-stable regimes that resemble nation states. These regimes have access to trained engineers and process equipment within their borders. State-like terrorist regimes have become skilled in cyber-security, partly as a consequence of the policies implemented by Western governments. Police and intelligence agencies have denied them access to conventional social media. These regimes have responded by developing cryptographic skills and peer-to-peer networking using techniques originating with the deep or dark web. This has connected terrorist groups with strong political, religious and ideological motivation to the semi-commercial hackers who already sell zero-day exploits, malware libraries and root kits.

A key theme in this paper is that neither government nor private industry has moved at the rate required to maintain our defences against the growing threats from cyber-criminals and terrorist states. Public-private partnerships have failed to deliver regulatory guidance and appropriate audit mechanisms. The fiscal crisis prevented many safety regulators from recruiting and retaining staff with sufficient expertise in both cyber-security and safety-applications. Cultural and organizational barriers between safety and security have compounded this. It takes time before someone with a deep knowledge of conventional cyber-security can also gain an understanding of the concerns that arise, for instance in the nuclear or aviation industries. This is important when cyber-security techniques cannot simply be transferred from more conventional office based systems to safety-critical environments.

Political and organizational barriers also help to explain our limited progress in securing national critical infrastructures. These barriers arise because different regulators are responsible for the cyber-security of national data networks and for the safety of particular industries. In the UK, this is illustrated by the distinction between OFCOM, the Centre for the Protection of National Infrastructures and the Civil Aviation Authority or the Office for Nuclear Regulation. In the United States, similar distinctions arise between the Federal Communications Commission, the Department for Homeland Security and the Federal Aviation Administration or the Nuclear Regulatory Commission, not to forget the National Institute of Standards and Technology (NIST) as the body responsible for the Federal cyber-security provisions within the Federal Information Security Management Act (FISMA), or the host of local and State organizations that may also be included as stakeholders. These organizational and political distinctions create significant practical consequences. Companies often do not understand their reporting obligations for cyber incidents across national and international agencies, including Computer Emergency Response Teams (CERTS), police, intelligence and critical infrastructure organizations, as well as telecoms and industry regulators. This situation is compounded when superficial similarities between safety and security have led to the development of inappropriate policies that cannot be sustained using existing engineering practice. The following pages focus on two classes of concern. Firstly, there are situations in which cyber-security concerns undermine existing safety practices:

- Conventional safety risk assessments cannot be sustained when systems might be exposed to coordinated and malicious attacks;

- Existing safety-management systems offer limited support for cyber-security – especially given differences in incident reporting and root cause analysis between these two areas;

- Cyber-security concerns challenge many existing safety-related software engineering techniques, for instance, the use of software diversity and N-version programming lead to extended supply chains that are difficult if not impossible to secure.

The second, inverse set of concerns arise when safety issues complicate the application of existing cyber-security techniques:

- The limitations of conventional intrusion detection systems. White list enumerations of permitted processes cannot easily be applied to complex legacy systems and there are dangers when valid, safety-related processes are denied necessary resource. In contrast, black list enumerations of malware do not work because of the failure of cyber incident reporting in safety-related systems, noted in the previous section;

- The limitations of conventional forensic techniques. Existing support tends to focus on Internet Protocol related systems rather than on industrial Supervisory Control and Data Acquisition (SCADA) infrastructures using very different Programmable Logic Controllers (PLCs) and protocols. Further concerns stem from the competing risks that arise when deciding to either immediate isolate a compromised system leaving application processes in a potentially unsafe state or in continuing to operate until shut-down but with the risk of over-writing critical evidence;

- The limitations of conventional cyber-security policies for air-gapped SCADA systems. For control systems, where many devices are not networked the implementation of conventional security patching policies may arguably increase rather than decrease potential vulnerabilities.

2 The Failure of Safety Critical Techniques in Cyber Security

The introduction has argued that we are ill-prepared to face a growing range of threats against the COTS infrastructures that support many safety-critical industries. Later sections will explain why a range of existing cyber-security techniques cannot easily be applied in safety-related domains. In contrast, this section explains why safety-techniques are often compromised by cyber-security concerns.

2.1 Cyber-threats Undermine Safety Risk Assessments

The European Network and Information Security Agency (ENISA, 2006) and the European Air Traffic Management Organisation (EUROCONTROL, 2006) advocate the use of safety management concepts to support the cyber-security of critical infrastructures (Johnson, 2015). In conventional applications, safety management systems use incident reporting and other forms of operational monitoring to determine whether an implemented system meets the safety requirements derived form an initial risk assessment. If new forms of hazard emerge, or if the system failed to adequately mitigate an identified risk, then further development is required. In other words, risk assessment, design and operation, monitoring and incident reporting form a virtuous circle.

These components of safety management systems also provide the foundations of information security management systems. Risk assessment helps to identify threats and vulnerabilities. Appropriate design and operating procedures help to ensure that the threats are mitigated. Incident reporting and audit provide the feedback necessary to revise the initial risk assessments when new threats emerge and to identify any situations in which operations fail to meet the requirements derived from an initial risk assessment. In theory, the use of similar concepts should support the integration of safety and information security management systems. Unfortunately, these superficial similarities hide a host of differences that undermine attempts to transfer the benefits of safety management systems into the security domain, revealing the lack of engineering expertise and operational experience that has informed much previous guidance (Johnson, 2015a). For example, the presence of an intelligent adversary undermines independence assumptions in conventional safety assessments. Blended attacks are timed to co-incide with routine component failures. Similarly, if the symptoms of one form of cyber attack are identified then it is very likely that a system may have been compromised in other ways.

Cyber-security concerns undermine existing safety engineering practices in other ways. Not only do they challenge the probabilistic components of risk assessment, cyber-threats also undermine safety-related consequence assessments. One reason is that we have limited experience of the new forms of advanced persistent threat, such as the state machine that varied the behaviour of Stuxnet or the use of Command and Control servers to hide Duqu from intrusion detection systems. This makes it very dangerous to predict the potential outcome from future modes of attack. The growing interconnection of critical infrastructure leads to hidden interdependencies. There are also concerns over 'cyber-storms' where a single attack brings down many different infrastructures – for instance when critical systems run under the same variant of Linux or where multiple services depend on timing information from the same satellite infrastructures.

None of these caveats would be significant if we had a range of tools and techniques that could be used to combine conventional safety risk assessments and cyber-threat analysis. Fault trees have superficial similarities to attack trees but the underlying semantics are different. In consequence, many organisations end up with parallel systems that are incapable of transferring lessons between safety and security. There are some notable exceptions (Piètre-Cambacédès and Bouissou 2010, Johnson, 2015). However, there are few published case studies in integrated approaches to safety and security and even less agreement over the general utility of these tools across different industries.

2.2 Cyber-threats Challenge Safety Incident Reporting

It has taken many years to establish strong incident reporting cultures across safety-critical industries. In contrast, very few companies have the same security reporting culture. One reason for this is that employees reporting safety concerns are typically protected by a 'no blame' or 'proportionate blame' environment. In contrast, security violations can trigger disciplinary or legal action against other employees. A mismatch between security policy and practices can undermine a nascent reporting culture. Management implicitly approve of many security violations, for example the use of USB devices by sub-contractors, because they are anxious to maintain operations. Such incidents are seldom reported.

Many companies are reluctant to report incidents to Computer Emergency Response Teams (CERTs), regulatory agencies or industry associations. The loss of control and the reliance on external agencies can also compromise intellectual property where investigators must be familiar with commercially sensitive information in order to diagnose the causes of an attack. There are also political concerns over the exchange of information about cyber-security incidents across national borders, even to otherwise friendly states.

Further barriers prevent the use of incident reporting to support safety and cyber-security. Lessons learned applications ensure that safety recommendations are disseminated as widely as possible. The aim is to avoid any recurrence of potential accidents. However, the disclosure of information about a cyber-incident might encourage future attacks. It can undermine market confidence; it can trigger regulatory action and litigation.

While there are well-established reporting mechanisms for safety concerns, cyber-incident reporting has been undermined by a series of 'turf wars' across Europe and North America. For many companies, it can be unclear whether reports should be sent to an industry regulator, such as the US Federal Aviation Administration or Nuclear Regulatory Commission, to a security agency, such as the Department of Homeland Security, or the US CERT, or to telecoms regulators who have responsibility for collating information about wider cyber-security concerns, such as the Federal Communications Commission. In some cases, a single incident must be reported to more than one agency. For example, in the UK a cyber-attack with safety related consequences must be reported to the national industry safety regulator and potentially also to a subset of the National Crime Agency, the National Cyber Crime Unit, GOVCERT, the UK Information Commissioner as well as the CESG/Centre for the Protection of National Critical Infrastructure via providers registered under the Cyber Incident Response (CIR) or the Cyber Security Incident Response Scheme (CSIR).

2.3 Cyber-security Undermines Safety-Critical Development

The tensions between safety and security extend across the engineering lifecycle, from risk assessment to detailed development practices. For instance, redundancy is typically used to increase the dependability of critical systems. If one component fails then a backup can maintain operation. However, this provides few benefits in software related systems without some level of diversity. Two redundant versions of the same code are likely to contain the same bugs and hence will fail in the same way, even if they are provided with slightly different data. In consequence, N-version programming techniques rely on using two or more contractors to develop multiple versions of the same program. In the event that one fails, it is intended that the other will not. The use of a diverse supply chain helps to ensure that both programs do not share common bugs, assuming that their requirements are correct.

Unfortunately, software diversity creates immediate problems for security management. The end user must secure two or more different supply chains – in other words, redundant diversity opens up multiple routes through which compromised code might be integrated into a safety-critical system. The customer must audit multiple sub-contractors to ensure that they meet agreed cyber-security requirements. These security concerns are seldom considered in safety-critical software development. Customers have few guarantees that suppliers have vetted their staff, have prevented the introduction of code from untrusted sources, etc.

The meta-level point is that integrating safety and security reveals a host of tensions, which can only be addressed through integration. Without this, we cannot assume that the two isolated communities will deliver viable solutions to the problems identified in this paper. For example, a small number of safety-critical companies are now offering diverse supply chains from within their own organization. In other words, they will provide customers with two pieces of software each performing very similar functions but with assurances that they were implemented by different teams of employees using diverse development methods. This simplifies the supply chain but relies upon a range of innovative software management and development practices. It remains to be seen whether such practices are strong enough to address the natural safety concerns that arise when redundant software comes from the same supplier.

3 The Failures of Safety Techniques in Cyber-Security

The previous section has argued that there is an urgent need for integrated tools and expertise because safety-techniques are often compromised by the introduction of cyber-security concerns. In contrast, the following paragraphs argue that existing cyber-security policies cannot easily be applied in safety-related domains. We focus on three examples:

1. Conventional intrusion detection systems undermine the safety of complex applications;
2. Secondly, existing forensic guidance for conventional office based systems would lead to loss of life in safety-critical systems;
3. Finally, the air-gapped architectures of many SCADA environments undermine existing principles of security management.

3.1 Safety Concerns Limit Cyber-Intrusion Detection Systems

NIST (2012) advocate the use of several different intrusion detection systems (IDS) within critical applications. This raises significant concerns when an IDS might erroneously block critical processes in safety-related applications. There are two main approaches to intrusion detection. Blacklisting relies on detecting the characteristics of malware. Whitelisting is discussed in subsequent sections and relies on recognizing approved code.

Most blacklist IDS are designed to protect office-based systems. The signatures and symptoms of malware are compiled from evidence about incidents reported through a range of mechanisms, including honey pots (Spitzner, 2002) but also through confidential reporting to the major security companies.   Section 2.2 summarized the barriers that prevent the exchange of information about cyber incidents in safety-critical systems.   Unless we can identify the malware signatures that characterize the growing threat to industrial control systems then attempts to develop blacklist IDS will provide very limited protection for SCADA applications (Naedele, 2007).

In order to protect a system, it is important to update a blacklist as soon as a malware signature has been identified.   However, this creates problems in safety-related applications where there are requirements to conduct exhaustive tests prior to any software modifications.   Uploading a corrupted blacklist could also cause the failure of a detection system with knock-on consequences for safety-related processes.   Safety engineers would face a difficult decision between the competing requirement to update blacklists as soon as a new signature was identified and the requirement to ensure that the new signature did not undermine the safety of application processes.

In contrast to blacklisting, whitelist IDS ensure that only approved programs can be executed.   They profile and report any deviations from 'normal behavior'. This can be implemented by creating a hash digest of all software applications. If the hash of an executable does not match anything in the list, it will trigger a security event.   It is also important to prevent unauthorized users from changing the lists indicating which files can be executed.

Whitelisting offers benefits for safety-critical applications.   The focus on identifying 'normal processes' eliminates the need to continually update malware signatures.   Whitelists provide some protection against zero-day exploits – even if the signature of an attack is unknown, the malicious code will not be included on the approved hash list.   However, the application of this approach raises a number of concerns.   For instance, the same attack across multiple instances of a control system will simultaneously lead to a large number of distributed security events. These can overwhelm an organization's ability to respond in a timely fashion and may also be triggered by non-malicious causes. Software updates that are not reflected by changes in the whitelist can lead to a large number of false positives. Safety-critical processes could be denied computational resources.   It, therefore, becomes imperative that staff and sub-contractors follow agreed security update procedures during all software installations.   In some safety-critical applications this is relatively straightforward – for instance in long-lived SCADA systems where software updates on PLCs are relatively rare.   In other contexts, such as air traffic management, where tens of sub-contractors each have intellectual property concerns, it can be very hard to determine what is and what is not a 'normal process'.

Data diodes ensure that information can only travel in one direction; for instance, by removing the send and receive transceivers from one direction of a fiber-optic cable.   They can be used so that process data only flows from an operational zone to business systems but not vice versa.   These devices can also isolate IDS from critical processes. The uni-directional flow of data reduces concerns that the detection system will have an adverse effect on application safety. Unfortunately, greater levels of monitoring lead to an increasing number of false alarms. This can undermine cyber situation awareness and can lead to denial of service when operators incorrectly halt an application that they fear has been compromised.   In contrast, raising IDS tolerance thresholds increase the potential for missed positives.   In safety-critical systems this leads to the possibility that the over-tolerant configuration of an IDS allows companies to continue operating with malware inside critical applications.   In conventional office-based systems, machine-learning techniques have been successfully deployed with threat visualization to integrate automated intrusion detection with human decision-making. Further work is required to determine whether these approaches might also be adapted to address the false-positive/false-negative concerns that undermine cyber-situation awareness in safety-critical systems.

3.2 Safety in Air Gapped Systems Undermines Cyber Policies

Most existing cyber-security tools and techniques focus on distributed architectures that are based around the conventional IP-stack.   In contrast, many safety-related applications rely on computational devices such as PLCs that are isolated even from local area networks. The behavior of monitoring and control applications may not change for over a decade. The 'air gap' between the device and any network improves the cyber-security of safety-critical applications because it limits the opportunities for remote attacks.   However, the 'air gap' also limits opportunities to use blacklist IDS.   There is no easy way for system administrators to automatically update nodes with malware signatures. Operators must manually install any updates on each isolated device across the plant.   This leads to a paradox.   It is hard for any attacker to compromise a stand-alone PLC unless it is hooked to another device – for example to install a patch or update the IDS.   The more often these updates occur then the greater the risk of cross-contamination. Systems managers of safety-critical systems, therefore, often deliberately ignore conventional cyber-security advice, preferring to leave isolated devices unpatched. Other problems limit the application of whitelisting in air-gapped systems. Without network access, it may be weeks or months before an engineer can examine the logs in sufficient detail to note an infection on a remote device.

3.3 Safety Concerns Undermine Conventional Cyber-Forensics

Detailed guidelines cover the forensic analysis of cyber-security incidents. For example, the US Department of Justice (2004, 2008) suggest that forensic investigators must preserve the 'chain of evidence':

- "Immediately secure all electronic devices, including personal or portable devices.
- Ensure that no unauthorized person has access to any electronic devices at the crime scene.
- Refuse offers of help or technical assistance from any unauthorized persons.
- Remove all persons from the crime scene or the immediate area from which evidence is to be collected.
- Ensure that the condition of any electronic device is not altered.
- STOP! Leave a computer or electronic device off if it is already turned off".

These principles support cyber forensics in office-based systems.   They also illustrate the problems of integrating existing security practices into safety-critical systems.   It is hard to envisage how any investigatory agency could immediately secure "all electronic devices" distributed across a compromised process control system, or indeed how to enumerate all of the devices connected to a modern, national air traffic management system. Many safety-critical companies have minimal access control policies, so that it is not always clear to external agencies who exactly has authorized access to the devices at a crime scene.   Typically, there are strong forms of perimeter access – where only staff and authorized sub-contractors can gain access to a facility or machine room but once inside they have wide-ranging access to racks and network components. This is a strong contrast with financial institutions and even web service providers where it is normal to have fine grained access control policies that prohibit software engineers from accessing a machine room.   Removing "all persons from the crime scene" could be catastrophic in a crowded Air Traffic Control centre or nuclear control room where operations, engineering and safety management must cooperate during contingency operations, including the aftermath of a cyber-attack. To "leave a computer or electronic device off if it is already turned off" would prevent the use of redundant protection systems.

  The Department of Justice guidelines aim to preserve evidence by urging investigators to turn off any compromised systems. Continued operation may overwrite valuable data or enable attackers to disguise the manner in which a system was compromised. However immediately isolating a safety-critical system might endanger the lives of the public and of operators. Starting a fallback system can reduce this risk until an application reaches a safe state. However, this increases the risk of cross-contamination. In other words, halting a primary application to preserve forensic data can lead to the infection of the secondary system at a time when engineers and investigators are unlikely to know the mechanisms by which an attack was originally propagated. Without some form of integration, it is impossible for operational staff, senior management and investigatory agencies to balance the risks between the safety of application processes, the potential for cross-contamination and the legal requirements to preserve the evidence necessary for prosecutions in the aftermath of an attack.

4 Conclusions and Further Work

We face a growing range of cyber-threats to safety-critical systems. State-like terrorist groups have access to significant finance and engineering resources. The threats to safety-critical systems also stem from commercial markets in malware through the peer-to-peer networks of the Dark/Deep web where zero day exploits can be bought by those lacking the technical skills necessary to develop them. At the same time our vulnerabilities are increasing – through the integration of COTS applications including Linux, Voice over IP (VOIP) and Satellite Based Augmentation Systems into safety critical applications. The public-private partnerships established to enhance cyber-security have done little to address these concerns partly because the fiscal crisis and organizational barriers have left us without regulators who are competent in the cyber-security of safety-critical applications.

Superficial similarities between safety and security have led to the development of policies that cannot be sustained using existing engineering techniques. There are unique concerns for safety that prevent us from simply re-using existing guidance from office based systems in the aftermath of cyber-attacks. We cannot immediately isolate safety-related processes during forensic investigations without risking the lives of those who depend on critical infrastructures. Similarly, we cannot reuse convention Intrusion Detection Systems if these applications could block critical processes or if updates to malware signatures inadvertently bring down safety-related systems.

There are many areas for further work, including the causal analysis of cyber-incidents in safety-critical systems. Systemic factors help create the context in which an incident or accident is likely to undermine the safety of application processes. In contrast, security investigations tend to focus more on deliberate or unwitting violations. This focus on the direct human causes of a security incident is similar to the 'perfective approach' that characterized safety-related reporting more than a decade ago (Johnson, 2003). We might, therefore, expect that the focus of security investigations to shift towards systemic factors in the future. For this to happen it seems likely that we will need a new generation of root cause analysis techniques. Most existing approaches use counter-factual reasoning in the aftermath of safety-related incidents. Recommendations are derived by identifying causes, which had they been prevented then the incident would not have occurred. Such reasoning cannot easily be applied to cyber-attacks. It is hard to argue that a security incident would have been prevented given that adversaries launch multiple, simultaneous attacks, some of which go undetected.

References

ENISA, (2006), Risk Management: Implementation principles and Inventories for Risk Management/Risk Assessment methods and tools, Conducted by the Technical Department of ENISA, Section Risk Management, Heraklion, Greece, June 2006.

EUROCONTROL (2006), Safety Case Development Manual, Technical report DAP/SSH/091, Brussels, Belgium.

Naedele, M. (2007), Addressing IT Security for Critical Control Systems, Proceedings of the 40th Hawaii International Conference on System Sciences, IEEE Computer Society.

Johnson, C.W., (2003), Failure in Safety-Critical Systems: A Handbook of Accident and Incident Reporting, University of Glasgow Press, Glasgow, Scotland, ISBN 0-85261-784-4, available in in electronic form.

Johnson, C.W., (2015), Contrasting Approaches to Incident Reporting in the Development of Security and Safety-Critical Software. In F. Koorneef and C. van Gulijk (eds.), SAFECOMP 2015, Springer Verlag, Heidelberg, Germany, 400-409, LNCS 9337, ISBN 978-3-319-24254-5.

Johnson, C.W., (2015a), Barriers to the Use of Intrusion Detection Systems in Safety-Critical Applications. In F. Koorneef and C. van Gulijk (eds.), SAFECOMP 2015, Springer Verlag, Heidelberg, Germany, 375-384, LNCS 9337, ISBN 978-3-319-24254-5.

Piètre-Cambacédès, L. and Bouissou, M., (2010), Modelling Safety and Security Interdependencies with BDMP (Boolean logic Driven Markov Processes). IEEE International Conference on Systems Man and Cybernetics (SMC), 10-13 Oct. 2010, 2852 – 2861.

Spitzner, L. (2002), Honeypots tracking hackers. Addison-Wesley. pp. 68–70. ISBN 0-321-10895-7.

U.S. Department of Justice (2004), Forensic Examination of Digital Evidence: A Guide for Law Enforcement.

U.S. Department of Justice (2008), Office of Justice Programs, Electronic Crime Scene Investigation: A Guide for First Responders, Second Edition, Washington DC, 2008. http://www.nij.gov/publications/ecrime-guide-219941/

U.S. National Institute of Standards and Technology (NIST, 2012), Computer Security Incident Handling Guide (Draft), Special Publication 800-61 Revision 2 (Draft), Gaithersburg, Maryland.

Human error in safety-critical programming

Harold Thimbleby

Swansea University

Wales, UK

Abstract *It is self-evident that we need an effective safety culture to avoid human error (and its consequences) in programming, yet many of us program as if safety is trivial, and if we just use the right tools it should be even easier. Although it is an unwelcome message, we are deceiving ourselves about how easy safety is, and this deception is self-serving, achieving nothing other than entrenching ignorance of error and its influence over us. The solution is called "resilience" and of the various techniques of resilience, mathematics and Formal Methods are basic tools for safety-critical programming — but they are not sufficient without a proactive commitment to be resilient. We provide a worked example that helps show that error is not "out there" as an abstract concept but deep inside us. Error is an unavoidable companion to our programming that we urgently need to master.*

1 Introduction

Human error can be defined informally as "not doing what we wanted to" but, unfortunately, we only realise we have made an error in hindsight — that is, *after* something has obviously gone wrong. If we noticed we were making an error, we would correct it and there would be no unwanted consequences: there would in fact be no error.

In programming there is often a long separation, in both time and space, between committing an error (e.g., a typo on line 21,045) and discovering its consequences (e.g., a user in another country loses control of an industrial process). This separation makes it hard to recognize errors contemporaneously with any opportunity to mitigate them. Indeed, there is a natural drive for product delivery as soon as possible at the expense of saving up errors for the future — a process dignified with the name *technical debt* (Allman, 2012). What programmers save today in lower development costs, users (if they are still alive) pay back in the future, effectively paying the interest on the technical debt the programmers borrowed.

Error is not just something to study in the abstract, but it affects our own lives and work whatever we do, and by its very nature we under-estimate its impact. Often when unwanted things happen, they have many causes, and we can often overlook or deny our own contributions.

In programming, error takes on a dark character: our errors as programmers may affect very large numbers of things and people later, much later, if — when! — the systems we build fail to achieve their objectives in a safe and dependable way.

We do not just program for end users, but we often program tools for programmers. Some errors we make affect all programmers using our programs, and all users using their programs. There is therefore an enormous leverage for amplifying error in programming, an enormous multiplying factor that scales up the human impact of our errors.

Unless we are incompetent, "our errors as programmers" may not seem to make sense, or if it makes sense it is unwelcome. If we knew we were making errors we would certainly correct them, and the only cost would be a delayed project, if that. We do not think we make *real* errors, because we correct them, and the ones we do not notice, we do not even realise we have made. Many errors slip past our awareness until, later, they become only too visible.

Only when NASA's $125 million Mars Orbiter spacecraft failed to complete its first orbit of Mars did anyone suspect an error had occurred (Lloyd, 1999). Lockheed Martin software had used pound second units while NASA had specified Newton seconds, resulting in a conversion error, and the spacecraft flying too low and burning up. As NASA said:

> A single error like this should not have caused the loss of Climate Orbiter. Something went wrong in our system processes in checks and balances that we have that should have caught this and fixed it.

In other words, there were meta-level errors — also not noticed until too late.

There are many other examples of critical programming errors, notably the classic Therac-25, London Ambulance Service, and rather complex errors such as the UK's National Programme for IT. Of course, these aren't *just* programming errors, but in every case the programmed systems failed.

Although it was not a programming problem — elsewhere I have argued that poor understanding and use of Human Factors in engineering design contributed to the loss of a world war (Thimbleby, 2015a). There were thousands of critical engineered products (in this case, the German WWII Enigma) all with the same preventable design problems that the designers had overlooked — for Human Factors reasons identical to those we expand on in the present paper. The harsh lessons of the Enigma apply directly to programming design problems today: the lessons are not just of historical interest, for the Human Factors problems are exactly the same today as they were then. People are the same. NASA could have been speaking for them: human error is a fact of life, so if things go wrong because of "human error" then this implies that there are oversights in the systems and processes.

2 Perspectives on human error

Human error has of course been widely studied in psychology and in relation to accidents, patient harm, financial errors, aviation, environmental and other man-made disasters. There has been relatively little discussion of programming error, not least because the investigation or research tends to stop at the programming error, rather than digging into why it happened — or more usefully — how it might have been avoided or mitigated. Also, of course, few psychologists can program well, so the conceptually hard issues of programming tend to be glossed in the research literature.

Human error happens because we do not notice the sequence of events leading up to the visible error until (if ever) it is too late; if we had noticed it, we would have modified the steps we made. Of course, we know errors happen frequently, particularly with other people, yet we are largely unaware of our own errors — we were as they happened, and we may remain oblivious subsequently and we may even deny we made any mistakes.

We are more aware of other people's errors than our own, so we often think we are privileged: we make few errors (we think), so why bother to use protective techniques (e.g., Formal Methods) that are hard work?

Similarly, people working on building construction sites are reluctant to wear protective gear because they have never in their lives needed it! (If they had ever needed it, they would have injured themselves and probably would not have returned to the industry.) Thus people, everyone you know, and in particular you yourself, are lucky survivors — healthy, working humans therefore have a low awareness of error and its consequences. We do not know how lucky we are, and we do not know how lucky our colleagues are! If any of us had made big programming errors, we would not be high-flying programmers working in a safety-critical industry; but we are. What if it was merely chance that we have been successful, and our error-free performance to date is more luck than skill?

This is called *observer bias*: the very fact that we are observing something influences what we can see. The very fact we have got so far in our careers means we have already been successful as programmers; conversely, the "bad" programmers whose bugs somebody noticed have been moved out of the way — these people are not around to observe anything they can tell us. An extreme example is (for the sake of argument) that some programmers have written code that has killed them; but it is impossible that we (who are alive) have made such a mistake. Therefore our personal estimate of the probability of our programming killing ourselves is too low.

So we think we are better than we are because we are, and can only be surrounded by, successful survivors. Observer bias in programming thus encourages complacency — which may then be much further exacerbated by accountants seeking ways to lower costs, which means the next error may be catastrophic as we have the "costly" slack to manage it trimmed away in the call for efficiency.

2.1 Examples of real errors

Many familiar examples of human error thankfully dissect other people's errors, preferably big disasters, like the Mars Orbiter. Unfortunately for us, human error strikes closer to home.

We know human error happens regularly and affects everything *we* do, when we think about it. Stage magicians are adept at fooling us: their "magic" is just cleverly presented error in our perception: what our eyes see cannot be true, therefore it must be magic. Despite us all agreeing magicians are fun, it is hard to take seriously the blatant exploitation of our senses as evidence that we do make errors, so predictably in fact that a stage magician can get an entire audience fooled over the same errors.

Here is a predictable error you may be familiar with — which, if so, unfortunately makes it easy to dismiss. *Except that* it is but one example from a whole class of error (Kahneman, 2012): and you will be susceptible to others in this class, even if you are not fooled with this example:

If a baseball bat and a ball cost a total of $1.10, and the bat costs $1 more than the ball, then how much does the ball cost?

The ball costs $0.10, right?

Wrong!

Instead, just do the maths. Call the cost of the bat x and the cost of the ball y. We are told $x + y = 1.1$ and that $x - y = 1$; use the latter to substitute for x in the first equation to get $1 + 2y = 1.1$, so $y = 0.1/2$ or 0.05. So, the ball costs 5 cents, not 10 cents.

Almost certainly, if you got the right answer to Kahneman's problem, you did something very close to this mathematics in your head: *you used mathematics to think clearly*. However, if you (or your friend) didn't get the right answer: you should be worried that most of computer programs have *far* more complicated specifications, and how likely, then, is it that they have errors in them — and errors the programmers are unaware of?

Donald Rumsfeld, twice US Secretary of Defense, is perhaps most famous for his February 12, 2002, Department of Defense news briefing where he talked about unknown unknowns:

As we know, there are known knowns; there are things we know we know. We also know there are known unknowns; that is to say we know there are some things we do not know. But there are also unknown unknowns — the ones we don't know we don't know.

This may be thought-provoking, but he missed out another possibility: "unknown knowns." These must be the things that we don't know that other people do know.

Ironically, we know something Rumsfeld himself didn't know: there are 2^2 tuples of two items (here, *known* and *unknown*) and Rumsfeld only enumerated 3 of them, and of course $2^2 - 3 = 1$. So elementary formal reasoning exposes one missing case; surely a basic oversight in a very high profile statement!

The moral is that it is very easy to make errors you do not notice, and that formal reasoning — at least, once it is appropriately applied to the problem — provides another perspective that can help detect and solve problems.

Rumsfeld's missing case is particularly interesting. Your unknown knowns (what you do not know that others do know) is a problem that Rumsfeld overlooked, yet uniquely it is the one case that is very easy to rectify. You just need to find those people or tools that provide the missing points of view to turn your unknowns that are knowns that *others* know into knowns that *you* know.

2.2 Human Factors

Human Factors is a speciality combining psychology and engineering: it is concerned with how people work with engineered systems. Just as computer architecture affects performance, and strategies for improving performance, human mental and physiological architecture influences how humans perform. Combined with the varieties of computer hardware, such as keyboards and displays, when we interact with computers there are many ways to improve performance and dependability — as well as Human Factors such as satisfaction.

Human Factors, then, is concerned with the human aspects of the operation of systems, often emphasising the possible failings of humans. Blaming the "human factor" or saying the cause of an incident was "human error" has become just misdirecting jargon to blame a human operator for a mistake. Properly understood, however, *human factors* (HF or HFE for Human Factors Engineering) is the scientific study of human behaviour with particular reference to work, such as operating machinery, flying aircraft or programming — demanding tasks that involve attention and multi-tasking, phenomena that are known to lower human performance in predictable ways. Human error, for instance, can be induced by maintaining attention for a long period (causing fatigue) or multi-tasking (which induces omission errors). Human Factors is not the study of human "foibles" but of the consistent and predictable factors of performance and behaviour that are common across individuals.

Human Factors is particularly interesting because everyone, the people using, not using, or studying it, and we who are reading about it, are humans and therefore subject to the same biases and errors that are the very concern of the field. It is easy for people to ignore Human Factors precisely because they are human! Whereas bad engineering (like the Mars Orbiter) fails, and cannot easily be ignored, people routinely make decisions without the benefit of Human Factors insights and therefore rarely wonder about trying to make decisions that could be or could have been better informed by Human Factors.

Much work in Human Factors concentrates on operators or users performing work-related tasks, such as airline pilots or nurses. Very little has been studied about programmers (or designers and engineers more generally), the people who make the systems such users have to use. Even less has been studied about how programmers create languages or compilers that enable other programmers to create systems.

Certainly, the popular culture of languages such as PHP, Perl, JavaScript, Mathematica (and many others) creates the unspoken impression that "real programmers" can handle this. Somehow these languages live in a world where the solution is to make them ever more complicated, rather than safer, probably because their designers have a weak hold on the theory of language design, and one choice seems as good as another — hence features proliferate, interact badly with each other, and orthogonal features are rare.

The evidence is that it would be much better to improve the systems that are used to reduce error, rather than continually try to avoid error: a mature attitude is human error is the symptom of bad systems. Hence languages with strong typing catch errors that programmers are known to readily make.

These issues are well known background in the safety-critical community (e.g., Dale and Anderson, 2014) but they are rarely explicit, and even more rarely generalized to include *your own* hands-on programming practice when you are actually doing it.

2.3 Causes of error

The Human Factors that concern us in this paper affect programming (and in a more sophisticated way than keyboard design); for our purposes, Human Factors can be divided into two categories:

- how and why we make errors, *and*
- how and why we fail to notice or take our errors seriously.

My lists are not exhaustive.

Tunnel vision, formally known as loss of *situational awareness*. Our brains are finite, so when we concentrate on a task, inevitably there are fewer cognitive resources available for other activities. If this matters, it is called tunnel vision —

we can see what we think are the important part of the tasks, but the rest of the situation has gone blank. If the wider situation matters, then errors can occur. For example, programming is very complex and hard work: just getting a program to work at all is a minor miracle. The effort of getting the program to work means we *will* inevitably lose awareness of other issues. The cat can wait; we just let the cat out later, which will not matter much, but we may also lose awareness of critical issues, such as safety properties, usability and other issues that do matter. We have no idea what we are ignoring.

Thus explained, tunnel vision is obvious, but the problem is that when tunnel vision occurs you do not know it is occurring. You are not just ignoring a distraction, but your perceptual system is blocking your awareness even that there *is* a distraction. This may be efficient if you are doing the right thing, but how do you know you *are* doing the right thing?

Confirmation bias (Reason, 1990). The scientific method has rightly adopted falsification as its key method, but in everyday thinking we look for confirmation rather than falsification. We want to know why we are right, and we seek out confirmation that we are right. For example, when testing programs it is easier to devise tests that check the program works than devise tests that ought to crash it. Because we wrote the program, we know how it works, and this tends to limit our thinking to variations on how it works — rather than exploring the many ways it may not work.

Fundamentally, and perhaps understandably, we do not like being wrong: it is uncomfortable to be wrong. So we seek reasons why we are right. This is confirmation bias. We may end up with many reasons why we feel right — yet logically we need only one reason to be wrong, but we dare not look for it.

Attribute substitution (Kahneman, 2012). Many problems are hard, and rather than think about them, we may substitute some easier criterion. If the easier problem is unrelated to the original problem, we may end up solving the wrong problem. The classic case when attribute substitution arises is in interviewing. It is hard to choose candidates correctly, but cute candidates look good. Attribute substitution would mean thinking if they *look* good, they *are* good.

In programming, attribute substitution can occur when we go after visual effects and cute animations rather than dependability, compositionality, and many other properties that are much harder to think about. In fact, we cannot see these properties, so we may not notice we are not thinking about them — particularly if we like what we *do* see.

We live in a consumer world, and manufacturers are aware of this weakness. We are happily seduced to want products that look beautiful; if beauty is relevant (say, for a wedding ring or a movie) that is fine, but if we want a program to be safe, how nice it looks may well be a distraction.

And since we've already decided that the cute things are good, our eyes quickly glaze over when somebody tries to explain the real drawbacks.

Illusion of transparency (Gilovich *et al*, 1998). I have spent my lifetime living with myself, and I think I am and my programs are easy to understand. In fact, I even think that you think I am easy to see through. You know that I can't remember the reference to the illusion of transparency, don't you?

This is the illusion of transparency: we all think we are easier to understand than we are. In programming it could mean that when I implement a program I naturally think it is easier for you to understand it than it really is. How *could* you not know you press W, X and then Y to do Z? This is such elementary, simple knowledge, you must be stupid that you do not understand my program!

The illusion of transparency is one reason we find commenting our own code so hard; it is obvious what it does, so why waste time explaining it?

Fixation (Woods *et al*, 2010). Failing to revise one's current model of a situation, despite the evidence, is fixation error. It always seem easier to add new conditions to a program (e.g., to program around bugs) than to start again — it ends up getting more and more complicated, but being fixated on the original model is never questioned. Fixation has been argued to be the chief problem of error: we keep on doing the same thing, despite any evidence (which we are probably ignoring). Here is an example of fixation error:

> Insurers have estimated that such operator errors [fixation] account for about 80 percent of crane accidents. This incident repeats a familiar scenario: continued operation of a technological system despite its misbehaving. (Petroski, 2012, p317)

The chief solution to fixation error is to have more points of view (e.g., an *effective* team) and an authority gradient to enable diverse views to play a part. Fortunately we have "refactoring" as a word to remember to jog us out of some of our fixations.

2.4 Denial of error

There are lots of reasons (some are listed above) why error occurs; but once an error occurs we tend to deny it. Worse, in denying it, the subjective frequency of error seems lower than it is objectively, and thus it becomes even harder to admit to it. If we work in an organization, who wants to be the first person to admit making an error?

Denial. In the 1840s, when the doctor Ignaz Semmelweis discovered that washing hands reduced mortality from around 10–35% to about 1% (Best and Neuhauser, 2004) — two decades before Pasteur and Lister introduced antiseptic principles (Lister, 1867) — he was reviled. His blunt message was that doctors' *failures* were causing peuperal fever. What caring person (which doctors are!) wants to think that their routine behaviour is responsible for death?

Although we now understand about infection control, the cultural issues of denial of error persist. Today, in the USA about 750,000 people a year suffer a cardiac arrest in hospital, and are often treated with a defibrillator. Yet 30% of responders use defibrillators incorrectly. They are not redesigned to be easier and safer to use *because no doctor wants to admit they do not know how to use them correctly* (Pronovost and Vohr, 2011).

"Denial" may seem immature, but it is rational: if you would lose your job by admitting a problem, denial at least means you keep your job, and you keep your head down and hope nobody spots your bugs or, worse, traces them back to you.

But denial can happen for unconscious reasons too ...

Cognitive dissonance. You've spent a lot of time and effort programming the system, and a lot of time understanding it, so you are going to think it is great (Tavris and Aronson, 2007). Otherwise you would have to admit you wasted all that time and effort! And you are not stupid; therefore you did not waste that time. Ironically, the worse the system *really* is, the more the cognitive dissonance will mislead you. Cognitive dissonance, then, is a reason to be in denial.

Success bias. When your program works, it is really nice and you think it is wonderful, but the bugs in your program that you do not understand you will tend not to experience. Therefore, your perception of your program weighs success highly and tends to disregard problems.

Over-confidence. Our unawareness of these issues ensures we are also poor at estimating the effect of our unawareness, and hence we are likely to over-estimate our competence at (for example) thinking we are good at programming. Worse, if we have good professional feedback about our lack of programming skills, we are unlikely to learn anything because we do not understand the feedback: it is easy to think "I learned nothing from that professor, so I must know a lot." This is called the *Dunning-Kruger Effect* (Kruger and Dunning, 1999).

3 Some solutions for safety-critical programming

All of us are ignorant about programming; it is one of the fastest-developing areas of human endeavor, and none of us can keep up. What is unknown today is news tomorrow, and widely adopted the next day: consider QR codes as a blatant example (Masahiro *et al*, 1998). We are adopting new technologies faster than completing existing projects. *Nobody* is a skilled functional programmer, skilled parallel programmer, skilled logic programmer, skilled object-oriented programmer, skilled database programmer, skilled web programmer, skilled at Formal Methods, skilled graphics programmer *and* a skilled symbolic mathematics programmer — let alone skilled in all the APIs and other resources one needs to be remotely effective. There are things we do not know, and a lot of things we do not know we do not know. We all have tunnel vision about programming.

3.1 Mathematics and science: reliable reasoning

In contrast to our normal error-prone behaviour, mathematics and science are the two key human inventions that enable us to spot error, correct error, reason reliably and reach consensus on what should be valid knowledge. If we disagree about any issue, turning it into a mathematical or scientific question in principle provides a standard methodology that allows us to jointly reach the truth, agree it, and know the limitations of that truth.

The reason we need mathematics and science is because we make errors, including errors we do not notice or even know about. We once thought the Earth was flat, the stars were fixed, that the continents did not move, and so on. We had centuries of strange ideas about disease until Koch, Pasteur and Lister showed the world that diseases are caused by invisible germs. There are many similar major errors that persisted in various forms and then were radically corrected; Lightman and Gingerich (1992) present a wonderful discussion.

Mathematics as a subject less obviously struggles with error; instead, mathematics continually expands, reducing the unknown unknowns, decreasing ignorance. In fact an advantage of mathematics is we often know what we do not know, whereas in science we more often have no idea.

"Individual" mathematics (such as what we do when we calculate), though, is notorious for error. Mathematics has developed many methods for detecting error rapidly — double entry book-keeping being a prime example, and error-correcting codes the generalization of that to all forms of communication, not just accountants communicating reliable financial information.

Both science and mathematics value multiple ways of thinking. In science, an idea that cannot be replicated or duplicated has little significance. In mathematics, abstraction and generalization are fundamental values. That 17 has no factors (other than 1 and 17) is not interesting; but that there are an infinity of such numbers is so interesting it had already been written up around 300BC in Euclid's *Elements*. A theorem is of course a concise way of saying a lot of truths, and thus establishing at once many things that do not need to be individually checked.

One of the interesting features of mathematical reasoning is that, at least in principle, everything needed for reasoning is visible on the page. In almost all other subjects, much of the reasoning and evidence is beyond the paper and out in the world — no scientific paper for instance can say everything about the issue: the paper has to abstract away from reality and take a lot on trust. And by doing so it may make errors that cannot be seen in the paper itself. Unsurprisingly, there are plenty of fraudulent scientific papers, but hardly any fraudulent mathematics papers, where it is not so easy to get away with error, accidental or deliberate. There are, unfortunately, lots of fraudulent programming papers (Thimbleby, 2004; 2012).

The reason we use Formal Methods in programming, then, is because humans make errors, and Formal Methods provides other point of views, primarily based in mathematical reasoning, which itself "is designed to" pick up issues that normal, error-prone human thinking misses.

Nevertheless, there are still residual errors: programmers may think some programming is so obvious that Formal Methods are not needed, programmers may correctly specify the wrong thing, and programmers may fail to specify all of the intended behaviour of the program. Ironically, user error is rarely handled using Formal Methods — it is messy and hard to formalize. In some super-human way we have to overcome our own error-prone blinkers to properly consider the errors users will make.

3.2 Iterative design

The international standard ISO 9241 proposes *iterative design*: build a system, try it, then improve it, and repeat. Iterative design is needed because in programming we are generally building systems that have never been built before and we do not really know everything about the systems that we need to implement them correctly. We should therefore — following ISO 9241 — build prototypes, evaluate them and iteratively converge on a better system. Iterative design is a process that helps find and eliminate errors.

Iterative design is best practice, but it faces many hurdles. To most people who build systems, how they work is so obvious that iterative design seems like an unnecessary expense. (Whether the systems are obvious to anyone else is a matter that needs careful empirical evaluation.) Designers get fixation, confirmation bias and cognitive dissonance: whatever they have designed is obviously good, they think. Marketing people who often drive product design have no idea how complicated design is and they do not want to know how much iterative design could improve their ideas. And with the usual rush to market, iterative design seems like an unnecessary delay.

3.3 Multidisciplinary teamwork

Conventional wisdom is that resilience to error is achieved by using teams effectively, particularly multidisciplinary teams. Different people in a team can see things differently and can point out error trajectories earlier so preventative steps can more often be taken. Formal Methods may therefore be thought of as a "team player" that brings a new perspective, that thinks differently, and hence exposes the errors in our normal error-prone thinking.

3.4 Checklists

Checklists are so simple they are easy to ignore, yet they are an excellent way to overcome tunnel vision and manage complexity. Gawande has written a marvellous book on checklists that I very strongly recommend (Gawande, 2011).

The following checklist summarises a few of my points:

- Use Formal Methods
- Prove theorems
- Involve people from other disciplines
- Try to make your program fail
- Use iterative design
- Use tools to help find and track errors
- Read Atul Gawande's *Checklist Manifesto* ...

This is far too obvious to be needed, right? One of the problems with checklists, particularly to anyone who has the authority to get them used, is that it seems like a waste of time just stating the blindingly obvious and in such a simple way! But the point is that when people are doing a demanding task (such as programming) the bigger picture will be lost to tunnel vision. When you are *not doing* the task, you have spare mental capacity and things are relatively obvious, so write a checklist to help for when you *are doing* the task — when things will not be so obvious. Note that checklists should be developed as rigorously as programs — they are of course a sort of program for programmers — and, in particular, they should be reviewed by multidisciplinary teams and subject to iterative design.

I think in our worked example, next, you will soon find that you very easily get tunnel vision, etc, and a good checklist (that was followed!) might have helped.

3.5 This list is not complete ...

Reason (1990) is the classic reference on human error; although he recently refined his perspective: how people are *resilient* to errors is more interesting than the errors themselves (Reason, 2008). Alamerti (2012) is an excellent, insightful and brief introduction to the wider complexity of the field. Kahneman (2012) and Woods *et al* (2010) are also highly recommended. However, none of these authors discuss programming.

Within software engineering there are many techniques for resilience beyond those we have space to explore in this paper that (to varying degrees) help safety-critical programming. Open source, for example, helps by engaging a wider perspective from many other programmers.

4 A worked programming example

I want to illustrate the issues — the clash of Human Factors and safety-critical programming — by way of a very small example. If you appreciate this little example once you have worked through it, I'd like you to think how much worse real life must be.

Consider, then, a simple case where the requirement is to read a number from a user; to be more specific, to read (say, from the keyboard) a non-negative numeral in conventional base-10 Arabic notation, and convert it to a floating point number.

Suppose the sequence of keystrokes the user presses is s_1, s_2, ... $s_{|s|}$ and the software should return a floating point number $f(s)$.

At this point, many people think the problem is so obvious they would go straight into coding it. Or you would use some standard function of your favourite language and hope it has the same specification as what is wanted (or you might renegotiate the specification).

Alternatively, a moment's thought creates a complete specification ...

$$f(s) = \sum_{i=1}^{d(s)-1} s_i 10^{d(s)-i-1} + \sum_{i=d(s)}^{|s|} s_i 10^{d(s)-i}$$

where $d(s)$ is the position of the decimal point (if s_p is the decimal point, then $d(s) = p$). If we further define $d(s)$ to be $|s| + 1$ in the case there is no decimal point, then the formula above for f also works correctly with numbers with no decimal point, since the second sum (which calculates the fraction value) will be vacuously zero.

This is not a complicated problem, but in general we may want to prove some theorems about our specification. Theorems, by the way, are a way to think explicitly about a problem to provide another perspective that (by convention) we feel obliged to prove.

Here, the simplest theorems might be along these lines:
$$\begin{aligned} f(st) &= f(t) + f(s) \times 10^{|t|} \\ f(s.t) &= f(s) + f(t) \times 10^{-|t|} \end{aligned}$$

These are the sorts of property we want our specification to have. (I think they happen to be elegant as well, which makes them more plausible to me!)

We could also list 20 basic rules (though we don't need all of them thanks to theorems like those above):
$$\begin{aligned} f(s0) &= 10f(s) + 0 \\ f(s1) &= 10f(s) + 1 \\ &\vdots \\ f(s.t8) &= f(s.t) + 8 \times 10^{-|t|-1} \\ f(s.t9) &= f(s.t) + 9 \times 10^{-|t|-1} \end{aligned}$$

This case analysis (which isn't yet complete without the base cases) might immediately suggest an implementation strategy, particularly if you use a programming language with pattern matching.

Regardless ... I would now like you to *please* go ahead and refine and implement this specification.

Actually, it is *so* easy, we can code it directly, say in JavaScript, without much thought.

I gave in and did it for you, but please check my code carefully, which is shown below.

```
// given a string s return index position of decimal point,
// or length of string if none
function d(s)
{       for( i = 0; i < s.length; i++ )
            if( s[i] == '.' ) return i;
        return s.length;
}
```

```
// given a string s return the floating point value of it
// considered as a base-10 Arabic numeral
function ReadNumber(s)
{       var dlocal = d(s), f = 0;
        // integer part
        for( i = 0; i < dlocal; i++ )
            f = f+(s[i]-"0")*Math.pow(10, dlocal-i-1);
        // fractional part
        for( i = dlocal+1; i < s.length; i++ )
            f = f+(s[i]-"0")*Math.pow(10, dlocal-i);
        return f;
}
```

It goes without saying that it is good practice that what was f in the specification has become `ReadNumber` in the program, since single-letter names like f are commonly over-used.

It takes a little further thought to map the specification to JavaScript's array base of zero and to convert characters to digit values, but it's not difficult (I was briefly tempted to use `s[1]` as the base and ignore `s[0]`).

One also notices that the specification conveniently takes digits to denote their own values, though of course in standard character codes, the digit character 0 may have a code like 48, and a calculation is then required to convert from the character's code to the decimal digit value. We assume (since we happen to know this is a design principle of Unicode) that the digits 0123456789 have consecutive numerical codes, so the correct numerical value of a digit character c (say "2") would be $c-$"0", where we assume "0" gives 48 or whatever the code value of the character 0 is.

You may also notice that there are many optimizations possible; for example,

```
f = f + (s[i] - "0")*Math.pow(10, dlocal-i-1)
```

can be rewritten

```
f = 10*f + s[i] - "0"
```

thus avoiding some nasty and slow JavaScript arithmetic.

Finally, it is worth checking a few cases like ReadNumber("12345"), Read-Number("123.45") and so on — all of which should work nicely. Please do that, or at least double check I have not made a typo in copying my code into this paper while reformatting it to make it look nicely typeset.

You may have noticed how trivial this exercise seems to be, and it might be tempting to think that it therefore deserves very little attention.

How much simpler could it possibly get? It is practically the identity function:

```
ReadNumber("12345") = 12345
ReadNumber("123.45") = 123.45
```

In JavaScript, it is tempting not even to bother, since the language *itself* guarantees "12345"==12345 without even needing to implement any functions like ReadNumber!

After all these deliberate distractions (which were to try to make the experience of the problem realistic — though your real programming problems will surely be even more complex), which of the following design issues did you fail to think about?

1. The requirements do not discuss error. (Even after my quoting NASA as saying not managing error was an error!)
2. The specification gives zero if the string is empty (which the program correctly implements). Was this special case intended?
3. The specification gives 0 if the "number" is just a decimal point (an isolated decimal point should surely be a syntax error).
4. The specification does not handle more than one decimal point; instead, it gives an undefined numerical result. The definition of $d(s)$ is not even correctly defined if there is more than one decimal point.
5. The specification is undefined if anything other than digits or decimal point is entered. However, being JavaScript, the program will execute "correctly" but will return unexpected values, both numbers and NaN ("not a number").
6. There are many invisible characters (such as null and space, end of file and others) that probably ought to be ignored.
7. The program may have undetected number overflow or underflow.
8. The program (being written in JavaScript) approximates numbers with values larger than $2^{53}-1$, which limits its correctness to values no larger than 9,007,199,254,740,991: with numbers larger than this, it will return imprecise values (only returning NaN when the number is larger than 10^{308}). If we had taken the specification literally, we should have used a big number implementation, or we should have revised the specification

to handle over and underflow or, for instance, to put bounds on $|s|$, such as $0 < |s| \le 10$.

9. If the user enters s typed into a text field, it is undefined what happens when the field becomes full (in fact, we did not define anything about the user interface). In particular, in many implementations what the user sees will be different from s. Trivial? Thimbleby (2015b) reports on a case where a user lost \$100,000 over this design error.

10. The specification is not internationalized. Indeed, there is an international standard on number representations, and this specification is not compatible with it. ISO-31 requires numbers less than zero to be preceded with a 0 digit (e.g., 0.5, not .5); ISO 31 also allows spaces between groups of 3 digits and permits comma and dot as the decimal point. (We needed some such numbers above, yet they cannot be handled by our program!)

11. If there is an error (as there evidently can be) handling the error is undefined.

12. Stylistically, defining a function called d (used to return the decimal point index) in JavaScript is stupid: our definition may override this common name and change functionality of the program elsewhere.

13. The requirements do not say what to do with character coding — for example, Unicode has several sets of digits in languages like Thai and Javanese, as well as multiple Arabic digit forms (e.g., bold digits, sans serif digits, etc). There are also digit-like forms, such as letter O which looks like zero.

14. If you believed my explanation earlier that we needed to subtract the character code of 0 from each character in `s`, you made an error. You were fooled! In fact, when JavaScript parses `s[i]-"0"` it sees the subtraction as arithmetic and therefore (if that is the right word for JavaScript) coerces both `s[i]` and `"0"` to be numbers, so it actually does the conversion of `s[i]` for us, then subtracts zero (i.e., the coerced value of `"0"`).

15. Having read the last point, you might want to try a test like `ReadNumber("12 345")` and you will find it goes wrong (but in a strangely nice way!): in this example, JavaScript takes the numerical value of `" "` to be zero rather than being a syntax error. In other words, our program accidentally inherits a bug (a misfeature) of JavaScript.

16. Many numbers have a purpose, and "just" implementing numbers may conflict with those real purposes. Very simple examples include that credit card numbers and NHS numbers are both numbers, but they require spaces in interesting ways that our blind implementation of generic numbers would fail on. For example, my credit card number is 4821 8197 1724 3995, and you are welcome to try typing that *with the spaces* into any web site; you may also like to find out what our JavaScript code above does with it.

17. Did you notice my deliberate programming bugs? For example, the controlled variables of the several `for` loops were not declared as local variables, so any global variable `i` will be corrupted.

18. Did you write any test code? If you did, did you only check your code works (confirmation bias: such as trying just the examples I gave above) or did you try to find ways it does *not* work — such as corrupting global variables (which it does).

19. The requirement was to return the floating point value corresponding to the Arabic numeral *s*. The program code above may return an integer instead, which is not a problem in JavaScript (because integers are represented as floats anyway), but it could be a problem in other programming languages.

20. JavaScript has a lot of peculiarities like `ReadNumber(".1") * Read-Number(".1")` will not be 0.01 as you might expect (it is 0.010000000000000002 because of the way JavaScript's IEEE floating point works). The fact that the function implements the specification may not be sufficient for the program it is used in to have the expected properties implied by the specification.

21. Finally, the requirement was to read a number from the user. Did you think about how to handle user error? *Users will make errors*. How will the user be able to detect and correct errors? Why doesn't the specification discuss the delete key (to correct errors) and why doesn't the specification discuss syntax (to define erroneous input, such as missing numbers or numbers with incorrect syntax) and how to handle it?

4.1 Discussion

Reading a number is trivial and our exercise was a simplification of a trivial exercise. Yet I hope it was still complicated enough to illustrate common Human Factors problems.

Worryingly, reading a number is a critical application in many safety-critical systems (such as radiotherapy machines) — and surprisingly many systems get reading numbers wrong (Thimbleby, 2015b). In fact, a considerable part of failing to read numbers correctly (as shown by Thimbleby, 2015b) is failure to handle user error correctly: an astonishing number of everyday user interfaces fail to handle the delete key correctly during number entry.

It is very likely that if you seriously played along with the game of thinking about the implementation (section 4 above) that you failed to notice many of the potential errors listed. (Even better, you may have found errors I overlooked in my list. I am not immune from error!)

The lesson, I hope is, even using Formal Methods (albeit, only sketched in this paper) cannot make us immune from many routine errors such as those listed above. Not only do we make errors but we are unaware of our making errors: as our workload or concentration increases, our awareness of design issues — including safety issues — outside of our immediate focus necessarily decreases. Hopefully the pressure and complexity I created around formalizing parsing a number created a bit of realistic tunnel vision, so you lost sight of bigger issues.

Along with tunnel vision, perhaps you also experienced confirmation bias and probably denial — like: you would not have made any errors if you had actually taken the problem seriously, would you?! You are in good company: Apple, Casio, Hewlett Packard and others make all the same mistakes (Thimbleby, Oladimeji and Cairns, 2015; Thimbleby 2015b).

5 Conclusions

The programming task is so cognitively demanding that it is practically inevitable that there will be oversights in the requirements, specification, implementation, and testing.

Hopefully, this paper has provided a perspective to start thinking strategically about human error. We tried to explain the reason why Formal Methods helps — it provides rigorous methods for expanding our thinking and enabling discussion and debate (with yourself and with colleagues and even users) — but also why Formal Methods are not sufficient. Human error cannot be avoided. Nobody is immune from error.

Formal Methods thinks differently, it thinks differently to your normal thinking, and therefore using it helps reduce error. But Formal Methods is not sufficient, so always ask: "What *other* ways of thinking, what other points of view can be used?"

This will take other experts with the other points of view to contribute. Crucially, these other points of view have to be integrated with the team: simply having people tell you things you ought to think about increases your workload and may deepen any problems with tunnel vision. People need to be involved and share the work.

- **Avoiding error is the most strategic thing to do**. Put effort (and this paper argued: much more effort, and more informed and more diverse effort, than you thought) into programming safely.
- **We under-estimate error**. Our own psychology misses error and under-reports it — to the extent of denial — and we are distracted by blaming people rather than fixing systems. We do not hear enough from users about error: few users will provide feedback after an error (if they go out of business they will give no feedback at all).
- **Even with precautions, error still happens**. Saving lives after accidents is as important as avoiding accidents. In medicine not having a plan for after an incident is called *failure to rescue* (Schmid *et al*, 2007); or as

Glegg (1969) put it nicely: if you do not design for error, error will still happen … and then you will not know what to do.

We as programmers are human and subject to error; one of the most pernicious types of error is failing to appreciate and mitigate human error in the use of our programs. I do not know what errors you made in our simple case study, but even if you got into the spirit of the exercise: did you think about problem 21?

Finally: humans, both programmers and users, will always make errors. You are not immune. A specification cannot be error-free until both it and the human process to implement it handles error appropriately.

Acknowledgement. Ross Koppel, Michael Parsons and Michael Harrison made very helpful comments. This research was funded by UK EPSRC Grant No. [EP/L019272].

References

R. Amalberti, *Navigating Safety: Necessary Compromises and Trade-Offs — Theory and Practice*, Springer, 2013.

E. Allman, "Managing Technical Debt," *Communications of the ACM*, **55**(5):50–55, 2012. doi 10.1145/2160718.2160733D

M. Best and D. Neuhauser, "Ignaz Semmelweis and the birth of infection control," *Journal of Quality & Safety in Health Care*, **13**(3):233–234, 2004. doi 10.1136/qshc.2004.010918

C. Dale and T. Anderson, *Addressing Systems Safety Challenges*, Proceedings of the Twenty-second Safety-critical Systems Symposium, SCSC, 2014.

A. Gawande, *The Checklist Manifesto: How To Get Things Right*, Profile Books, 2011.

T. Gilovich, K. Savitsky and V. H. Medvec, "The Illusion of Transparency: Biased Assessments of Others' Ability to Read One's Emotional States," *Journal of Personality and Social Psychology*, **75**(2):332–346, 1998. doi: 10.1037/0022-3514.75.2.332.

G. L. Glegg, *The Design of Design*, Cambridge University Press, 1969.

D. Kahneman, *Thinking, Fast and Slow*, Penguin, 2012.

J. Kruger and D. Dunning, "Unskilled and Unaware of It: How Difficulties in Recognizing One's Own Incompetence Lead to Inflated Self-Assessments," *Journal of Personality and Social Psychology*, **77**(6):1121–1134, 1999. doi 10.1037/0022-3514.77.6.1121.

A. Lightman and O. Gingerich, "When do anomalies begin?" *Science*, **255**(5045):690–695, 1992.

R. Lloyd, "Metric mishap caused loss of NASA Orbiter," CNN, http://edition.cnn.com/TECH/space/9909/30/mars.metric.02, September 30, 1999.

H. Masahiro, W. Motoaki, N. Tadao, N. Takayuki and U. Yuji, *Optically readable two-dimensional code and method and apparatus using the same*, US Patent US5726435 (A) — 1998-03-10, 1998.

H. Petroski, *To Forgive Design*, Belknap Press of Harvard University Press, 2012.

P. Pronovost and E. Vohr, *Safe Patients, Smart Hospitals: How One Doctor's Checklist Can Help Us Change Health Care from the Inside Out*, Plume, 2011.

J. Reason, *Human Error*, Cambridge University Press, 1990.

J. Reason, *The Human Contribution: Unsafe Acts, Accidents and Heroic Recoveries*, Ashgate, 2008.

A Schmid, L Hoffman, M. B. Happ, G. A. Wolf and M DeVita, "Failure to rescue: A literature review," *Journal of Nursing Administration*, **37**(4):188–198, 2007.

C. Tavris and E. Aronson, *Mistakes Were Made (but not by me)*, Harcourt Inc., 2007.

H. Thimbleby, "Give your computer's IQ a boost," *Times Higher Education Supplement*, 9 May, 2004.

H. Thimbleby, "Heedless Programming: Ignoring Detectable Error is a Widespread Hazard," *Software — Practice & Experience*, **42**(11):1393–1407, 2012. doi 10.1002/spe.1141

H. Thimbleby, "Human factors and missed solutions to Enigma design weaknesses," *Cryptologia*, 2015a. doi 10.1080/01611194.2015.1028680

H. Thimbleby, "Safer User Interfaces: A Case Study in Improving Number Entry," *IEEE Transactions on Software Engineering*, **41**(7):711–729, 2015b. doi 10.1109/TSE.2014.2383396

H. Thimbleby, P. Oladimeji and P. Cairns, "Unreliable numbers: Error and harm induced by bad design can be reduced by better design," *Journal of the Royal Society Interface*, **12**(110):20150685, 2015. doi 10.1098/rsif.2015.0685

D. D. Woods, S. Dekker, L. Johannesen and N. Sarter, *Behind Human Error*, 2nd ed., Ashgate, 2010.

Competence Considerations for Systems Safety

Carl Sandom

iSys Integrity Limited

Sherborne, UK

Abstract *People often use the word 'competence' without understanding what it means even when it is vital for safety. This paper examines common definitions of competence to identify the individual components and understand the principles underlying the specification and assessment process. Safety management must facilitate the achievement and maintenance of competence for those developing and operating safety-related systems. The paper examines the theory and principles underlying the attainment and maintenance of competence providing a framework for a discussion on competence assurance. The specification of competence criteria is an important safety management activity and these are unique for different systems. This paper describes how competence criteria can be specified and assessed for safety-related systems. Safety assurance is ultimately based upon the competence of the people involved hence competence evidence is essential for the validity of any safety claim. The paper examines common safety assurance issues associated with competence and some suggestions are made on how to improve the validity of safety claims based upon competence.*

1 Introduction

Competence for any professional is a desirable attribute but it is an essential requirement for those involved with the development, maintenance and operation of safety-related systems. The international safety standard IEC 61508 (IEC 2010) now has normative requirements for demonstrating the competence of those involved in safety-related systems activities across all safety lifecycle phases; however, the standard lacks guidance on how to fulfil that requirement.

New technologies, particularly those containing software, have enabled systems to function more effectively and allowed more sophisticated ways to make them safe. Paradoxically, new technology has also brought its own challenges such as increased design complexity. The accelerated use of new technologies and the associated complexity increases the importance of the activities undertaken by people engaged in the design, development, maintenance and use of safety-related systems. The achievement of sufficiently low levels of risk is critically dependent on individual and team competence.

In parallel, the pace of change in industry continues to accelerate, with frequent restructuring and much movement of people between roles, between companies and even between sectors. Ever newer technology requires new skills. Even if new staff possess these skills, they may be unfamiliar with the organisational culture and more importantly the safety culture. Long term familiarity of managers with the capabilities of their staff can no longer be assumed, so increasingly organisations need to establish Competence Management Systems (CMS) in order to satisfy themselves, their customers and regulators that their staff are competent for the tasks to which they are assigned.

Competence is a vital issue for those involved in hazardous systems, not just for the system developers, operators and maintainers but also for those providing safety assurance based upon expert opinion and judgement.

2 Understanding Competence

This section examines the meaning of competence and identifies its main components in order to understand the underlying principles associated with the process of attaining and maintaining competence (section 3) and also specifying and assessing competence (section 4).

2.1 Competence Definitions

The term *competence* generally means the ability to do something successfully or efficiently and many synonyms are used including: capability; ability; proficiency; expertise and skill. It is useful here to differentiate between the closely related (and oft confused) terms *capability* and *competence* as there are important but subtle difference between the two concepts. Capability describes the ability of an organization while competence describes the ability of a person to do something (Holt and Perry 2011).

The UK Engineering Council (EC 2013) provides the following definition of competence:

> The ability to carry out a task to an effective standard. Its achievement requires the right level of knowledge, understanding and skill, as well as a professional attitude.

The definition of competence given by UK Office of Rail Regulation (ORR 2007) is similar:

> The ability to undertake responsibilities and to perform activities to a recognised standard on a regular basis. Competence is a combination of practical and thinking skills, experience and knowledge, and may include a willingness to undertake work activities in accordance with agreed standards, rules and procedures.

The UK Health & Safety Executive (HSE 2007a) defines competence as:

> The ability to undertake responsibilities and perform activities to a recognised standard on a regular basis.

The HSE also assert that to be competent an organisation or person must have:

- Sufficient knowledge of the tasks to be undertaken and the risks involved.
- The experience and ability to carry out duties in relation to the project, to recognise limitations and to take appropriate action to prevent harm to those carrying out or affected by work.

Competence can develop (and decay) over time and it is vital in abnormal and emergency situations. Generally, people develop competence through a progressive mix of initial training, on-the-job learning, instruction, assessment and formal qualification. In the early stages of training and gaining experience, people should be closely supervised and as competence develops, the need for direct supervision should be reduced (HSE 2007b).

2.2 Competence Components

In the sciences methodological reductionism provides explanations of concepts in terms of their individual, constituent parts and their interactions. Similarly, while general definitions of competence are helpful, a more detailed examination of its separate elements can provide a better understanding of how it may be acquired and maintained.

Many different explanations of the elements of competence exist; however, an examination of the similarities between the EC, ORR and HSE definitions reveal three main components of competence which are shown in Figure 1.

1. **Knowledge** which is acquired through training, both formal and on-the-job, and is required to enable people to formulate a plan of action to undertake an activity.
2. **Skills** which are the things that experienced people often do subconsciously. Skills can be thought of as the execution part of a plan of ac-

tion. Skills are an observable act or behaviour (sometimes referred to as ability) exhibited while undertaking an activity.

3. **Attributes** are associated with personal qualities such as determination, integrity, effective communication etc.

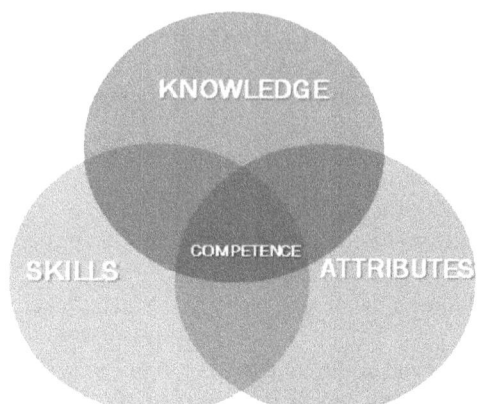

Fig. 1. Competence Components

It should be understood that the proportions of each component contributing to competence as depicted in Figure 1 is entirely context dependent and it is not fixed. Competence involves much more than technical training, it includes a person's attitude and behaviour as well as experience and knowledge of the application domain (HSE 2007a).

IET publications on competence (IET 2007, 2016) suggest that competence consists of: technical skills; behavioural skills; underpinning knowledge and understanding. A distinction is made between technical and behavioural skills. Technical skills can be thought of as those vocational skills learned for a specific role (e.g. an aircraft pilot's motor skills or their ability to interpret meteorology reports) while personal behavioural skills are more general (e.g. the ability to communicate effectively or problem solving ability).

Competence might be transferable from one work situation to another, but the extent to which this is possible depends very much on the *context* in which apparently similar competence is required. For example, a person considered competent to develop software for an aircraft In-Flight Entertainment system will almost certainly not be considered competent to undertake the development of the Flight Management System (FMS) for that aircraft without having the experience and detailed knowledge of FMS functionality, standards and, importantly, how the FMS is used operationally.

3 Attaining and Maintaining Competence

This section examines the theory and principles associated with the process of attaining and maintaining competence; providing a foundation for an examination of the specification and assessment of competence (section 4) and a framework for a discussion on competence assurance (section 5).

3.1 Conscious Competence Model

Much of the literature on competence comes from the teaching profession and is encapsulated in various theories of learning. One prevalent model of learning, and change management in general, is the Conscious Competence Model (CCM) (Robinson 1974) which will be examined here as it is useful to frame the discussion presented later on organizational and individual safety competence.

Notwithstanding the various claims to original authorship, many of the proponents of the CCM consistently advocate the separate stages of learning (or change) as shown in Figure 2.

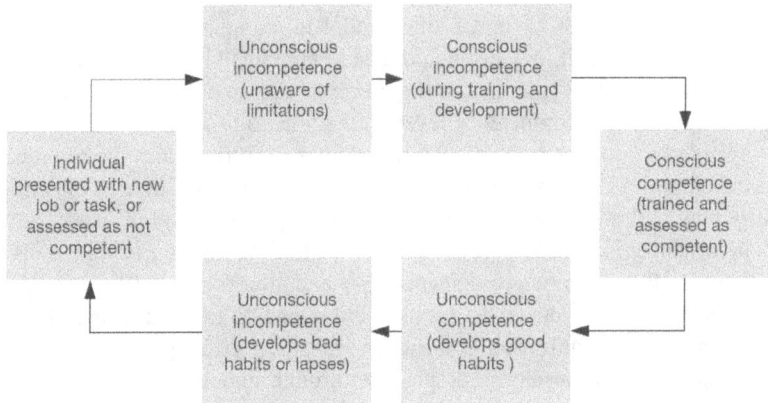

Fig. 2. Conscious Competence Model (adapted from Robinson 1974)

Unconscious Incompetence is a state when a person or organization is blissfully unaware of their lack of a specific skill, knowledge or attitude required for a given task.

Conscious Incompetence is a state when a person or organization becomes aware of their lack of a specific skill, knowledge or attitude required for a given task.

Conscious Competence is when a person or organization has consciously attained a degree of skill, knowledge or attitude required for a given task but it requires conscious effort to complete.

Unconscious Competence is when a person or organization has attained a high degree of automatic skill, knowledge and attitude required for a given task and it requires minimal or no conscious effort to complete.

Unconscious Incompetence is when an unconsciously competent person can regress to unconscious incompetence due to changing environmental factors or the erosion of competence through the development of bad habits.

Once a state of unconscious competence is attained, proactive measures must be taken to maintain that state and avoid unconscious incompetence; for example professional or chartered engineers are required to undertake Continuing Professional Development (CPD) to maintain their knowledge, experience, skills and personal qualities. CPD encompasses both the acquisition of new skills to broaden competence and the enhancement of existing skills to keep up to date with evolving knowledge.

3.2 Unconscious Competence

The name of the CCM model may not be appropriate as the name implies that conscious competence is the aim when in fact the ultimate aim is the attainment and maintenance of *unconscious competence*.

Nonetheless, the CCM is a useful model to frame any discussion of safety competence and how it may be acquired and maintained both at the organizational and individual level. Typically, an organization can be characterized as operating at the unconscious incompetence level until some point in time when they are either awarded a contract with safety requirements to fulfil or they simply recognize that developing safety-assured systems is a good business strategy and from necessity they will transition to the conscious incompetence level.

The organization could then take action to initiate training to take individuals and teams to the conscious competence state and after time (and perhaps some on-the-job training) individuals and teams could transition to the unconscious competence state. The organization could even transition directly from the conscious incompetence to unconscious competence through judicious recruitment of competent people.

The aim for any organization dealing with safety should be to facilitate the transition of the organization and individuals from conscious incompetence to unconscious competence and to maintain that level through the formal development and implementation of a CMS (see HSE 2007a, 2007b) and the specification and assessment of suitable competence criteria for the activities they undertake (see IET 2007, 2016).

4 Specifying and Assessing Competence

The international safety standard IEC 61508 (IEC 2010) now has normative requirements for demonstrating the competence of those involved in safety-related systems activities across all safety lifecycle phases; however, the standard lacks guidance on how to fulfil that requirement. It is useful to examine the general principles of competence management and identify where the detailed specification of criteria are required for the achievement and maintenance of safety competence.

The specification of safety competence is unique for different systems and operational domains and it is an important element of safety management. The requirement to demonstrate safety competence involves the identification of safety-related activities and their associated tasks each at a specified level of competence for a given system; these are referred to collectively as competence criteria. Competence criteria for safety-related systems developers are significantly different to those for developers of non-hazardous systems because different technical skills, knowledge and personal attributes are usually required. However, the specification of competence criteria must be based on a coherent model of competence to provide a common reference framework.

4.1 Competence Model

A competence model sets out the relationships between various concepts used when evaluating competence, in particular the relationships between roles, activities, tasks, attributes, competence criteria, levels of competence and activity or personal competence profiles. Discussions within this paper are based on the general competence model shown in Figure 3.

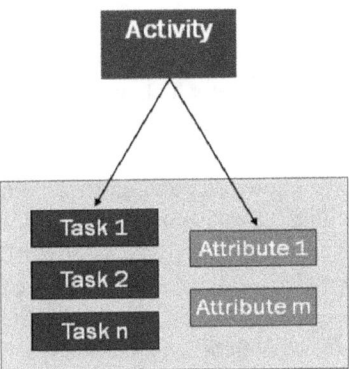

Fig. 3. Competence Model (adapted from IET 2007)

A person working either alone or in a team, performs an *activity*. Figure 3 shows that each activity is subdivided into a set of *tasks* each of which require particular technical skills and knowledge. All the tasks in an activity also require behavioural skills and underpinning knowledge and understanding which are expressed as a set of *attributes*.

The relationship between a role and an activity can be confusing and needs to be clarified. An activity may be undertaken by a person or by a team. When working in a team, each person may contribute to the completion of the activity by performing a role within the team and carrying out part of the activity. A role can therefore be equivalent to an activity or it may be a part of an activity.

An activity can be thought of as the high-level things that an organization needs to do to achieve a specified goal; for example, from the activities associated with achieving the goal of providing functional safety assurance. Each activity is decomposed into a set of tasks each of which require the necessary technical skills and knowledge to be defined. In addition, each activity is decomposed into a set of attributes each of which require the necessary behavioural skills and knowledge to be defined.

An example of this hierarchy for a safety-related activity, using the competence model in Figure 3, is shown in Table 1.

Table 1. Competence Hierarchy

Criteria	Safety Example
Activity	Independent Safety Assessment
Task	Safety Auditing
Attribute	Effective Communication

4.2 Competence Levels

The tasks and attributes defined for a specific activity are typically specified at three different levels of competence which are summarised as follows (IET 2007, IET 2016):

Supervised: has sufficient knowledge and understanding of good practice, within the organization or within the relevant industry sector, to be able to work on the tasks associated with the overall activity without placing an excessive burden on the practitioner or expert who is responsible for checking their work.

Practitioner: has sufficient knowledge and understanding of good practice, and sufficient demonstrated experience, to be able to work on tasks without the need for detailed supervision.

Expert: has sufficient understanding of why things are done in certain ways, and sufficient demonstrated managerial skills, to be able to undertake overall responsibility for the performance of a task or activity.

For any CMS established and operated within a safety-related domain, the validity of the competence criteria will have a critical influence on the efficiency and effectiveness of the CMS and in turn on the ability of an organization or individual to provide safety assurance for safety-related systems.

4.3 Competence Criteria

Competence criteria must be specified for the activities, tasks and attributes for which competence is required in an organization or individual. For an organization developing safety-related systems an example activity may be to provide *Functional Safety Management*; associated tasks may include: *Define Safety Management Policy*; *Allocate Safety Responsibilities* and *Promote Safety Culture*.

Typical attributes for the individual or team undertaking the activity may include: *Effective Communication*; *Professional Standing* and *Personal Integrity*.

Task or Attribute Title		
Description of the task or attribute		
Supervised Practitioner	**Practitioner**	**Expert**
Competence criteria for supervised practitioner level.	*Competence criteria for practitioner level.*	*Competence criteria for expert level.*
Competence criteria for both supervised practitioner level and practitioner level.		*Explanation for non-relevance of level to task or attribute*

Fig. 4. Generic Task/Attribute Criteria

A generic task or attribute criteria specification is shown in Figure 4. Each task or attribute has a set of criteria specified for it which state the competencies required to fulfil the task or attribute at any or all of the 3 competence levels discussed in 4.2.

For each of the competence levels appropriate to a given task or attribute, competence criteria will be specified. For example, for the activity *Functional Safety*

Management, the Expert level competence criteria for the task *Define Safety Management Policy* could be expressed as (IET 2007, IET 2016):

> Has developed at least one corporate safety management policy and has been involved in the development/ review of others. Can identify organization methods and procedures, which have had to be updated to meet new standards in functional safety assurance, and show how the updated methods and procedures fit within the organization's safety management system.

Similarly, the Expert level competence criteria for the attribute *Effective Communication* may be expressed as (ref. 8-9):

> Is acknowledged as proficient in communicating information orally in all situations. Has established effective liaison with the organization's management such that safety issues are raised at the highest level. Has effective relationships with relevant external organisations, such as regulatory bodies.

Figure 5 shows how a competence profile can be specified for a specific activity where Tx and Ay depicts Tasks and Attributes for the specified activity.

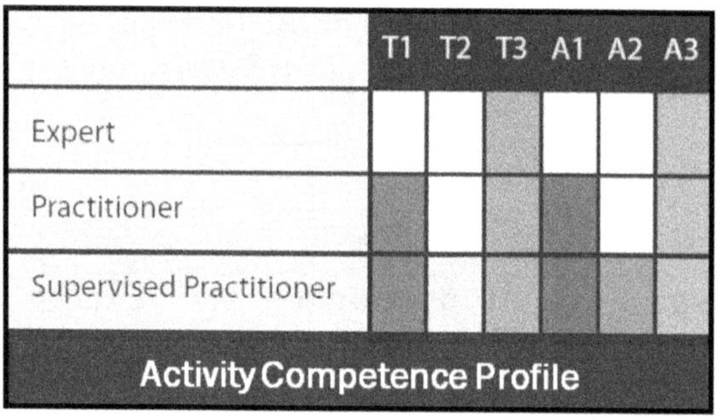

Fig. 5. Activity Competence Profile

An activity competence profile is specified in terms of an appropriate competence level for each task and attribute (note: the column colours have no significance here). The competence profile is specified for each task and attribute in terms of the three competence levels resulting in a profile for all tasks and attributes comprising a given activity. This gives a minimum activity competence profile with differing levels of expertise required for the different tasks and attributes.

4.4 Competence Criteria Process

An explicit requirement of the HSE competence management principles (HSE 2007a, 2007b) is for the specification of competence criteria and an assessment process to give confidence that all people undertaking safety-related roles are competent to perform specific work activities.

Table 2. Competence Criteria Process

Competence Criteria Process	Step
Define (Activities, Tasks, Attributes, Criteria)	1
Specify (Activity Competence Profile)	2
Assess (Individual Competence)	3
Compare (Activity Competence Profile v Individual Competence)	4

Hence processes are required to achieve the requirement. Table 2 (adapted from IET 2016) outlines a four step competence achievement process. Each step of the competence achievement process is briefly summarised here:

Step 1: Definition. Addresses the definition of the relevant activities and associated tasks and attributes applicable to an organisation implementing a competence management scheme. In addition, for each task and attribute the associated competence criteria need to be defined at each competence level.

Step 2: Specification. Addresses the specification of the minimum competence profile required for a given activity or role (but not for an individual).

Step 3: Assessment. Addresses the assessing of the current competence of an individual or individuals against predefined criteria for all activities, tasks and attributes relevant to a specific assignment. The result is recorded as the individual's assessed competence profile, together with the validity period for the assessment.

Step 4: Comparison. Addresses the comparison of an individual's competence profile against the competence profile required for a specified activity or activities to determine the suitability of the individual for that activity or activities.

A detailed explanation of the competence criteria specification and assessment process along with guidance and example competence criteria for safety-related activities can be fond in the IET Code of Practice: Specifying and Assessing Competence for Safety-Related Practitioners (IET 2016).

5 Competent Safety Assurance

The discussion so far has focused on the theoretical aspects of competence: what competence is; how it may be attained, defined and assessed. The remainder of the paper will examine some common issues related to organisations and individuals for whom competence is a crucial element for the provision of safety assurance.

5.1 Organisational Safety Issues

If an organization does not have competent people working on the development, operation or maintenance of safety-related systems then the organization is unlikely to produce tolerably safe systems. Organizations that are new to developing safety-related systems don't usually have safety competent staff and often the approach to competence will be reactive and programme dependent rather than the implementation and operation of a defined corporate-level CMS. In the absence of a CMS, typical organizational competence deficiencies can be categorised as: Distributed Competence; Limited Competence or False Competence which can be summarised with reference to the CCM in section 3.

Distributed Competence. When an organization predominantly (or exclusively) outsources the responsibility for the safety engineering process to specialist safety consultants while their own staff provide the system and domain knowledge. When distributed competence is the norm an organization can at best be operating at the *unconscious competence* level and will remain so unless staff with competence in general safety processes are trained or recruited.

Limited Competence. When an organization that does not have core safety competence is awarded a contract to develop, operate or maintain a safety-related system. Typically, they select existing employee(s) to undertake the safety roles with minimal training. These organizations have limited competence and the responsible person undertaking the safety role will quickly reach the conscious incompetence stage while the organization itself can blissfully remain at the *unconscious competence* level.

False Competence. When an organization has a false or misleading view of its safety competence; this can occur when safety staff are out of date with changes in safety knowledge or have inadequate skills. An organization such as this may consider itself to be operating at the unconscious competence level when it may in fact have regressed to a level of *unconscious incompetence*.

An organization could exhibit one or more of the above competence limitations discussed above and these are generalizations drawn from many existing organisations, even including some that possess mature safety management systems. Organizations that exhibit characteristics of Distributed Competence; Limited Competence or False Competence can often have serious problems providing adequate safety assurance as they will be operating at either the unconscious competence or unconscious incompetence states and will be unaware of the safety-related skills, knowledge and attitudes necessary to competently undertake safety-related activities.

In addition to addressing safety competence deficiencies at the organizational level, a CMS must also consider potential deficiencies related to the competence of individuals involved in the provision of safety assurance.

5.2 Individual Safety Issues

Safety assurance is ultimately based upon the competence of the people involved in the safety process and individual competence is a vital requirement for assessing the validity of any safety claims. Individual competence can have a significant influence on the safety engineering process; particularly where professional judgement is applied and there is a critical relationship between safety competence and the application of sound professional judgement for safety-related systems developers.

Professional judgement (or expert opinion or engineering judgement) can be defined as the ability of a person or group to draw conclusions, give opinions and make interpretations based on a combination of evidence from diverse sources such as experiments, measurements, observations, knowledge and experience (McKenna and Mitchell 2006). Professional judgement is frequently used by systems developers of all disciplines and it relies upon a combination of impartial and biased facts and opinions and, for anything but simple scenarios, subjectivity can be hard to discriminate from objectivity. The problems of objectivity and perception when applying professional judgement to decisions on risk have been well documented (Adams 1995).

Professional judgement is often used when an expert doesn't have any accurate or statistically significant data and the order of magnitude required for the solution to be acceptable is estimated by applying judgement gained through a combination of: academic training; experience and professional development - in other words competence. Professional judgement can be considered poor if highly subjective evidence is accepted as fact without consideration of where or how the evidence is derived and without an appreciation of when it is overstated or simply invalid. Safety assurance claims are always founded to some extent upon professional judgement and unless the person (or group) making those judgements are competent to do so, conclusions, opinions or interpretations may be derived from incomplete or inadequate evidence (Sandom 2011).

Safety assurance is ultimately a matter of professional judgement and professional judgment is based upon competence. Safety-related practitioners have a responsibility to show where professional judgement has been applied and, for safety assurance claims, how that judgement is defensible. The application of professional judgement is a necessity for any systems development; however, it remains problematic; particularly for safety-related systems development. Safety assurance evidence can be deficient due to safety competence limitations and also safety claims may be over-reliant on professional judgement.

5.3 Competence Evidence

At both organizational and individual levels it has been argued here that competence is a critical element for all safety assurance claims. Regardless of the specific method used for demonstrating safety assurance, it is also asserted that an essential goal for safety assurance must be to demonstrate competence validity at both organizational and individual levels. A claim of competence validity must be supported by comprehensive and compelling competence evidence which should be routinely sought to support an overall safety assurance claim.

All safety evidence must be both comprehensive and compelling and to demonstrate that both direct evidence and meta-evidence (i.e. evidence about evidence) should routinely be sought (Sandom 2011) to underpin safety assurance arguments. If it is accepted that the validity of a safety claim is critically dependent upon both organizational and individual competence then compelling evidence and associated meta-evidence related to competence must also be provided. Competence evidence must be presented to support both organizational level claims of competence and meta-evidence must be presented to support individual level competence-based claims.

Claims of Competence. An explicit claim must be made based upon the presentation by an organization of compelling evidence of the existence of a proportional CMS with adequately defined safety competence criteria that together enable an organization to effectively specify and assess the competence of the individuals involved in the safety assurance process. If safety competence cannot be managed then no claim can be made about the validity of either claims to fulfil explicit safety competence criteria like those of IEC 61508 (IEC 2010) or the implicit requirement for competence where professional judgement is applied to other safety assurance activities.

Competence-Based Claims. These claims are made to provide assurance that other safety claims, arguments and evidence are based upon sound professional judgements made by competent people. For example, a claim could be made that a safety-related system can achieve a certain failure rate and evidence would be presented to support the failure rate claim; perhaps from a Human Error Analysis (HEA) of the system. At some level the claimed system failure rate will be underpinned by the application of professional judgement (e.g. human failure rates in the HEA); therefore, meta-evidence is required relating to the identification of where the judgement was applied and for each instance the competence of the individuals providing that judgement.

6 Conclusions

Competence has been defined as the ability to carry out a task to an effective standard; its achievement requires appropriate technical knowledge, skills and personal attributes. The aim of safety-related organisations and individuals should be to transition from unconscious incompetence to unconscious competence and to maintain that state through effective competence achievement processes and through the specification and assessment of suitable competence criteria.

Organizations with significant competence deficiencies cannot provide adequate safety assurance as they are simply unaware that a problem even exists let alone have any awareness of the detailed competence criteria necessary to undertake safety assurance activities.

Professional judgment is applied by safety engineers during the safety assurance process and the validity of that judgement is critically dependent upon individual competence. Safety assurance therefore relies fundamentally upon the competence of all those contributing to the development, operation and maintenance of safe systems.

The validity of any safety claim is critically dependent upon organizational and individual safety competence therefore compelling evidence and associated meta-evidence must be provided to support both *claims of competence* whereby safety competence can be assessed and managed and *competence-based claims* when competent judgements are made.

Acknowledgments   The author acknowledges all contributors past and present to the IET *Blue Book* upon which some of the ideas in this paper were founded.

References

Adams J (1995) Risk. Routledge, London
Holt J, Perry S (2011) A Pragmatic Guide to Competency: Tools, Frameworks and Assessment. British Computer Society
IET (2007) Institute of Engineering and Technology Competence Criteria for Safety-Related System Practitioners. IET Publications
IET (2016) Institute of Engineering and Technology Code of Practice: Competence for Safety-Related Systems Practitioners, IET Publications
IEC (2010) International Electrotechnical Commission 61508, Functional Safety of Electrical/ Electronic/ Programmable Electronic Safety Related Systems. Ed 2
ORR (2007) Office of Rail Regulation Developing and Maintaining Staff Competence. Railway Safety Publication 1
McKenna S, Mitchell J (2006) Professional Judgment in Vocational Education and Training: A Set of Resources. 2nd Ed. Commonwealth of Australia, Department of Education, Science and Training
Robinson W L (1974) Conscious Competency - The Mark of a Competent Instructor. The Personnel Journal - Baltimore , Vol. 53, pp538-539
Sandom C (2011) Safety Assurance: Fact or Fiction? In: Tony Cant (ed.) proceeding of the Australian System Safety Conference, Melbourne, 25-27 May 2011, Conferences in Research and Practice in Information Technology (CRPIT), Vol. 133
Sandom C (2015) Unconscious Competence and Safety Assurance. proceedings of 33rd International Systems Safety Conference, 24 - 28 August 2015, San Diego, USA
UK EC (2013) UK Engineering Council Standard for Professional Engineering Competence (UK-SPEC). Third Edition
UK HSE (2007a): UK Health & Safety Executive Managing Competence for Safety-related Systems, Part 1: Key Guidance, Crown Copyright
UK HSE (2007b): UK Health & Safety Executive Managing Competence for Safety-related Systems, Part 2: Supplementary Material, Crown Copyright

Confirmation Bias within Safety Case Arguments

Chris Hobbs

QNX Software Systems

Ottawa, Canada

Abstract The preparation of a Safety Assurance Case has been an integral part of the development of railway systems for many years, being one of the requirements of EN 50129. For automotive systems, ISO 26262 also mandates the creation of a Safety Case. Increasingly, Safety Cases are also being required for the certification of medical devices. Researchers have demonstrated that producing a Safety Case untainted by confirmation bias is extremely difficult, or even impossible. This seriously affects the level of confidence that can be placed in the Safety Case argument. This paper describes the results of an experiment to determine whether the notation used to represent the Safety Case argument influences the structure of that argument.

1 Introduction

A Safety Case consists of three parts: a claim, an argument for the correctness of the claim, and the evidence to back the argument. For a safety-critical embedded system, the claim may assert compliance with a standard (e.g., "We claim that the system, when operated in accordance with the associated Safety Manual, meets the requirements of IEC61508 for continuous operation at safety integrity level (SIL) 2.") or may be more general and claim that, subject to certain conditions of use, the product is adequately safe.

In her widely circulated, but not published, 2012 paper "White Paper on the Use of Safety Cases in Certification and Regulation", Nancy Leveson points out one potential flaw in the use of Safety Cases to argue the correctness of the claim:

> An important component of mindset is the concept of confirmation bias. Confirmation bias is a tendency for people to favor information that confirms their preconceptions or hypotheses regardless of whether the information is true. People will focus on and interpret evidence in a way that confirms the goal they have set for themselves. If the goal is to prove the system is safe, they will focus on the evidence that shows it is safe and create an argument for safety.

Confirmation bias is present in all of us and this paper describes an experiment to determine whether the form in which the argument is expressed can affect the amount of confirmation bias introduced.

2 Confirmation Bias

Although the term "confirmation bias" was coined by Peter Wason in (Wason 1960), the concept has been understood at least since the time of Thucydides (4th century BCE), who wrote "For it is habit of mankind to entrust to careless hope what they long for, and to use sovereign reason to thrust aside what they do not desire" (translation from (Strassler 1996)).

A simple experiment, a variant of Wason's original experiment, can be used to demonstrate confirmation bias. The sequence 2, 4, 6, 8 is written up on a whiteboard and a group of people is told that these numbers are being generated in accordance with a "secret" rule. The group is invited to guess numbers that might continue the sequence, the objective being to uncover the rule. I have tried this dozens of times with groups of people and almost always the group guesses 10, 12, 14, 16. I accept these and write them onto the end of the sequence. Then someone will take a random guess and suggest a number such as 153. I accept that and write it onto the end of the sequence. Suddenly the group sits up, and perhaps tries 150, which I reject.

Eventually the group works out that the rule is: "each number must be greater than the previous one". The important observation is how much more information the guesses 153 and 150 provided than the 10, 12, 14, 16. The group believed that the rule was "next even number" and looked only for confirmation of that belief.

If confirmation bias affects the ability of a group to determine a simple numerical rule irrelevant to safety, how much more difficult is it to avoid confirmation bias when presented with the challenge to "demonstrate that this system is safe"?

3 The Safety Case Argument

A safety case is defined in section 9.1 of [00-56 2007] to be:

"...a structured argument, supported by a body of evidence that provides a compelling, comprehensive and valid case that a system is safe for a given application in a given operating environment."

The critical term here is "structured"; an unstructured pile of paper with the argument for the safety of the system buried somewhere within it, is worthless.

In order to structure the argument, it is common to use a semi-formal notation (i.e., a notation where the syntax is formal, but the semantics are informal). Popular notations are the Structured Assurance Case Metamodel (SACM), Goal Structuring Notation (GSN) and Bayesian Belief Networks (BBNs). SACM is defined by the Object Management Group and GSN is defined in the GSN Community Standard Version 1. There is no standard for how a safety case argument should be expressed in BBN, but the present study adapts the descriptions given in reference (Fenton and Neil 2013).

This study compares the GSN and the BBN notations. The primary difference between these notations is that BBNs are quantitative whereas a GSN diagram is deliberately qualitative. One justification for not quantifying the level of confidence in an argument is that it is very easy to start to believe the resulting numbers. For example, if three or four inputs, with estimated confidence levels of "roughly" 60%, 70%, 75% and 45%, are combined, these can be used by a BBN tool to provide an overall confidence of 62.35424412442%, an obviously unreasonable precision, given the lack of precision of the inputs.

However, BBNs provide a very flexible and rich environment for expressing different levels of confidence and combining evidence in different forms. In particular the so-called "noisy" conjunctions (e.g., "noisy-or") can be very expressive when working with inexact data.

The purpose of the experiment described below was to see whether either of these notations provided a better protection against confirmation bias on the part of the person producing a Safety Case. In the event, although one strong indication did appear during the experiment, it was not possible to attribute it directly to confirmation bias.

4 Experimental Procedure

4.1 Goal of the Experiment

The experiment was performed to see whether the level of confirmation bias introduced into an argument structure was correlated with the notation used. To determine this, two groups of people were asked to work on an incomplete draft of the same argument, but annotated using either GSN or a BBN.

4.2 General

All employees located at QNX Software Systems' head-office in Ottawa, Canada, were invited to participate in the experiment. In total 40 people volunteered from a range of departments including Engineering Development, Verification, Technical Publications, Marketing, Sales, Legal, Training, and Support. None of the volunteers was informed about the purpose of the experiment and each signed a paper promising not to converse with other participants until the study was complete.

Initially, each volunteer was individually presented with a statement chosen so that its correctness would to be difficult to assess objectively. The statement was:

"Within the next 25 years, fully autonomous cars will become commonplace in Ottawa."[1]

One unforeseen advantage of using this statement was that, because QNX Software Systems is involved in the development of automobile systems, the experimental participants assumed that the experiment was actually about the feasibility and acceptability of autonomous cars, thus providing a smoke-screen for its real purpose. Participants were generally surprised when, after the experiment, they were told that it had nothing to do with autonomous cars.

Each volunteer was individually asked to place a mark on a line ranging from "I strongly disagree" to "I strongly agree" to indicate his or her strength of belief in the statement. To avoid any tendency amongst the volunteers unconsciously to prefer a higher score, the line itself was neither graduated nor marked:

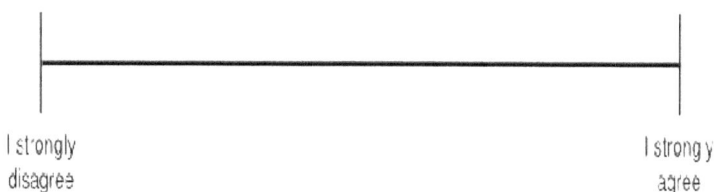

I strongly
disagree

I strongly
agree

After each volunteer had made a mark, the position of the mark was privately measured by the experimenter on a scale of 0.0 ("I strongly disagree") to 1.0 ("I strongly agree"). Those volunteers with scores between 0.1 and 0.4 and between 0.6 and 0.9 were invited to take part in the experiment. These ranges were chosen not only to exclude those people with fixed (and presumably immutable) views (< 0.1 and > 0.9), but also those without any strong view (0.4 to 0.6). Note that these

---

[1] Ottawa was chosen because it was the city best known to the participants and also because it is not California --- it has long winters that make driving difficult by reducing visibility in snowstorms, producing slippery roads, and obscuring road signs and road markings. City driving in Ottawa can be considered a far more hostile environment than, for example, highway driving in California.

ranges were defined as part of the experimental procedure, hidden from the subjects, but agreed by the experimenters, before any results were obtained.

The remaining volunteers were randomly divided into two equally-sized groups of 12 people.

Each participant was individually interviewed as described below and asked whether he or she was familiar with BBNs or the GSN; none was. Each was then introduced to the BBN (Group A) or GSN (Group B) notation using a trivial example: the main claim being that "Canada is in North America" and the evidence being Google maps (which are known to contain errors), a Wikipedia article (not always correct) listing the members of the North American Free Trade Agreement and newspaper articles about North American games (baseball, ice hockey) being played in Canada. Figure 1 illustrates the GSN argument although it should be noted that it was not exposed to the participants in this form: rather it was incrementally drawn on a whiteboard with explanations. Figure 2 illustrates the same argument as it was drawn on the whiteboard for members of group A.

Group A participants were then individually exposed to an incomplete BBN argument regarding the ubiquity of autonomous cars in Ottawa, which they were asked to extend and enhance to provide a more complete argument. Group B members were asked to do the same, but starting with an equivalent, incomplete argument expressed in the GSN. Incomplete arguments were prepared because, while it was felt unreasonable to expect non-technical staff to prepare a full GSN or BBN, it was considered reasonable for them to extend and complete an existing one. Both formats of incomplete argument contained precisely the same argument points relying on the same evidence.

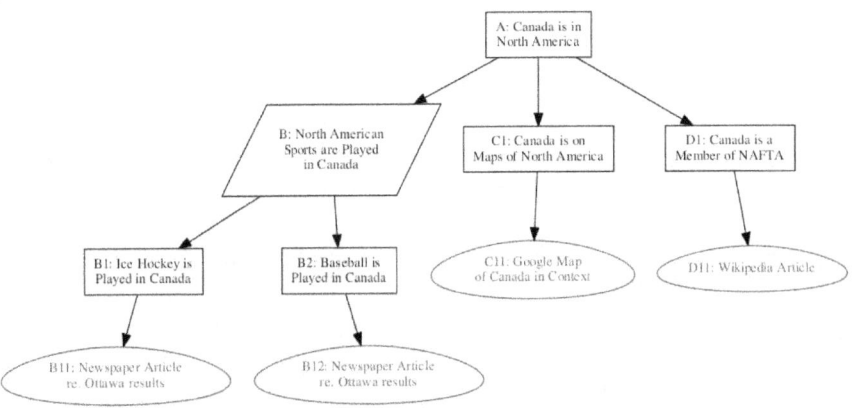

Fig. 1. Trivial GSN used for education

The BBN was presented using the AgenaRisk tool; the GSN using QNX Software System's inhouse-developed tool. Bias could be introduced by the choice of tool as AgenaRisk is more visually pleasing than the GSN tool developed by QNX Software Systems, but each participant was only exposed to one tool and so there was no way that they could be influenced by the (lack of) sophistication of the other.

Following the work with the BBN or GSN, any additional arguments presented by the participants were saved and, without sight of the level of belief they had indicated at least 5 days previously, the participants were again asked to mark a line indicating their strength of belief in the above statement. If confirmation bias were present, it was assumed that this would be revealed by the choice of additional arguments and in a strengthening of the opinion away from the neutral point of 0.5 (perhaps moving from 0.3 to 0.25 or from 0.7 to 0.8).

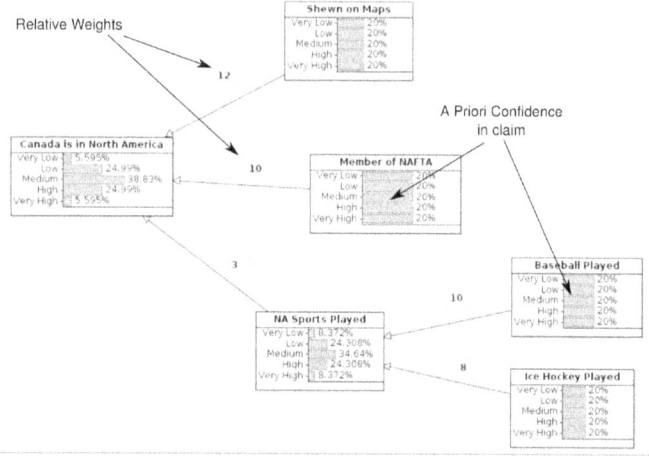

Fig. 2. Trivial BBN used for education

4.3 Group A: BBN

The BBN for this experiment was created using a 5-level degree of confidence (very low, low, medium, high, very high) and values were combined using a normal distribution (truncated at 0 and 1) with the mean being the weighted means of the inputs and the variance 1E-5. The weights represented the strength of a particular part of the argument.

Each participant was individually exposed to the same incomplete BBN, prepared to justify the statement about autonomous cars given above. With the experimenter actually making the changes to the model, each participant was encouraged to add new argument structure and to change the weights that had been assigned to the argument strengths. The experimenter refused to contribute to the discussion, even when directly asked, "what do you think about X?".

Once a participant was comfortable with the structure, weights and evidence of the BBN, he or she was asked to place a level of confidence on each "leaf" node. Having completed this, the participant was again asked to review the whole BBN, the computation was performed and the result displayed on the screen.

Without either the experimenter or the participant seeing what strength of belief had been chosen in the first part of the experiment, the participant was finally asked to mark another line (of a different length) with his or her current strength of belief and was reminded not to discuss the experiment with anyone else.

4.4 Group B: GSN

Each participant was exposed to an incomplete GSN argument that had been prepared to justify the participant's belief for or against the statement about autonomous cars according to his or her initial response. The argument was presented on paper (to allow the participant to scribble on it) and displayed on a screen. As the argument was described, the participant was encouraged to add new argument structure and evidence. When the participant suggested changes, these were immediately added to the diagram on the screen by the experimenter. As with Group A, the experimenter declined to discuss any aspect of the argument, simply acting as scribe to edit the model.

Once the participant was comfortable with the argument structure and evidence, he or she was again asked to review the whole GSN argument.

As with Group A, without either the experimenter or the participant seeing what strength of belief had been chosen in the first part of the experiment, the participant was finally asked to mark another line (of a different length) with his or her current strength of belief and was reminded not to discuss the experiment with anyone else.

5. Creating the Incomplete Arguments

One problem encountered almost immediately, while the experiment was being set up, was the realisation that the GSN is not rich enough to express evidence that tends to counter the goal being justified. Appendix C2 of the GSN standard does contain a symbol (a coloured rectangle) for expressing a "problem", counter-evidence for a goal, but this is not explored elsewhere in the standard.

As an example, consider the claim that autonomous cars may become commonplace because people, in general, would prefer to be driven than to drive themselves. This is true for many, but there are also people who actually enjoy driving, particularly for leisure rather than commuting.

For the BBN, this presented no problem: it was sufficient to make the claim that "People prefer to be driven" and allow the experimental subjects to (strongly) agree or (strongly) disagree with this claim. Those who believed that most people actually prefer to drive themselves simply disagreed strongly with the statement.

For the GSN, it was necessary to create two GSNs: one including the goal that "people prefer to be driven" and one with the goal that "people prefer to drive themselves". This immediately introduces confirmation bias into the structure of the argument: given the statement that "people would prefer to drive themselves", it is only natural to seek evidence that that is the case: "yes, I know a lot of people who really enjoy driving; they wouldn't want an autonomous car".

Reference (Goodenough, et al 2013) provides a different approach to this problem: that of "eliminative induction". With this technique, a claim is justified only insofar as reasons for doubting its truth can be eliminated by evidence or argument. Confidence then depends on how many doubts have been identified and removed. It would be interesting to repeat the experiment described in this paper using eliminative induction.

6 Experimental Results

6.1 Candidate Selection

A total of 40 people volunteered for the experiment and were subjected to the first stage of selection described above. The results were as shown in table 1.

Table 1. Candidate Selection

Confidence Range	Quantity	Note
0.0 - 0.1	1	Rejected from study
0.1 - 0.2	3	
0.2 - 0.3	2	
0.3 - 0.4	3	
0.4 - 0.5	2	Rejected from study
0.5 - 0.6	5	Rejected from study
0.6 - 0.7	2	
0.7 - 0.8	7	
0.8 - 0.9	7	
0.9 - 1.0	8	Rejected from study

For the purposes of the study, the initial opinions of the volunteers were irrelevant, other than to exclude those with extremely strong or indifferent views; this was not an experiment about autonomous cars![1]

Of the 40 volunteers, 16 were rejected at this first stage of selection as listed in table 1. This exclusion was asymmetric, with 13 people being rejected who believed in the statement, while only 3 people were rejected who doubted it. However, the purpose of the experiment was not to determine whether autonomous cars will become commonplace in Ottawa. The remaining 24 volunteers were assigned randomly to group A or group B: twelve members to each group.

6.2 Experience During BBN and GSN Completion

During the discussion of the BBN or GSN diagram, almost every participant added at least one, and in many cases many, additional elements to the argument structure. One interesting observation was that many participants, apparently unconsciously, added evidence that contradicted their initial assessment in contradiction to what would be expected by the confirmation bias.

Although it is not relevant for the experiment, the experimeter realised that a new argument structure combining all the factors added by participants would be a much stronger argument: a reason for ensuring that all safety case arguments are

---

[1]   However, it is interesting to note that, amongst an engineering community focused largely on automobiles, a majority of people felt confident that autonomous cars would be commonplace, even in Ottawa's hostile environment. This was particularly noticeable amongst those working in Engineering Development.

reviewed, discussed and dissected by as many people with relevant experience as possible.[1]

6.3 Confirmation Bias in the Investigator

Confirmation bias is everywhere and it would be naïve to assume that the investigator was immune to it during this experiment: the experiment was performed to demonstrate a view held by the investigator and, however much he tried to hold to the null hypothesis that there was no difference between GSN and BBN, it could be thought inevitable that evidence would be sought to substantiate the preconceived view. This being so, the experiment described here should be considered only an initial exploration of the concept of confirmation bias. The experimenter took care during the interviews to remain neutral, while adapting the GSN or BBN in accordance with the participants' input.

When the investigator's view was explicitly sought by the participant ("What do you think about the feasibility of X?"), the investigator deliberately but politely refused to respond until after the session was complete.

6.4 Summary of Findings

Group A (BBN) Results

Table 2 summarises the changes in opinions from those participants in Group A (subject to the BBN argument). It can be seen that, of the 12 participants, 7 weakened their view (i.e., their score moved towards 0.5, and in 2 cases crossed 0.5 to indicate a complete change of view), 1 view was unchanged and 4 participants strengthened their view, albeit in one case to a very small degree.

---

[1]   Many factors positively and negatively affecting the feasibility of autonomous cars were raised by participants during the experiment, but the one that seemed to resonate most was whether a solution would be found to the moral problems associated with autonomous driving: the car detects something moving into the road ahead which it assesses as 83.4% probable as being a child (but might be a blown plastic bag or large raccoon) and has to make a choice of running the obstruction over, swerving the car off the road into a tree or changing lanes into the path of an oncoming truck (which, as several participants pointed out, might also be autonomous).

Table 2. Group A Results

Participant	Change in Belief	(S)trengthing or (W)eakening of belief	Movement	New Arguments
1	0.75 → 0.71	W	+0.04	-
2	0.39 → 0.34	S	-0.05	Negative
3	0.19 → 0.31	W	+0.12	Negative
4	0.75 → 0.38	W	+0.37	-
5	0.77 → 0.74	W	+0.03	Positive
6	0.27 → 0.28	W	+0.01	-
7	0.80 → 0.86	S	-0.06	-
8	0.35 → 0.61	W	+0.26	Positive
9	0.39 → 0.30	S	-0.09	Positive
10	0.80 -> 0.82	S	-0.02	Positive
11	0.68 -> 0.68	-	0.0	-
12	0.86 → 0.82	W	+0.04	-

Taking movement away from 0.5 as negative and movement towards 0.5 as positive, the total movement amongst the 12 participants in group A is 0.65.

The final column in table 2 indicates whether the arguments that the participant added to the argument were predominantly positive, negative or neutral to the idea that autonomous cars would be commonplace in Ottawa in 25 years. In most cases the participants introduced several new arguments and the entry in the final column of table 2 is a summary, prepared without a knowledge of the participant's belief.

Of particular interest are those two participants who changed (rather than simply weakened or strengthened) their view. Participant 8 moved from disagreeing with the proposition to agreeing with it. The evidence that this participant added to the BBN was a combination of positive (e.g., there will be no age limit on driving), neutral (e.g., car ownership will disappear) and negative (e.g., there will be substantial transition problems to overcome when the autonomous cars share the road with conventional ones), yet the participant's view changed from disagreement to agreement.

Group B (GSN) Results

Table 3 summarises the changes in views from those participants in Group B (subject to the GSN argument). It can be seen that, of the 12 participants, 2 weakened their view (i.e., their score moved towards 0.5) and 10 participants strengthened their view.

Table 3. Group B Results

Participant	Change in Belief	(S)trengthing or (W)eakening of belief	Movement	New Arguments
13	0.82 → 0.87	S	-0.05	-
14	0.72 → 0.81	S	-0.09	Negative
15	0.82 → 0.83	S	-0.01	Positive
16	0.19 →0.16	S	-0.03	Negative
17	0.71 →0.72	S	-0.01	Negative
18	0.72 → 0.75	S	-0.03	Negative
19	0.67 → 0.65	W	+0.02	-
20	0.71→ 0.92	S	-0.21	Negative
21	0.27→0.00	S	-0.27	Positive
22	0.87→0.84	W	+0.03	Negative
23	0.19 →0.17	S	-0.02	Negative
24	0.88 → 0.91	S	-0.03	Negative

Taking movement away from 0.5 as negative and movement towards 0.5 as positive, the total movement amongst the 12 participants in group B is -0.7.

As with table 3, the final column of table 4 indicates whether, in general, the additional arguments added to the model by the participant were positive or negative.

The strongest movement in this group was that of NV who moved from being mildly unconvinced about the ubiquity of autonomous cars to being totally unconvinced. Some of the additional evidence that NV added was positive (e.g., there would be energy savings from the more conservative autonomous cars, making them more acceptable to governments and drivers) and some negative (e.g., the security problems are unlikely to be solved). Overall the majority of arguments that NV added were positive to the idea of autonomous cars being commonplace, yet his level of belief dropped.

6.5 Debriefing Session

Once the experiment was complete, a debriefing session was held to which all of the volunteers, whether rejected at the first phase or not, were invited. For the first time the volunteers knew who else had been part of the experiment and had the purpose of the experiment explained.

Several participants said that following the BBN or GSN session, they had tried to mark the line in the same place as previously to indicate that their views had not changed, but had been tricked by the different length of the line. After discussion, however, there was agreement that, as both groups had been given the same lines to mark, this did not affect the significant difference in the results of the two groups. It was also noted that the changes were generally very small, but again the correlation between direction of movement and group (BBN or GSN) could not be explained by random fluctuation.

One useful suggestion was that some people should have been chosen as a control group. These would be exposed neither to the BBN nor the GSN argument, but would simply have been asked to mark a line again, after an interval of a couple of weeks.

8. Summary and Conclusions

8.1 A Clear Distinction

The difference between the changed views of people exposed to the arguments expressed in BBN and GSN form is clear: 64% of the participants exposed to the BBN weakened their view while 83% of those exposed to the GSN strengthened their view.

This would seem to indicate a statistically-significant difference1 between the two notations when expressing an argument.

8.2 Confirmation Bias?

The question of whether this significant difference arose from confirmation bias is less clear, as can be seen from the rightmost columns of tables 2 and 3. One flaw with using the statement about autonomous cars is that many people have never really thought deeply about them and, when asked to think about the concept, negative arguments tend to arise: "I'd never really thought about the problems associated with liability, licencing, manual override, the transition period,

---

1   A Kolmogorov-Smirnov test for comparing two populations reduced to Weakening/Strengthening gives $D_{max} = 0.6$ while the critical value for rejecting the null hypothesis that the two samples are from the same distribution is $D_{24;0.01} = 0.333$. Thus the null hypothesis can be rejected at the 99% level.

etc." The predominant category in tables 2 and 3 is of those people who believed in the concept of the autonomous car, presented negative evidence (against confirmation bias), and then strengthened their belief in the autonomous car! It seems easier to find arguments as to why something should not be true, than to find arguments as to why it should be true. This would emphasise the possible efficacy of the eliminative induction technique described in (Goodenough et al 2013); see below.

9. Further Study

The data collected during the experiment could also provide fertile ground for further mining: e.g., were participants from different departments (engineering, marketing, etc.) differently influenced?

Performing the experiment has provided a much deeper understanding of how evidence can be manipulated in relation to confirmation bias. During the experiment, the author was made aware, through the GSN Google Group, of the approach proposed in (Goodenough et al 2013) and this would intuitively seem to be an approach less susceptible to confirmation bias. The "eliminative induction" described in that paper inverts the normal GSN approach of making a claim and then documenting the evidence that the claim is justified. Instead, given the claim, the reasons for doubting it, the so-called "defeaters", are documented and, through evidence, successively eliminated. Three types of defeater are considered: rebutting defeaters that provide a counter-example, undermining defeaters that raise doubts about the validity of the evidence and undercutting defeaters that describe circumstances under which the conclusion is in doubt.

It would be interesting to repeat the experiment described in this paper, but using an eliminative induction approach instead. For those participants who initially believed that autonomous cars would be commonplace, this would mean identifying as many reasons as possible for doubting this. An incomplete list might include:

1. Insurance issues are intractable: who would pay in the event of an accident – the "driver", the car manufacturer, the software designer?
2. An autonomous car is software on wheels. The end-user licence for a simple application today is typically 20 pages long. It would be impossible to produce an intelligible end-user licence for an autonomous car.
3. The transition period with both autonomous and non-autonomous cars on the road would be too dangerous.
4. Although large companies are investing in the technology, there is no evidence that any tests have been performed in Ottawa winter conditions.
5. Drivers would find a car that kept to the speed limit and did not accelerate towards a traffic light turning red too conservative and frustrating.

Confirmation bias would help rather than hinder the production of such a list. Evidence would then have to be produced to eliminate or at least mitigate each of these issues.

Acknowledgements
The author would like to thank QNX Software Systems for allowing its employees to be sufficiently distracted from their day jobs to take part in this experiment.

References

[00-56 2007] Defence Standard 00-56 Issue 4 (Part 1): "Safety Management Requirements for Defence Systems", UK Ministry of Defence. p. 17

GSN COMMUNITY STANDARD VERSION 1, November 2011. Available from http://www.goalstructuringnotation.info/documents/GSN_Standard.pdf

OMG: Structured Assurance Case Metamodel (SACM), Version 1.1 available from http://www.omg.org/spec/SACM/1.1/

Norman Fenton and Martin Neil (2013), Risk Assessment and Decision Analysis with Bayesian Networks, CRC Press, 2013, ISBN 978-1-4398-0910-5

John B Goodenough, Charles B Weinstock and Ari Z Klein (2013), Eliminative Induction: A Basis for Arguing System Confidence, New Ideas and Emerging Results Workshop, International Conference on Software Engineering, 2013

Peter C Wason (1960), On the failure to eliminate hypotheses in a conceptual task, Quarterly Journal of Experimental Psychology, ISSN 1747-0226

Robert B Strassler (1996), The Landmark Thucydides, ISBN 0-684-82815-4

Beyond arrangements – making the link between safety management and safety culture

Rebecca Canham, Ben McCaulder & Shona Watson

Greenstreet Berman Ltd[1]

London, UK.

Abstract *High Reliability Organisations (HRO) rightly invest significant effort to ensure their Safety Management System (SMS) is effective – for example ensuring appropriate and reliable engineering along with robust and resilient organisational arrangements. The benefits of these arrangements go far beyond mere compliance and contribute to high performance. As well as the organisations themselves, regulators, such as the Office for Nuclear Regulation (ONR), also recognise the importance of these arrangements, actively seeking evidence of their effectiveness within such themes as Leadership, Capable Organisation, Decision-Making and Learning from Experience. Whilst approaches to describing and assessing the visible elements of the SMS can be readily identified, there needs also to be a focus on the cultural and behavioural elements of the organisation's commitment to safety and high performance. People are part of the system, rather than an unpredictable add-on. An effective SMS will guide and control behaviours with respect to safety. But how do you go beyond the arrangements in order to understand and be confident in the social and interpersonal influences – leadership and commitment, supervision, prioritisation of safety, and so forth? This paper addresses this need to understand and manage the link between the management system and the organisational safety culture, whilst recognising that neither can be considered independently, due to the interconnected nature of complex systems. Considerations are presented to prompt reflections of SMS and organisational culture with respect to compliance and performance.*

---

[1] 10 Fitzroy Square, Fitzrovia, London, W1T 5HP Tel: 020 3102 2110

1 Introduction

1.1 Background and context

Within UK safety legislation the goal of safety management is broadly to reduce levels of risk to as low as is reasonably practicable, or 'ALARP' or 'Broadly Acceptable'. An organisation's Safety Management System (SMS) provides a systematic and structured means of managing and controlling the safety risks apparent as a result of its operations, thereby supporting achievement of an ALARP position.

Safety culture has been defined as follows by the UK Health and Safety Commission, (1993):

"The safety culture of an organization is the product of the individual and group values, attitudes, competencies and patterns of behaviour that determine the commitment to, and the style and proficiency of, an organization's health and safety programmes."

Following the enquiry into the Ladbroke Grove Rail crash, the Cullen Report (2001), cites safety culture as: "the way we typically do things around here". In line with these definitions, it can be said that the safety culture within any organisation will be reflective of the attitudes and behaviours of the individuals and groups that make up the organisation. Furthermore, these attitudes and behaviours will undoubtedly influence the organisation's safety performance, be that positively or negatively.

If mature, an organisation's safety culture can drive positive behaviours and foster continuous reflection and improvement, thereby maintaining or enhancing an ALARP position. If safety culture is not mature, however, an organisation's safety culture can result in a downturn in safety performance, moving the organisation away from a demonstrable ALARP position. A successful SMS can positively influence safety culture by creating the environment in which the culture exists and develops. An ineffective SMS on the other hand is likely to produce discontent and increase the frequency and acceptance of poor safety practices.

It would therefore seem that the presence of an effective SMS or safety culture in isolation may fail to deliver the positive safety performance sought by both the organisation themselves and their regulator(s). Here, we explore the link between safety management and safety culture, what might be done to improve both aspects, and the value added when considering the two together.

2 Safety Management

2.1 What is a Safety Management System (SMS)?

Organisations have a legal duty, and often regulatory requirements, to put in place suitable arrangements for managing health and safety. These arrangements are often collectively termed the 'safety management system'. The SMS should include measures and processes to support, for example: safety leadership; risk profiling; quality documentation; competence management; worker involvement and being a learning organisation (HSE, 2013). Therefore, an organisation's SMS is the systematic structure and process by which it manages the safety of its operations.

2.2 The value of an effective SMS

Implementing a successful SMS should not only be a means to ensure regulatory compliance, but should also offer a variety of business benefits, if a suitable system is applied, including:

- Efficiency improvements that may actually reduce costs
- A demonstrable improvement in safety performance
- An opportunity to improve culture and morale amongst the workforce through sight of a clear commitment to safety
- An improved public perception of the organisation and its activities

A successful SMS however, must be tailored to the needs and expectations of the organisation it serves.

2.3 Developing an effective SMS

While the fundamental principles should apply to all SMS a one size fits all approach is unlikely to succeed. With this notion in mind an effective SMS is likely to encompass some core principles. The UK Health and Safety Executive (HSE), in its publication 'Managing for Health and Safety' HSG65, has aligned with the "Plan, Do, Check, Act" model, as made popular by Deming (The Deming Institute, 2015), as illustrated in Figure 1. In the context of safety management these core elements can be understood as follows:

- Plan – Determination of the overall safety policy and structure for the organisation, along with definition of the safety performance sought.
- Do – Undertaking of adequate risk assessment to understand the exact needs of the organisation, and activation of systems and processes to control and manage these.
- Check – Using the means and measures defined during the Plan activity performance is measured (both actively and reactively).
- Act – Review information gathered during the Check activity in order to iterate the SMS and organisational activity following enhanced understanding, in order to achieve ALARP.

As is perhaps evident by the core elements of Deming's model, the SMS should not be a static or linear process with a defined and absolute conclusion, rather a continuous and responsive activity. This is further evident within the cyclical illustration provided in Figure 1.

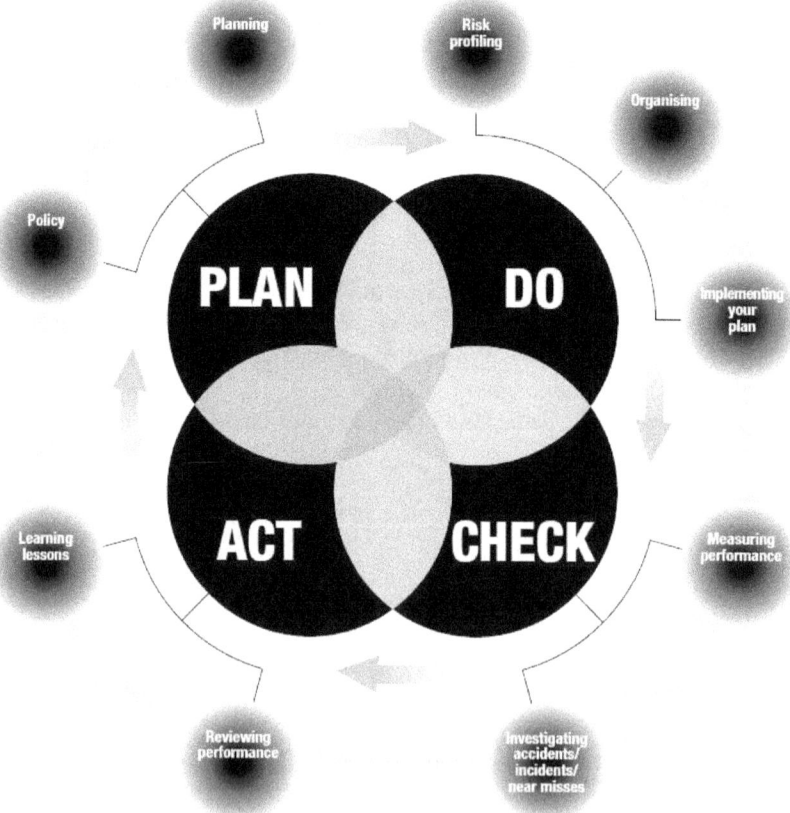

Fig 1: HSG65 Plan, Do, Check, Act model and process, as presented in HSE (2013) publication HSG65.

The exact form of an organisation's SMS should, as noted above, take account of the specific needs of the organisation. As such it is probable that the SMS of a large and complex organisation will, by necessity be itself complex. Similarly a smaller organisation with a lower risk profile is likely to require a much simpler SMS. Whilst the exact structure and content may differ in scope and extent between two such organisations the fundamental principles of what the SMS is required to do, and how it is to function remain the same.

2.4 SMS within the High Reliability Organisation (HRO)

With this notion in mind we can consider the High Reliability Organisation (HRO). A HRO is defined not by its size, but through its ability to function safely in challenging environments, and where it would not be unreasonable to expect accident and incidents to occur.

Specifically the HRO, as a concept, is considered to be an organisation that functions well in circumstances where the consequences of failure are real, significant and potentially widely understood. It is highly likely that an HRO will utilise technology and be engaged in activity that is notably risky with a definite possibility of failure. These factors present the circumstances in which an HRO functions, what ultimately however sets an HRO apart from other organisations is the organisation's ability to employ processes and systems well in order to manage and control the apparent risk to a demonstrably ALARP position.

A suitable and well-functioning SMS is therefore a key component of an HRO as it will provide positive features such as highly trained personnel, regular and appropriate monitoring and auditing of performance and opportunity for continuous improvement.

2.5 Maintaining an effective SMS - an iterative process

To ensure that a SMS is functioning correctly it is sensible to measure its performance and this is enshrined within the Plan, Do, Check, Act structure. There are typically three means of undertaking such measurement:

- investigation of the results (data),
- assessing compliance with the system, or
- a process based approach that gathers feedback internally on the various components of the system.

Using data gathered in the form of performance statistics, accident rates, etc. is a very clear and easily applied method that employs materials that are already gathered as part of the SMS. However, such measures are one-dimensional in only considering output and these do not necessarily have a direct correlation to the adequacy and functioning of the SMS. It may be the case that a low accident rate has been achieved due, not to the success of the SMS but due to a change in circumstances or the termination of a certain type of risky activity.

Contrary to investigating only the data output of the SMS, a compliance based approach considers the components of an SMS, typically via some form of audit function against a defined standard or other example of relevant good practice. Again, such an approach will provide useful information but is unlikely to show the complete picture in that face value compliance with SMS expectations does nothing to inform of how well the system works within a given organisation.

The process based approach takes the basic notion of examining the various components of the SMS as done by a compliance based approach and applies further specificity and user feedback through interviews and the like with different users of the system (e.g. managers who define policy and operators who work to procedures). Such feedback should provide useful insight to the actual functioning of the system and how it is used. It may not however address its ultimate suitability or compliance with requirements.

All three of the above approaches provide a means of assessing and measuring the success of a SMS and understanding the benefits and drawbacks to its use. The use of any one means of measurement and evaluation is therefore unlikely to provide a complete picture. It follows then that an evaluation of an SMS should employ a combination of all of approaches if it is to be successful. Typically this may be achieved by formal auditing practice, surveys and analysis of quantifiable data but it can also be beneficial to conduct activity that elicits such information on a more frequent and communicative basis rather than at a dedicated audit or review stage. It is this application of frequent, if not continuous, and multiple means of measurement that will be a likely feature of a successful HRO.

2.6 The boundaries of safety management

While it is possible that the implementation of a well-functioning SMS will positively influence workers in their performance and morale it is understandable that this is only going to be able to go so far. A well-functioning SMS will provide workers and operators with the means to function safely through accurate and usable procedures, adequate training and the opportunity to feedback on actual performance. Such aspects will be of benefit, of course, but they will not automatically deliver a positive safety culture across all aspects of the workforce, where deviation from safety procedures is not tolerated, constructive challenge is welcomed and clear safety values are cascaded from management such that these are both understood and shared by all employees. It is this gap that we consider and understand later.

3 Safety Culture

3.1 What is Safety Culture?

It is widely accepted that an effective safety culture is a pre-requisite and enabler of effective safety performance. The UK's Health and Safety Commission (HSC, 1993) states:

> "The safety culture of an organisation is the product of the individual and group values, attitudes, competencies and patterns of behaviour that determine the commitment to, and the style and proficiency of, an organisation's health and safety programs. Organisations with a positive safety culture are characterised by communications founded on mutual trust, by shared perceptions of the importance of safety, and by confidence in the efficacy of preventative measures."

Put simply, safety culture can be described as the way people think and behave regarding safety.

3.2 The importance of safety culture

What role does safety culture play in safety performance? A simple analogy can be made to driving a car – there being three distinct requirements for safe driving:

- The design of the car;
- Traffic management arrangements;
- The driver.

Some cars are inherently safer than others; some traffic management arrangements are safer than others. However, even with good systems, it is the attitudes and behaviours of the driver that will determine the safety performance of driving. For the highest levels of safety performance, we require all three elements – attitudes, behaviours and systems – to be as effective as possible, as shown in Figure 2.

The 'health' of safety in an organisation is primarily defined by the key day-to-day behaviours of frontline staff and management, and the extent to which these are supported by an effective and flexible safety management system. The shared belief in the importance of safety, and the degree to which this importance is communicated, owned and acted upon across the organisation is what defines a positive safety culture. This emphasises the importance of proactive risk management through organisational culture, to complement the focus on technical systems.

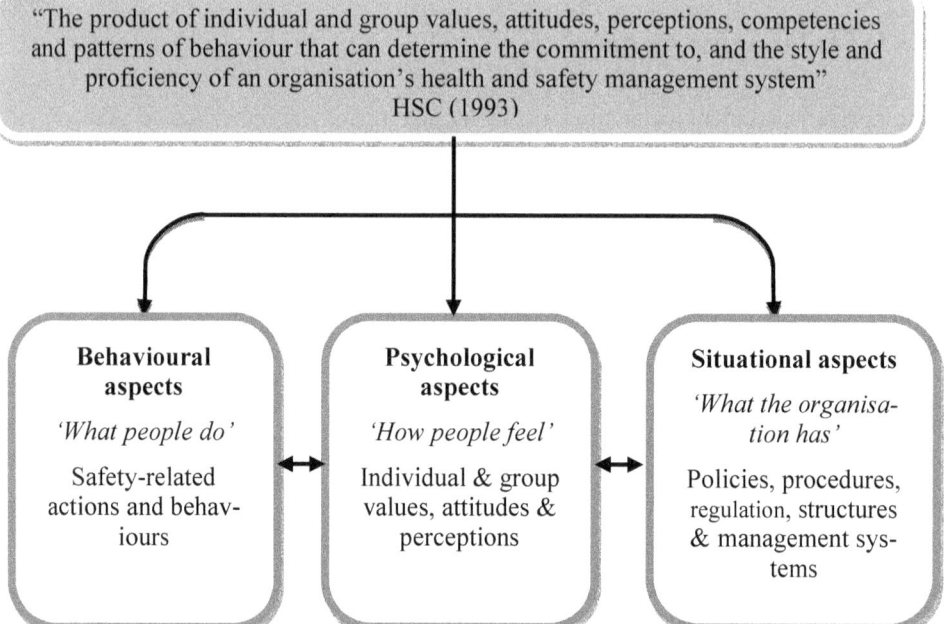

Fig 2: Overview of safety culture components

3.3 Safety culture maturity

Safety culture maturity provides an evaluative measure of the collective safety attitudes and behaviours within an organisation, usually indicated along a continuum or sliding scale. One such model is that of Patrick Hudson (2001), which has been used to good effect in safety-critical industries such as oil and gas, aviation, defence and healthcare. An adapted illustration of this model, is shown in Figure 3. It shows how an organisation can be categorised along the scale from counterproductive – caring less about safety than not being caught, through to championing – in which safety behaviour is fully integrated into everything the organisation does.

Models of safety culture maturity allow organisations to understand their cultural maturity by assessing the level of compliance with various key elements of safety culture, such as leadership, business communications, and degree to which unsafe practices are normalised, to name but a few. The adoption of a safety culture model allows an organisation to identify shortfalls and help address any behavioural or cultural issues they may have with a view to improving safety culture over time.

3.4 Developing and sustaining a mature safety culture

Safety culture cannot simply emerge and be self-sustaining. It has to be encouraged by senior managers, and then sustained with peer reinforcement and overt safety behaviour leadership. Leadership is important in the creation of a culture that supports and promotes high levels of safety performance. The management are vital in inspiring employees to a higher level of safety and productivity that means that they must apply good leadership on a daily basis.

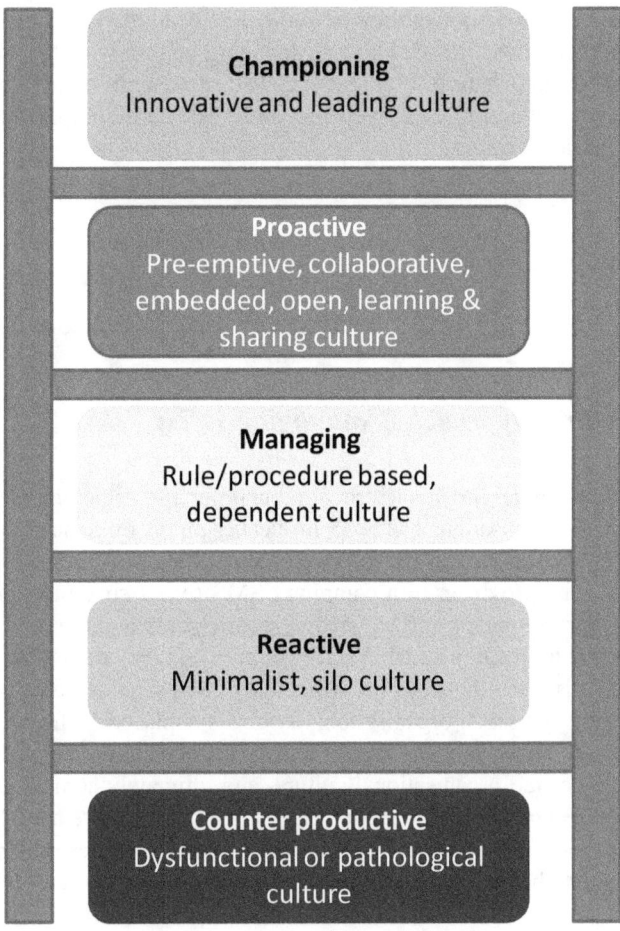

Fig 3: Greenstreet Berman Ltd adaptation of the Safety culture maturity scale

Such models of safety culture are not just about assessing culture in isolation; they suggest a strong relationship between the culture of an organisation and the development of a systems approach. It is critical to understand that systems can only facilitate progress to a certain point, in the absence of safety culture progressing in parallel (and vice versa). For example, if there is a safety management system but no real commitment or culture towards safety, then the management system will not be effective, as decisions will not prioritise safety or be used in earnest. Similarly, if there is a positive safety culture, but no management system, then the way that safety is organised may be inconsistent, under-resourced and difficult to channel through to front line operations. Both examples highlight a culture-systems mismatch.

Additionally, such a model neither assumes nor implies that an organisation should aim to achieve any particular level of safety culture. It is for the organisation to set their own vision for safety culture, relative to where they are now and commensurate to the activities they perform. For example, a business with less severe hazard activities may aspire to a Managing culture whilst a business with major hazards might aspire to be Proactive. The same maturity scale can then be used to measure cultural change over time and evaluate the impact of any initiative or organisational change.

4 Framing the issue

4.1 A structured approach to managing safety

As described above, an SMS is a structured approach to achieving and managing safety within an organisation. The SMS influences and is influenced by the organisation's safety culture as recognised by Reason (1998). It is therefore reasonable to assume that the provision of a compliant SMS (exact specification dependent on the particular regulatory and legislative requirements and the needs of the organisation) should proffer a suitable safety culture and by association good safety performance. It is often the case that when discussing safety culture as a concept, and how it exists within an organisation, that focus is placed on frontline workers. This can serve to reinforce the notion that workers need to be dictated to in order to achieve the desired organisational culture, and ultimately compliance. However, it is the focus and style of management that can easily have a more pervasive effect with aspects such as bias towards meeting performance and productivity targets over safety being key to shaping the overall culture.

4.2 Learning from experience

The accident which befell the BP Texas City plant in 2005 is a clear example of this. As identified by the US Chemical Safety Board (CSB, 2007) investigation, the plant's SMS focused on achieving basic safety requirements such as preventing slips, trips and falls without suitable levels of rigour applied to the significant hazards of the process. This was compounded by the lack of senior management consideration of cultural issues and a 'tick box' approach to compliance. Yet, on paper the SMS itself was likely to have been compliant with requirements such that it may have passed compliance based evaluation of its performance.

Instances such as Texas City demonstrate that a gap can develop between intended practice and reality and corporate understanding and on-site culture. In fact, it could be argued that safety culture and SMS can operate separately despite their apparent interrelation. In order to close such a gap it is necessary to understand how the two elements come about. An SMS is constructed as an entity with particular purposes, goals, needs and requirements (again as required by legislative requirements and the specific needs of the organisation). A safety culture, as we have seen, develops in a more organic manner as a result of various influences (only one of which is the SMS).

4.3 Considering the link between safety management and safety culture

If we consider the SMS and its development and construction; how might this negatively impact on an organisation's safety culture? Treated merely as a regulatory or legislative impediment to be addressed it is clear to see how this could be the case. Such goals, and their achievement, may not be negative but their application and influence can be so and can distract the development of the SMS away from more positive goals that may be of greater benefit to reinforcing a positive safety culture. The drivers that may have such an influence include:

- An overbearing focus on achieving regulatory compliance;
- An excessive desire to provide executive assurance above other requirements;
- Responding too carefully to protecting the organisation from a litigious society;
- Confusion brought about from trying to satisfy a diverse and varying group of stakeholders.

4.4 The potential for disconnect between people and systems

Developing an SMS with such drivers at the forefront may well deliver a complete and notionally compliant system with all the requisite components in place but it is likely to represent and foster a disconnect between the resulting 'perfect world' documentation and the 'actual world' operating environment. This issue can be defined succinctly as the Policy – Practice gap. It is necessary to understand why such a gap exists, what can exacerbate it and how this ultimately results in a poor safety culture.

Where the SMS is inappropriate and poorly aligned with the genuine needs of operations (e.g. through inaccurate or overbearing procedures) a lack of trust will result. Operatives may conclude that, as they cannot comply and get the job done, they need to find their own way of doing things. Often, line management are complicit in this with the result that operatives, line managers and sometimes even senior managers conclude that safety procedures are of no practical use and can be generally ignored. Such instances may reflect a failure of the SMS process to function in a continuous manner, particularly in its ability to monitor and respond to issues. Even where some monitoring does take place it may be simplistic, such as that undertaken at BP Texas City, which will at best hide genuine issues and possibly exacerbate them.

In such a situation, following extensive resources being applied to the development and implementation of the system, it is not surprising that there is a reluctance to challenge its accuracy and relevance. This in turn creates the situation where the SMS can become a self-serving entity that reinforces the status quo rather than a useful tool and an agent for change. It is plain how such an occurrence would negatively impact culture.

Once such a system is in place it is likely that necessity will force personnel to operate outside its bounds, as it simply does not align with their needs. Such activity is, of course, optimising violating behaviour where personnel are attempting to complete their goals despite the rules and processes under which they are required to operate. Their intent therefore is genuine, as staff may fail to see how a SMS relates to their needs (they are unaware of what it is trying to achieve) and it is not definite that their activity will be unsafe.

However, acting in such a way is a clear contravention of the rules as defined by the SMS and it is therefore understandable when a management response is to respond to such apparent transgressions with increased enforcement. Such enforcement may be formal disciplinary action or, at a lower level, minor corrective actions such as providing more training or communicating strongly the need to follow prescribed documentation. However, any response of that type clearly demonstrates a lack of understanding of the root cause of the issue. Such responses make no effort to understand and address the mis-alignment and can only serve to further reinforce the view that developers, managers and of the SMS do not understand the operating environment. Consequently they may struggle to produce a truly effective process or procedure. As a result, opportunities to rectify the issue become fewer over time as the safety culture degrades and the various stakeholders of the SMS drift further apart.

4.5 Cultural and behavioural elements of an effective SMS

It is clearly in the interest of those managing and implementing the SMS to ensure that it is responsive and changes to meet the needs of the operations at hand. However, this is often not the case. The unwillingness to change or develop the SMS such that it reflects reality often suggests a concern that doing so would actually challenge some of those extraneous drivers mentioned above.

Does the provision of excessive and overbearing control and performance recording satisfy an inflated need to document activity? If it can be demonstrated that it actually presents a burdensome regime, creating excessive workload and negatively impacting on operators' ability to undertake their tasks in a safe manner, would that need reduce? It is often the case that such systems are enforced under the misconception that by gathering extensive data in of itself this affords legal compliance; yet without responding to and addressing the findings of such data it can only become a stick with which to beat the organisation in the event of an accident should it highlight where problems had been previously encountered.

Notwithstanding such effects, the simple cost implications of maintaining a responsive SMS is an obvious driver to limit the amount of work dedicated to maintaining and improving an SMS on a frequent basis; or at least, this is the commonly held fear.

Current HSE guidance suggests that the development of safe working practice is more important than extensive documentation. The Construction, Design and Management (CDM) Regulations (2015) have been updated once again to address their regrettable impact as a driving force behind the generation of reams of generic risk assessment and method statements that had no impact on real world safety.

5 Aligning the SMS and Safety Culture

5.1 Closing the gap between people and systems

As mentioned previously a useful analogy for the interaction between a SMS and the organisation's Safety Culture is that of driving a car where the various elements of the overall system interact and contribute to an overall level of performance. In the case of both the SMS and the Safety Culture of an organisation the key common factor is the people within the organisation. The people within the organisation are responsible for developing and implementing the SMS and their behaviours both influence and are influenced by the Safety Culture.

To better align the SMS and safety culture for positive effect then the SMS must reflect and support activity not just tick boxes or define a perfect world. If we consider the development of such a system in the context of engineering a product the design phase is where we should seek to understand the genuine user requirements and not merely the designers expected model of the world.

If we assume that a gap already exists, then closing it fits within the context of modifications. Where modification is required, the motivation will come from the needs of users and clearly, without gathering and understanding what these are through relevant and accurate means of feedback, how can we expect to success-fully modify the product? The same principle applies to an SMS and therefore, where an inadequate and failing SMS exists that has fostered a poor safety culture it is necessary to go back and redesign. This must be an iterative process that uti-lises data and qualitative feedback gathered from measurement of the system (via a variety of means as discussed earlier) as to how the system is functioning, with particular emphasis placed on how personnel interact with it.

Simplistic monitoring and measurement, such as that undertaken at BP Texas City, will at best hide genuine issues and possibly exacerbate them. This may then result in an inappropriate response such as more training to reinforce inadequate arrangements. By its nature therefore such training will not recognise or address the underlying issues and will only seek to further damage the Safety Culture of the organisation through creating distrust.

5.2 Workforce engagement – the crucial link

Designing or redesigning the SMS so that it is relevant and effective requires that the drivers for the system are restated and the activity required to meet them is carefully planned. Clearly legal compliance is such a driver but it can be better stated as 'Creating a Safe Working Environment'. It becomes immediately appar-ent what is needed; the involvement of the people working in that environment, the knowledge of those who operate the process. A revised set of positive drivers may include:

- Creating a safe working environment;
- Meaningful Executive scorecards and Key Performance Indicators (KPI); and
- High levels of stakeholder engagement.

Informative and meaningful communications involving all strata of the organi-sation; 360 degree consultation and risk assessment teams involving end users; a lack of blame assignment; and genuine monitoring of real performance with accu-rate qualitative information, as well as basic quantitative statistics, will all con-tribute to the development of a genuinely credible SMS. Furthermore, such activi-ty will also help to foster a positive culture through engagement and mutual understanding that will help all parties, acting within the SMS, to move through the four stages of competence (unconscious incompetence, conscious incompetence, conscious competence, unconscious competence) to a shared position of unconscious competence that reflects simply 'the way we do things round here', when actions have become second nature.

Furthermore, as we have recognised, a SMS can only achieve its objectives if the organisation has a supportive and healthy culture that helps address safety issues at all stages of the management process (i.e. from the top to bottom of the organisation, for all activities carried out). It is essential therefore that to foster positive change, or reinforcement of positive aspects, the systems and processes must seek to, and demonstrably understand, the perceptions of the personnel operating within the system in relation to the social, environmental and physical environment in which they function. This is especially the case given the prevalence of change in most businesses at present. For an organisation to develop its culture, an organisation should have a means of:

- Defining the safety culture – norms and goals - which takes account of the opinions, perceptions and expectations of internal and external stakeholders;
- Communicating and demonstrating the organisation's commitment to these goals and norms, and maintaining this sense of commitment over time;
- Facilitating the achievement of stated goals and norms in a manner which recognises the perceptions of the system's users, such as workforce participation, empowerment, staff and management-contractor communications, training and resource management;
- Checking that the organisation's cultural goals and norms have been effectively achieved or at least the behaviour of people is consistent with these norms.

The genuine involvement of the workforce in developing a positive Safety Culture is essential. Any associated surveys and assessments should not be seen as an end to themselves but should be used to identify improvements and tangible benefits to safety performance.

5.3 Going beyond the arrangements

There are a range of organisational activities and processes that can influence achievement of the goals, particularly the perception of how the balance between safety and other priorities is to be struck and what level of safety performance (and hence behaviour) does the organisation seek to achieve. In turn, this is influenced by the messages communicated by influential people in an organisation, such as directors, senior managers and supervisors, how these messages are communicated, reinforced or contradicted by other messages and examples of behaviour. The extent to which behavioural expectations are reinforced by responses to examples of positive and negative behaviour will contribute to the setting of behavioural norms.

The extent to which people feel that they are responsible for managing and ensuring safety will influence the extent to which they become engaged in identifying and resolving issues, as well as the effort applied to meeting the behavioural expectations of the organisation. The extent to which people perceive that there is a significant risk will influence the willingness of people to commit time and effort to risk control procedures, assessing risks and resolving problems. Having a trusting and open relationship with your colleagues and other stakeholders will enable collaboration, collective learning and effective team working.

The willingness to report problems, assess their true root causes, develop effective responses and share the lessons will influence the extent to which the organisation can learn and improve over time. Having open, trusting, challenging cultures supports the process of organisational learning. A business which is overly rule-focused, is poor at team-work, task focused at the expense of working relationships may fail to apply lessons learnt from incidents.

6 Making the link between safety management and safety culture

In conclusion, it is clear that, while they are different entities, the link between the SMS and the Safety Culture of an organisation, is complex however, undoubtedly impacts upon safety performance. As a result it is imperative that they must be considered together as one will not excel without the other. It is the recognition of this and the willingness to learn and continually develop that distinguishes a HRO from an organisation which may have merely a nominally compliant and functioning SMS.

Whilst recognising this shared influence, it is clear that the culture of an organisation and the SMS are distinct and different. For example the culture of an organisation is, as discussed earlier, based largely on attitudes and beliefs, etc. which are somewhat nebulous and ill defined. Whereas the SMS, comprises more tangible and formalised structures, systems and components (such as, policies, procedures and reporting systems) which form part of the overall management system. These more concrete artefacts are, by their nature, easier to create, modify and change for the better. As such, it is most appropriate to apply effort to close the gap between policy and practice where we can directly influence change, whilst ensuring that doing so takes suitable heed of the factors which will reinforce a positive culture; and in reflecting the features of a HRO, we should strive to do this on a continual basis.

References

Construction (Design and Management) Regulations (2015) www.legislation.gov.uk/uksi/2015/51/pdfs/uksi_20150051_en.pdf (Accessed 19 Nov 2015)

Health and Safety Commission (1993). ACSNI Study Group on Human Factors. 3rd Report: Organising for Safety. (London: HMSO).

Health and Safety Commission (2001) The Ladbroke Grove Rail Inquiry. Retrieved from http://www.railwaysarchive.co.uk/documents/HSE_Lad_Cullen001.pdf. (Accessed 19 Nov 2015).

Health and Safety Executive (2013). HSG65 (3rd ed) Managing for health and safety. Retrieved from http://www.hse.gov.uk/pubns/priced/hsg65.pdf (Accessed 19 Nov 2015)

Health and Safety Executive (1993). ACSNI Human Factors Study Group: Third report – Organising for safety. HSE Books.

Hudson, P (2001). Safety Management and Safety Culture: The Long, Hard and Winding Road. In Occupational Health and Safety. Occupational Management Systems (ed. W. Pearse, C. Gallagher and L. Bluff). Crowncontent: Melbourne, Australia.

Reason, J. (1998). Achieving a Safe Culture: theory and practice. Work and Stress, 293-306.

The Deming Institute (2015) The Plan, Do, Study, Act (PDSA) Cycle https://www.deming.org/theman/theories/pdsacycle (Accessed 19 Nov 2015)

US Chemical Safety Board (2007). Investigation Report, Refinery Explosion and Fire, BP, Texas City, Texas, March 23rd 2005.

Development of an Adaptive Safety Monitoring Function

John Birch[1], Frederik Botes[2], Paul Darnell[3], David McGeoch[3]

[1] AVL Powertrain UK Ltd

[2] University of Bath

[3] Jaguar Land Rover Ltd

Coventry, UK

Abstract   *This paper describes the invention of an Adaptive Safety Monitoring Function at Jaguar Land Rover within the context of the development of a vehicle propulsion system according to the principles of ISO 26262. It outlines the way that conventional safety monitoring software is currently used to detect and mitigate faults in a propulsion control system. It then describes the typical challenges and drawbacks in developing such software. The paper then presents the theoretical principles behind how the proposed software algorithm could address these problems by adapting over time to the control software that it is monitoring. It explains the key challenge of ensuring safety by only adapting in a manner commensurate with the adaption of a driver's mental model to a change in relationship between vehicle acceleration and accelerator pedal input. It describes some of the practical problems encountered and solutions found during algorithm development before concluding with an outlook towards potential commercial applications.*

1 Introduction

1.1 Safety Mechanisms within the context of ISO 26262

The automotive functional safety standard, ISO 26262 [1], provides a framework for demonstrating that freedom from unreasonable risk has been achieved with regards to the malfunctioning behaviour of electric or electronic systems. With reference to Figure 1; in complying with this standard, 'hazardous events' associated with the malfunctioning behaviour of the system under consideration are identified, whose associated risk is addressed with the specification of one or more 'safety goals'. These safety goals are supported by the specification of one or more 'functional safety concepts' which, in turn, are further supported by the specification of one or more 'technical safety concepts'. Such safety concepts typically call upon the use of a number of 'safety measures'; activities or solutions to avoid, control or mitigate failures which can be implemented in the form of a variety of different technologies. 'Safety measures' may include 'safety mechanisms' which are technical solutions to detect faults or control failures in order to transition to or maintain a 'safe state'.

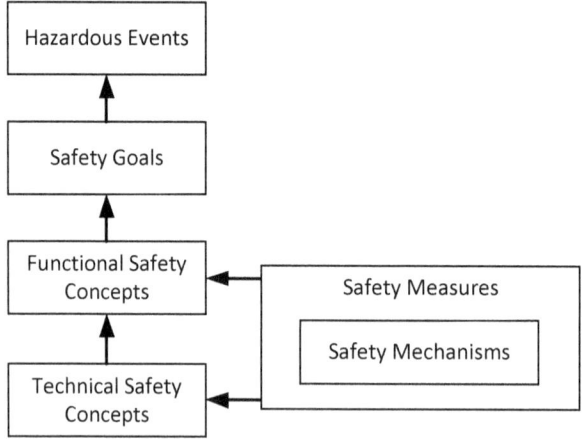

Fig. 1. Safety Mechanisms within the ISO 26262 Framework

1.2 Example Safety Mechanism: Software Monitoring Function

One example of a safety mechanism typically employed in the development of automotive control systems is that of a 'software monitoring function' (further referred to simply as a 'monitor').

Software control within complex safety-critical automotive systems has been in place since the 1980's in the form of ETC (Electronic Throttle Control) [2]. An activity to develop a standardised ETC safety concept, involving the use of software monitoring functions, was started by an alliance of German vehicle manufacturers through the 'EGAS working group' in the early 1990s [3]. The resulting EGAS standard was first publicised in 2002 and since then has served as a strong basis for arguing safety of much of today's ETC software [4]. The standard presents a '3-Level' safety concept, wherein:

'Level 1' contains the functional software
'Level 2' contains the functional monitoring software for detecting faults in the Level 1 software
'Level 3' contains software for monitoring the health of the controller on which the Level 1 and Level 2 software resides.

The purpose of such a monitor is to detect faults within the functional software to which it is assigned, and then to cause the system to transition to a 'safe state' should a fault be detected. The monitor is typically developed to a particular Automotive Safety Integrity Level' (ASIL) which is ultimately commensurate with the level of risk associated with the hazardous event that the monitor is put in place to mitigate. The higher the risk, the higher the ASIL, and therefore the greater the degree of rigour and effort associated with the development of the monitoring software.

1.3 Example Monitoring Function

An example (theoretical) monitor used within an automotive propulsion system is shown schematically in Figure 2. This particular monitor is designed to detect faults in the functional software that requests torque from a torque actuator (such as an engine or an electric motor) based on the acceleration that is demanded by the driver by pressing the accelerator pedal. It does this by:

1) Calculating an 'expected torque request', based on accelerator pedal position
2) Calculating (or simply detecting) the 'actual torque request' sent to the actuator

3) Calculating a 'torque error' based on the difference between the expected and actual torque requests.
4) Comparing the torque error with a pre-defined 'error threshold'
5) Taking an action to limit the torque request that reaches the actuator to a pre-defined (low) level if the torque error exceeds the threshold (note that this action ensures a safe vehicle reaction whilst still preserving a level of primary functionality, allowing the vehicle to continue to be driven.)

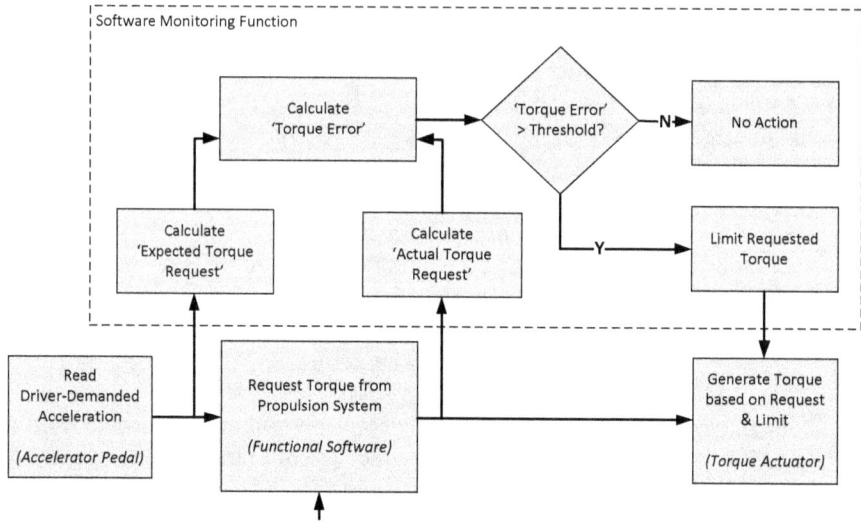

Fig. 2. Example Vehicle Propulsion System Controller Software Monitoring Function

1.4 Monitoring Function Principles

The principle attraction behind the use of the kind of monitor shown in Figure 2 is that because it is only a large torque error that would cause an 'unsafe' level of unintended vehicle acceleration, the software algorithm upon which this error is calculated can be relatively simple in comparison to the algorithm used for the corresponding functional software. Such simplification helps to limit the cost and effort associated with the monitoring algorithm which is particularly important due to the high ASIL (and therefore development rigour) typically assigned to it.

For example, to aid refinement of the propulsion system the functional software in Figure 2 may vary its torque request to the torque actuator a little dependent upon the temperature of the actuator (e.g. it may request more torque from a cold engine for a given accelerator pedal position that it would from a warm engine).

Because this temperature effect is relatively small it doesn't need to be modelled in the monitor algorithm; rather it is accepted that there will always be a correspondingly small torque error present; positive or negative, dependent upon actuator temperature.

In reality the functional software of an actual vehicle propulsion system will feature a large number of these dependencies in its transfer function from accelerator pedal position to torque request, such as:

- Driver mode selection
- Vehicle altitude
- Vehicle gradient
- Transient actuator response
- Battery state of charge
- Fuel level
- Etc.

These dependencies become 'noise factors' for the design of the monitoring function that will each introduce a variable torque error. When all of these torque errors are collectively considered over a vehicle drive-cycle they could give rise to a frequency-based distribution of combined torque error such as that shown in Figure 3.

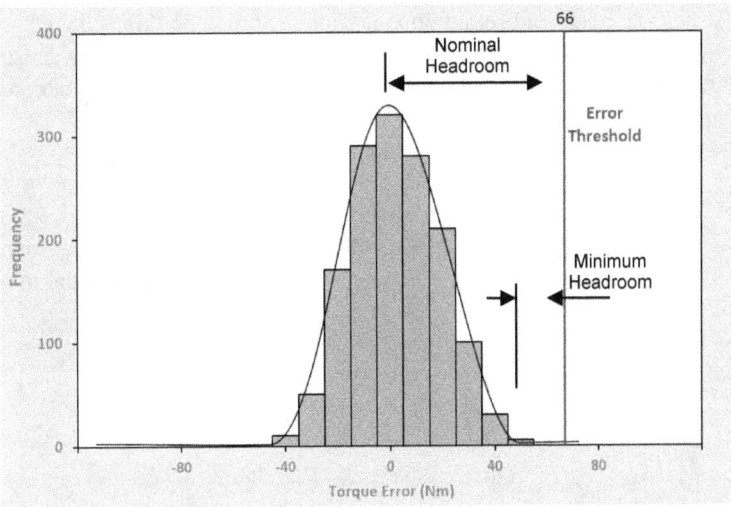

Fig. 3. Example Theoretical Torque Error Distribution

The error threshold such as the one indicated in Figure 3 is defined based on two separate and potentially conflicting requirements; *safety* and *robustness*.

On the one hand it has to be set sufficiently low such that in the case of a large, sudden torque error the monitor will react and limit the torque request to a level that is simply controllable by the driver – thus ensuring safety. On the other hand the error threshold has to be set sufficiently high such that with the worst-case combination of noise factors the monitor will not react and take action unnecessarily – thus ensuring robustness.

1.5 Monitoring Function Challenges and Drawbacks

The error threshold shown in Figure 3 is an example of one which is set with a minimum 'headroom' of around 15 Nm (the difference between the maximum positive torque error without faults and the error threshold) and a nominal headroom of 66 Nm (the actual value of the error threshold). For this example it may be that a 66 Nm positive torque error is simply controllable by the driver pressing the brake pedal, in which case, based on Figure 3, the monitor can be deemed to have met both safety and robustness requirements. However, in a real practical application, ensuring that both of these requirements are met can be a considerable challenge.

Consider the monitor shown in Figure 2. The sensor on which the actuator temperature measurement is based may exhibit a level of drift over its lifetime. This would result in a corresponding drift in the torque requested by the functional software. The distribution of torque error is likely to correspondingly drift over time and at some point would resemble Figure 4 rather than Figure 3, with an 'error distribution mean' of around 40Nm. Under certain conditions the error threshold will be exceeded solely due to the presence of a combination of noise factors without a large, sudden torque error being introduced - causing the monitor to fail to meet its robustness requirement.

The potential solutions to this problem include:

1) Using a sensor less prone to drift
2) Increasing the error threshold (without violating the safety requirement)
3) Taking the signal from the sensor as an additional input to the monitoring function, to allow the monitor to compensate

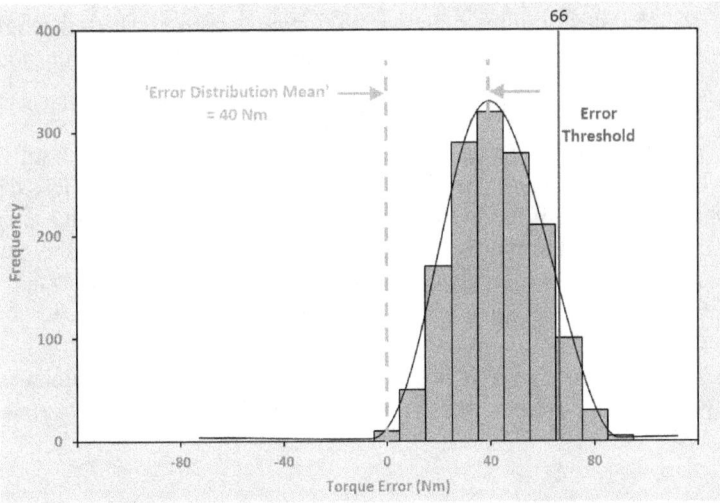

Fig. 4. Example Theoretical Torque Error Distribution with an 'Error Distribution Mean' of 40 Nm

In this case all three solutions potentially solve this problem. However, consider another example noise factor. The functional software may take a driver-mode selection as an additional input upon which to base its calculation of torque request. The selection of a 'sport' mode, say, might result in a significantly higher torque request for a given accelerator pedal position than that which would be requested if an 'economy' mode were selected. In this case it may be that the only option available would be 3).

Requiring the monitoring function to be based on additional inputs poses a number of drawbacks:

1) Monitoring function complexity
   The more inputs a monitoring function is based on the more complex it becomes, requiring additional cost, time and development effort

2) Monitoring function specificity
   The monitoring function becomes more specific to the functional software to which it is assigned, meaning that if a company has a number of different variants of functional software to monitor it must also develop and test a corresponding number of monitoring variants; again with cost and time implications.

3) Required ASIL of inputs
   In certain cases the additional inputs that are fed to the monitor would be required to be of at least the same ASIL as the monitoring function itself. Requiring these inputs to be of a high ASIL will likely increase the cost

and complexity of the functions or components from which they are generated, or it may be the case that such inputs cannot be practically sourced.

4) Lack of intellectual property (IP) protection
Consider the case that company X develops the functional software but employs Company Y to develop the monitoring software. Here the nature and operation of the functional software must be disclosed in sufficient detail to allow the monitoring software to approximate it closely enough to avoid robustness issues. Despite these limitations such a working arrangement is not uncommon in the automotive industry.

Making use of a monitor that has the ability to *adapt* to its functional software offers potential to overcome these drawbacks, as described in the next section of the paper.

2 Adaptive Monitoring Function: Principles

2.1 Overview

Shown in Figure 5 is a version of the monitoring function previously presented in Figure 1 but now with some additional 'adaptive' elements, indicated with dotted lines.

The monitor now additionally calculates the 'torque error distribution mean' (further referred to as the 'mean error') from the 'torque error' data recorded over a pre-defined time period. In order to calculate this mean error the monitor is effectively required to internally generate a real-time distribution such as that shown in Figure 4, which is updated with a pre-refined rate.

Based on this mean error the monitor calculates a 'torque error offset', which is constrained by two set limits; one for its maximum rate of change and the other for its maximum value.

A 'torque error with offset' is then calculated by subtracting this 'torque error offset' from the original 'torque error'. It is this 'torque error with offset' that is now compared to a threshold value to determine if the monitor is required to limit the requested torque.

These additional elements effectively allow the monitoring function to perform a level of closed-loop control; allowing it to maintain a low 'torque error with offset' even with changes in the transfer function of the functional software. This principle is illustrated in the example in section 2.2.

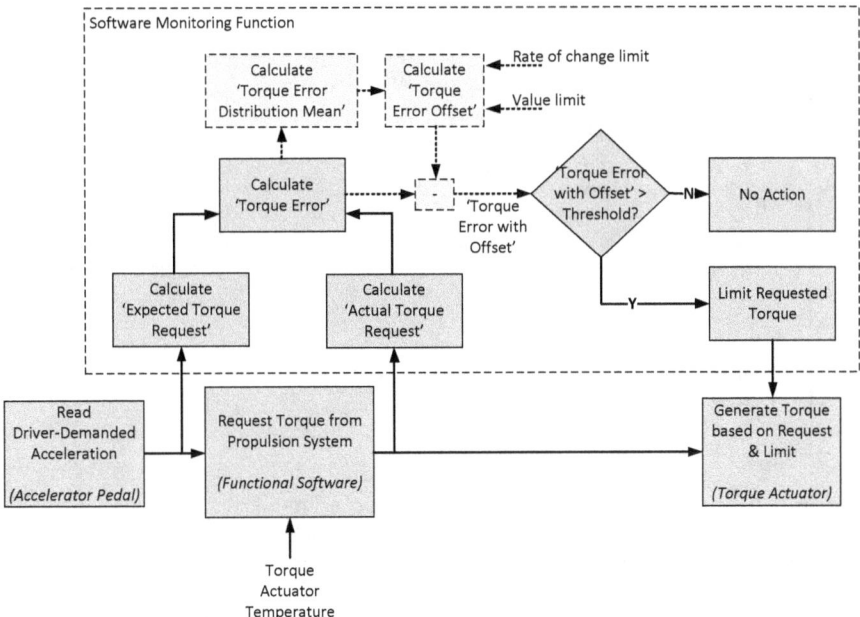

Fig. 5. Example Vehicle Propulsion System Controller Adaptive Software Monitoring Function

2.2 Example

Figure 6 shows the behaviour of an adaptive monitor of the type shown in Figure 5. The monitor calculates a torque error distribution from torque error data recorded over a 1 s window. At the start of the window the driver is demanding a level of vehicle acceleration that corresponds to a torque request of 50 Nm to the actuator. However, very early on in this window the functional software determines that an extra 10 Nm in torque request is required in order to continue to satisfy the driver acceleration demand. This is due to a measured change in actuator temperature, that the monitor doesn't directly read, which remains throughout the rest of the window. The mean error from the distribution over the 1 s is thus calculated at the end of the window to be ~10 Nm.

Ignoring the 'rate-of-change' and 'value' limits for now, the torque error offset at the end of this window will simply be calculated to be the same as the mean error, so ~10 Nm. This ~10 Nm is then subtracted from the 'torque error' at the start of the next 1 s window, so that this second window begins with a 'torque error with offset' of ~0 Nm. After four of these windows the 'torque error with offset' would be maintained at close to 0 Nm. The available monitoring headroom has therefore not been reduced by the 40 Nm which would have been the case with a conventional monitor.

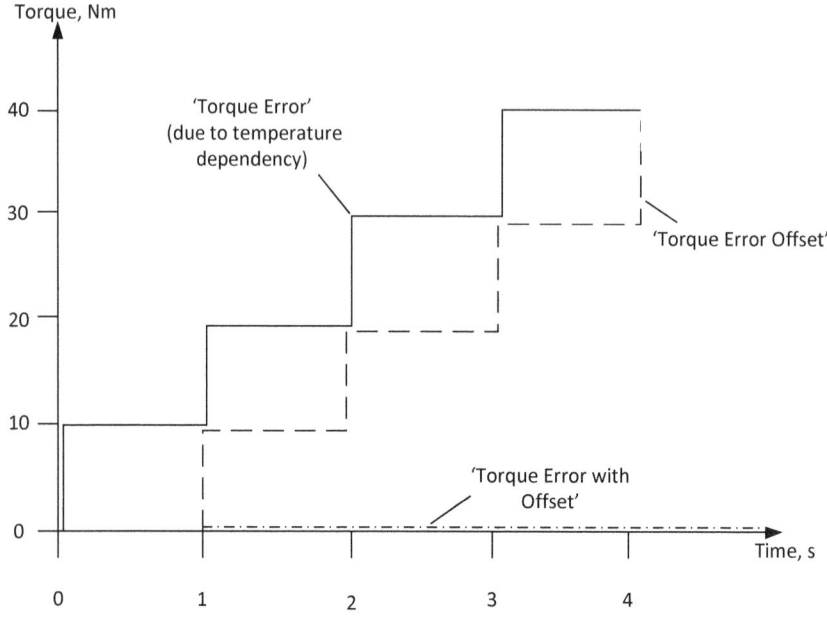

Fig. 6. Example Adaptive Monitor Behaviour

2.3 Ensuring Safety and Robustness

Consider the adaptive monitoring function behaviour described in section 2.2 in light of the requirements first presented at the end of section 1.4: the monitor must meet both safety and robustness requirements. The adaption of the monitor to the influence of a temperature dependency in the functional software is a means by which robustness is ensured. At the same time it also needs to ensure safety.

Consider a malfunction of the functional software such that instead of increasing the torque request by 10 Nm every second due to the influence of temperature it erroneously increases it by, say, 100 Nm as shown in Figure 7. If the monitor were to adapt to this behaviour then it would simply increase the torque error offset accordingly, and within 4 s the actuator would have been allowed to increase its torque output by 400 Nm with no change in the driver torque request. This could potentially result in a level of unintended vehicle acceleration that driver would not be able to simply control by pressing the brake pedal.

The adaptive monitor would mitigate this fault condition, thus ensuring safety, by the setting of the two limits shown in Figure 5:

- The rate of change of the torque error offset
- The maximum value of the torque error offset

If, by setting a limit on its maximum rate of change, the increase in offset is insufficient to allow full adaption to the increase in torque request then the 'torque error with offset' value will increase accordingly. This is shown in the example below in Figure 7. Once the 'torque error with offset' exceeds the predetermined error threshold value (shown as 150 Nm) then the monitor will take action to limit the torque request to the actuator.

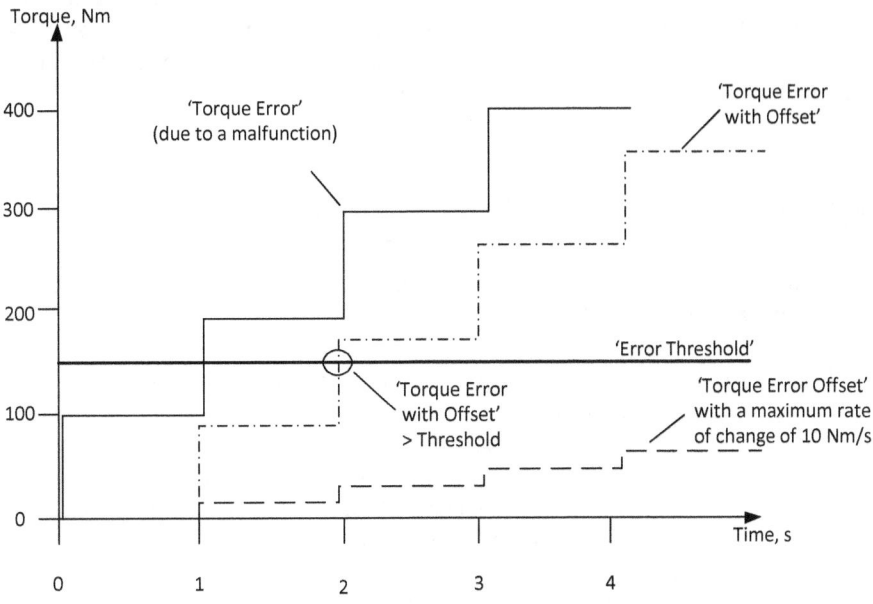

Fig. 7. Example Adaptive Monitor Behaviour with a Malfunction of the Functional Software and a Limit on Rate of Change of 'Torque Error Offset'

Similarly, it may be that the functional software requests a valid torque request increase of 10 Nm/s based on temperature, which the monitor would adapt to. However, instead of only operating within its design window of a maximum of 40 Nm for temperature compensation, it malfunctions and continues to increase its torque request indefinitely. This condition would be mitigated by a maximum value assigned to the torque error offset which in this case could be set to match that of the functional software; 40 Nm.

2.4 Key Enabler

The key enabler for monitor adaption is the fact that there is a driver in the loop, actively sensing vehicle acceleration and changing the accelerator pedal position accordingly. As long as vehicle acceleration does not deviate too quickly from that which the driver expects, according to his or her mental model – which is fluid, then the vehicle can be considered to be behaving in a safe manner. There is no need for a fixed absolute relationship between accelerator pedal position and vehicle acceleration. The limits for how quickly and by how much the monitor is allowed to adapt need to be constrained by the corresponding rate and extent to which the driver's mental model is able to adapt.

2.5 Key Challenges

The two key challenges in setting these limits, and in the general design of the monitoring algorithm, are therefore:

1) Ensuring safety by preventing the monitor from adapting to genuine faults in the functional software that would otherwise cause an uncontrollable level of unintended acceleration
2) Ensuring robustness by allowing the monitor to adapt quickly enough and sufficiently to ensure the avoidance of unnecessary interventions under a worst-case combination of expected noise factors

In addition, the monitoring algorithm itself must be shown to have been developed and verified to an appropriate level of integrity. If it is required to ultimately support an ASIL B safety goal, for example, then it will need to be subject to the ASIL B software development requirements contained within Part 6 of ISO 26262. However, because the adaptive aspect of the monitoring algorithm itself may be common across applications it may be possible to avoid having to demonstrate satisfaction of these software safety requirements every time the algorithm is employed.

2.6 Promise

The promise in the use of an adaptive monitor is that of overcoming the drawbacks associated with a conventional monitor identified at the end of section 1.5. By adapting to the functional software, whilst still ensuring safety, in theory it is possible to add a whole host of additional inputs to the functional software (noise factors for the monitor) that the monitor doesn't need to be made aware of or directly accommodate. This potentially allows the use of a single, functionally simple monitor, across a range of functional software designs, calibration releases and vehicle applications. It prevents the need for these extra inputs to be of a certain ASIL, and reduces the need for communicating the intricate design of the functional software to whoever is designing the monitoring software.

The extent to which the promise of the adaptive monitoring concept can be realised in practice is the subject on ongoing development, as described in the next section of the paper.

3 Adaptive Monitoring Function: Practice

A set of prototype algorithms and functions are being developed in the simulation environment for the adaptive elements of the monitoring function from Figure 5, extracted below in Figure 8.

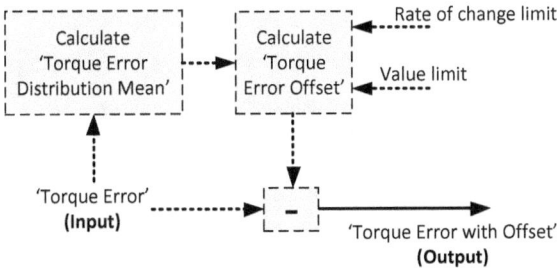

Fig.8: Adaptive Monitoring Elements Subject to Development

3.1 Calculation of 'Torque Error Distribution Mean'

A script has been developed which is capable of reading input error data ('Torque Error' in Figure 7), and using a histogram process to produce a probability density function (PDF) and distribution curve. From this process, the 'Torque Error Distribution Mean' is determined and initially used as the offset. The offset is then applied to the original 'Torque Error' input signal to create the 'Torque Error with Offset', subsequently referred to as the 'output'.

3.2 Model Input: Artificially Generated 'Torque Error'

An artificial 'Torque Error' signal has been generated as an input to the calculation of the 'Torque Error Distribution Mean'; as shown in Figure 9. The aim in generating this torque error signal is to be able to test the ability of the adaptive monitor to overcome the challenges identified in section 2.5; adapting to slow changes of torque error over time but reacting appropriately to torque error conditions that would cause a safety concern. As such, the error signal includes a range of features designed to simulate these different conditions; a number of ramps with various gradients along with several step changes of varying magnitude.

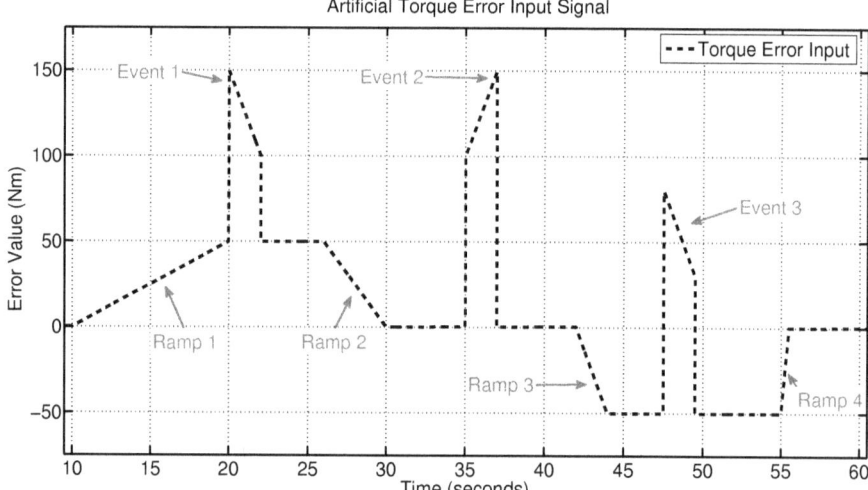

Fig.9: Artificially generated 'Torque Error' input signal.

3.3 Model Output: Initial Calculation of 'Torque Error Offset'

The first model developed for calculating the torque error offset was based directly on the schematic shown in Figure 5. The 'Torque Error Distribution Mean' is calculated and used as the basis for the 'Torque Error Offset'. An ideal adaptive monitor would be one that adapts to slow changes in error, but not to sudden step changes.

The only variable initially analysed in this model is the mean sample range (0.1 s, 1 s, and 2 s), i.e. the number of historical data points the function uses to evaluate the distribution mean. Maximum offset rate-of-change and maximum offset value limitations were not implemented initially. The resulting 'Torque Error with Offset' output using a 1 s mean sample range from this non-limited adaptive monitor is shown in Figure 10.

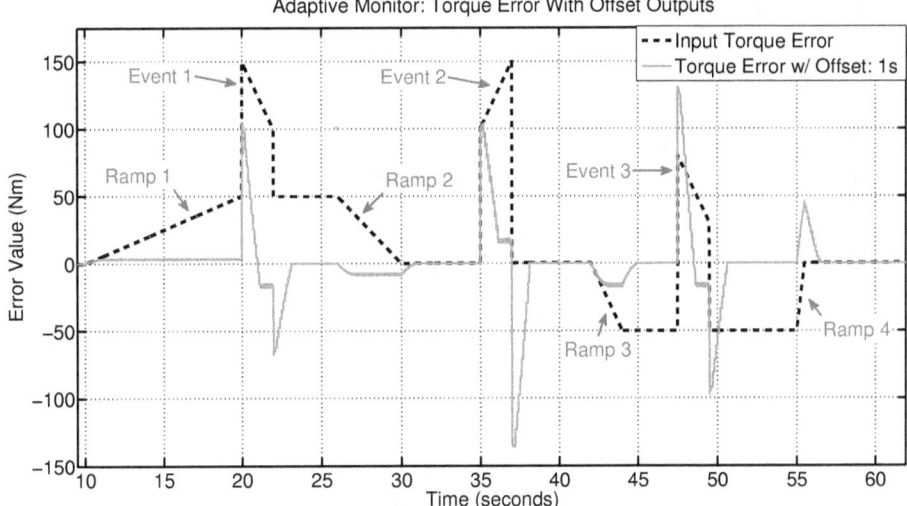

Fig. 10. Adaptive Monitor acting on an artificially generated 'Torque Error' input signal from Figure 9 with 1s mean sample range

Figure 10 shows that the adaptive monitor has adapted to the four 50 Nm input error ramps (ramps 1 – 4). What is apparent is that higher input ramp rates yield greater peak output errors.

At this point it is worth reconsidering the role of the driver described in section 2.4. The driver's mental model may be able to adapt to the input ramp 1 over the first 20 s. If so, the baseline from which a step change in error should be judged to be unsafe itself effectively rises with the error. With this in mind, examining Event 1 shows that the adaptive monitor has correctly reduced the absolute torque error spike as would be experienced by the driver from 150 Nm to 100 Nm; the driver would have adapted to the new 50 Nm baseline as a result of Ramp 1, and the error would only feel like 100 Nm to the driver. If there was an error threshold set at, say, 130 Nm then this would have been exceeded by a conventional monitor at this point, but not by the adaptive monitor.

Event 3 shows the opposite case, where the driver would have adapted to the -50 Nm ramp preceding it, but then experiences a 130 Nm step change. A conventional monitor would simply read the fault as an 80 Nm absolute fault, and could determine that the system is still in a safe state. However, relative to the driver's mental model, a 130 Nm fault has occurred which would feel less controllable. The adaptive monitor has successfully adapted to the drift and shows a 130 Nm absolute error output, which could trigger an appropriate fault reaction.

3.4 Model Output: Challenges in Initial Calculation of 'Torque Error Offset'

Figure 10 shows opposite overshooting at the end of each input event. The over-shoot occurs because the adaptive monitor's offset value after the initial step change in input is near the absolute torque error value. Therefore, when the input torque error signal suddenly drops back down towards zero (to 50 Nm, 0 Nm and -50 Nm for Events 1, 2 and 3 respectively) the output error is pushed by the same degree away from 0 in the opposite direction. At the end of Event 2 (37 s) the error offset would be about 145 Nm, which is acting on an input torque error of 150 Nm. This yields a 'Torque Error with Offset' of ~5 Nm. When the input torque error suddenly drops back down to zero, the monitor does not instantly adapt and is still applying a 145 Nm offset to what is now a 0 Nm input torque error, resulting in an undesirable, artificial, 'torque error with offset' of -145 Nm (shown in (3) in Figure 11).

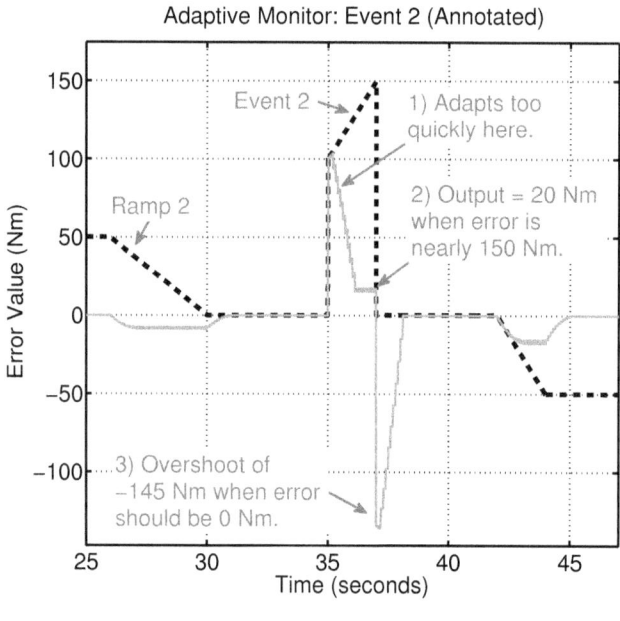

Fig. 11. Annotated Event 2 taken from Figure 10, highlighting some undesirable Torque Error with Offset output of initial model.

Figure 11 above focusses on Event 2, annotating where the initial model result-ed in undesirable output behaviour. After the initial 100 Nm step change at 35 secs, the Torque Error with Offset rapidly adapted to the Input Error (1), and thus failed to register the peak 150 Nm fault entirely (2). This shows the importance of implementing limits on offset rate-of-change and maximum offset, as without them, the adaptive monitor can compensate too quickly.

3.5 Model Output: Refinement in Calculation of 'Torque Error Offset'

This challenge and others that have been encountered are being addressed in on-going model development. A sample result of the latest model output at the time of writing is shown below in Figure 12, in which limits have been employed in the calculation of torque error offset.

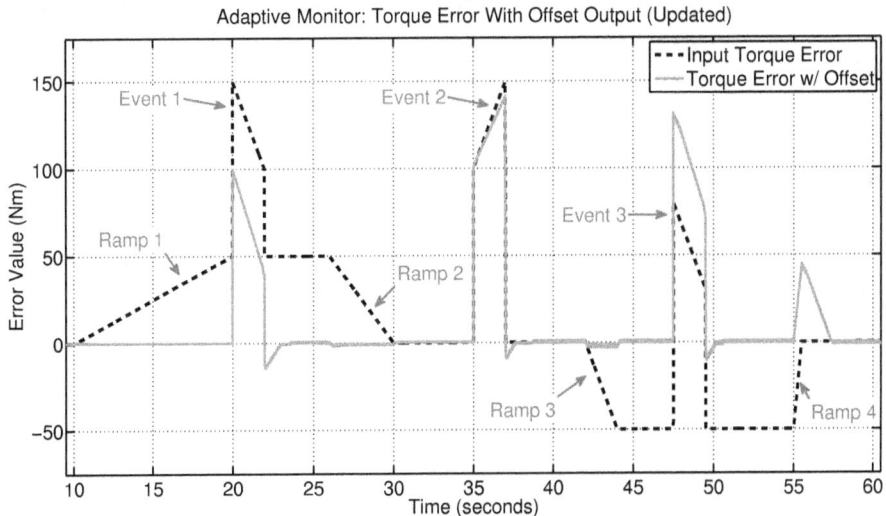

Fig. 12. Output of updated Adaptive Monitor model acting on an artificially generated 'Torque Error' input signal from Figure 9, showing updated 'Torque Error with Offset'

These latest results shown in Figure 12 are very promising. Around Event 1, the algorithm successfully adapts to the ramp in the input torque error, limiting the torque error with offset to only 100 Nm. Once the input torque error comes down to 50 Nm at the end of Event 1, there is only a very small opposite overshoot in comparison to that in Figure 8.

Event 2 is handled near-perfectly, showing very little variation from the input torque error, and peaking just under the 150 Nm input torque error. Again, there is an extremely small overshoot upon return; a large improvement over the initial model.

Finally, examining Event 3 shows the adaptive monitor has successfully adapted to the preceding -50 Nm ramp to bring the absolute error up to 130 Nm as the driver would feel it, with little subsequent overshoot at ~50s.

3.6 Future Work

From the desk-based work performed in the simulation environment it can sometimes be difficult to determine whether the monitor should be adapting to a particular event or not. Such determination requires a level of in-vehicle testing to be performed to understand how the different events 'feel' to a real driver, and how controllable they are. In effect vehicle testing of this sort will illicit properties of the driver's mental model to which monitor adaption can be tuned for compatibility. I.e. the testing will reveal what level of adaption, however small, a driver is able to comfortably accommodate.

4. Commercial Applications

The general example application of the adaptive monitor concept presented in this paper is that of the control of an automotive propulsion system. More specific examples of such controllers include:

- Internal Combustion Engine Controller
- Electric Motor Controller
- Hybrid Propulsion System Controller

Each of these types of controller make use of monitoring functions as safety measures, typically as part of an 'E-Gas' monitoring concept [2], and each are amenable to such monitoring functions being of the adaptive variety. It is worth noting that some monitoring functions are required to be robust to a significantly smaller set of noise factors than others, and for these monitors the potential benefits of being adaptive are less significant.

Each of these examples of propulsion system controller are required to operate as part of a larger control system that, at the very least, involves the driver adjusting his/her acceleration demand based on his/her perception of vehicle speed and acceleration. This larger control system could instead be a cruise control system, for example, which requests a level of torque from the propulsion system controller based on vehicle speed. In this case it could also be appropriate for the propulsion system controller to be adaptive. The cruise controller could adapt to a slow change in vehicle speed as a result of a slow change in actuator torque request that an adaptive monitor allows in a similar way that a driver would.

Following this line of reasoning it is therefore proposed that the concept of adaption could be applied to the monitoring of any safety-critical system within any industry domain in which the monitor is part of a larger control system that can accommodate such adaption.

References

[1] ISO 26262: 2011 Road vehicles -- Functional safety. Available at:
http://www.iso.org/iso/catalogue_detail?csnumber=43464
[2] D. McKay, G. Nichols, B.Schreurs (2000) Delphi Electronic Throttle Control Systems for Model Year 2000; Driver Features, System Security, and OEM Benefits. ETC for the Mass Market. Society of Automotive Engineers. Available at:
http://www.carprogrammer.com/Z28/PCM/FAQ/Delphi_Drive_by_wire_2000-01-0556.pdf
[3] EGAS Standard Monitoring Concept. Available at:
https://www.iav.com/sites/default/files/attachments/seite/ak-egas-v5-5-en-130705.pdf
[4] Li, S., Chang, C., and Zhao, H., "Functional Safety Development of E-motor Drive System for PHEV," SAE Technical Paper 2015-01-0261, 2015, doi:10.4271/2015-01-0261.

Managing the concept phase in the functional safety standard for automobiles

Masao Ito

Nil Software Corp.

Tokyo, Japan

Abstract *In the field of automobile development, the recent automotive function-al safety standard (ISO 26262) is now applicable. However, it is difficult to devel-op a system so as to comply with the standard. In this paper, I will focus on the concept phase of this standard because it includes the new ideas to keep a system safe. For example, the idea of "item" is one of them, and it means an abstraction of a system. It is important to think about safety in the very early phase because we can change the behaviour more readily than do after designing. On the other hand, it requires us to devise the new approach to calculating risk (a.k.a. ASIL) because we don't have detailed information on an item in the early phase devel-opment. In this paper, we introduce our approach CARDION, and we show how we deal with those characteristics of the standard.*

1 Introduction

There are many types of safety-critical systems; power plants, trains, airplanes and so on, and much experience has been accumulated in developing those systems. First we think of the characteristics of the automobile in comparison with other systems. We notice that the automobile has two features, and they are important to think about safety. The first feature is the movability of the system. In contrast, a power plant is stationary; the train moves from a station to the other station, but its path is restricted to the railway track. The airplane is freer to move, but its route is almost defined by the airway. On the contrary, the automobile can move around choosing roads.The second point is the diversity of users of a system. In a passenger car, most of the users (i.e. drivers) are ordinary people, not trained people (i.e. operators).

There is also a different feature that becomes conspicuous in recent years. The importance of the software is higher in the automobile field. The various systems known as advanced driver assistance system (ADAS) (Thalen 2006) have appeared on the market. Those are effective for safe driving, but the impact is large when an ADAS system fails. Because it automatically does braking / accelerating / steering a car in various situations with the ordinary driver. So, any trouble on these assisting systems might give us a fatal blow.

How should we keep the car safe in these circumstances? The most general approach is considering a safety mechanism and improving the reliability of the system so that the mechanism works correctly. We first focus on software, because ADAS is the software-intensive system.

As for software, we need a different perspective from the machinery and electronic equipment. Software doesn't change over time. It fails due to the error of logic or implementation. In the concept phase, the logic is the most important thing. That means we have to encompass fully the possibilities of the situation. We show an example of the recent Adaptive Cruise Control (ACC) problem that indicates the importance of the logic (CNN 2015). Here, milliwave radar of the ACC incorrectly responds to the metal surface of a tank truck car running the opposite lane. It falsely infers the existence of the forward car that doesn't exist and ACC suddenly slows down the car, and consequently increases the risk of rear-end collision. In recent years, this problem also occurred in Japan. If the designer does not consider the influence of reflection from the oncoming vehicle, even if we made the system reliable, safety cannot be ensured.

In this paper, for the software intensive system, we show how to ensure the safety in the concept phase and the correspondence between our approach and the ISO 26262 standard. First, we briefly introduce the ISO 26262 standard (chapter two), and we show our approach CARDION (chapter three). We applied our approach to many items. Through our experience, we consider the difficulties to conform to the standard (chapter four).

2 Concept Phase in ISO 26262

In this chapter, we briefly introduce the concept phase. This phase is the first phase in the lifecycle of functional safety process in the ISO 26262 standard (ISO 2011), and its goal is to define the functional safety concept of an item. The process is mainly divided into the four subphases:

(Subphase 1) Item definition
(Subphase 2) Initiation of the safety lifecycle
(Subphase 3) Hazard analysis and risk assessment
(Subphase 4) Functional safety concept

The term "item" means the abstraction of a system (10-4.2), that is, a system is realized from an item. For example, if a vehicle manufacturer would like to devise the new Adaptive Cruise Control (ACC) system, ACC is an item. And the ACC that is mounted on a specific model of the car is a system. In the item definition subphase, we have "to develop a description of the item about its functionality, interfaces, environmental conditions, legal requirements, known hazards, etc.". And it affects the behaviour of vehicle level, so the ACC is an item, but the milli-wave radar, which is one of the main parts of ACC, is not an item.

The item might be new one or revised one. For the latter case, we start the safety lifecycle by using the original one. In the initiation of the safety lifecycle subphase, we decide the starting point of development.

Before finding the hazards, we have to analyse "(t)he operational situations and operating modes in which an item's malfunctioning behaviour will result in a hazardous event". And we search the pair of operational situation and hazard. This is called hazardous events. Then we classify them from the viewpoints of severity, exposure and controllability. Finally, we calculate the automobile safety integrity level (ASIL) of the item by the ASIL determination table. A safety goal is set to each hazardous event.

After allocating the ASIL to the hazardous event, we define safety goals and write the functional safety requirements. In this requirement, we have to contain "safety measures, including the safety mechanisms, to be implemented in the item's architectural elements".

3 CARDION

3.1 Outline

Our approach consists of six process elements:

Step 1: Item sketching - sketching a system schematically
Step 2: Writing the top goal and decomposing it repeatedly
Step 3: Applying a guideword to each goal
Step 4: Finding the candidate hazards
Step 5: Assessing risk

Step 6: Creating the functional safety concept

In step 1, we depict the static structure and dynamic behaviour as the item sketch. The next step involves writing the goal model of an item (step 2). A goal tree is a tree structure in the goal model. Next, one applies guidewords of the hazard and operability study (HAZOP) (CEI/IEC 2001, Fenelon and Hebbron 1994, Redmill et al. 1999) to find out the candidate hazards (steps 3 and 4). In the risk assessment (step 5), we calculate the ASIL that is given by the class of the exposure, controllability and severity. Depending on the degree of risk we define the safety goals for hazards and then create the functional safety requirements to satisfy the safety goal (step 6).

The iterative execution of this process is essential. One might find the missing information from the previous loop. For example, when decomposing the goal, one might find a missing element of an original goal that helps locate other hazards.

It is also important to maintain the descriptions according to the level of abstraction of a goal and an item sketch until the end, respectively. This helps us in understanding the item, and it might be possible to reuse the item sketch.

3.2 Making an Item Sketch

The item sketch is the schematic description of a system, and it has a static and dynamic part. The schematic description provides a clue to finding hazards and helps assure the completeness of one's checking. The representation of the static structure of a system is shown in the diagram, for example, the class diagram of UML, the internal block diagram of SysML (OMG 2008), or the specification type representation of the CATALYSIS approach (D'Souza and Wills 1998). The finite state machine diagram is good for indicating system behaviour. Approximation is inevitable, however, because in the concept phase one cannot get the detailed information of the target system.

The static representation of the item sketch provides the relationship between the nouns in the description of a goal. Here, we use the cooperative adaptive cruise control (CACC) system (Naus et al. 2009) as an example. The function - communication between cars - is added to the adaptive cruise control system (ACC) (SAE International 2003). Information on the forward vehicle is acquired by communicating with the preceding vehicle as well as the information acquired from the images of the camera, radar, or laser imaging detection and ranging (LIDAR). Through communication, an action of the person driving the forward car ahead (for example, the degree of pushing the brake) will be transmitted immediately. The following car behind can react more quickly compared with a car using only a radar or a camera.

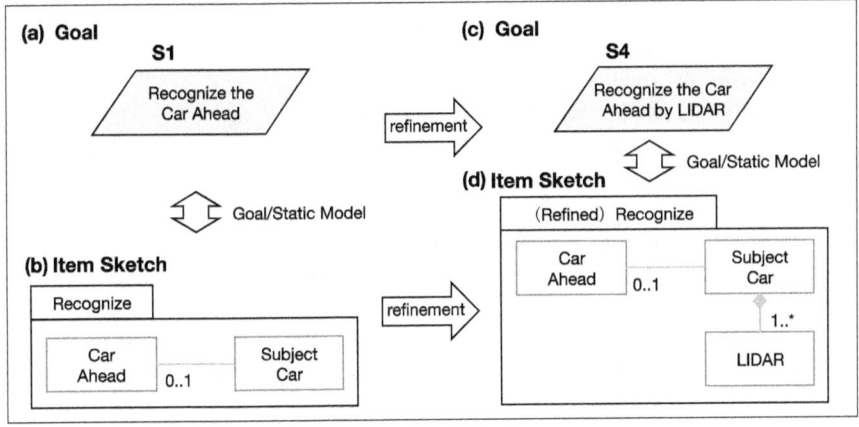

Fig. 1. Goal and Item Sketch (Static Model]

The sentence (S1): "(The subject car can) recognize the car ahead" can be described using Figure 1(a). This sentence includes two nouns: "car ahead" and "subject car," and a verb "recognize." With these morphemes, one can create a package diagram. This is a candidate of the static model of an item sketch (see Figure 1(b)). Here, we added the subject noun "subject car." In the case of omitting the subject word, one has to add it when creating an item sketch. And the verb "recognition" is used as a package name. If using the CATALYSIS approach, one can describe it as an action in the specification type, and it might be able to express a more clear meaning.

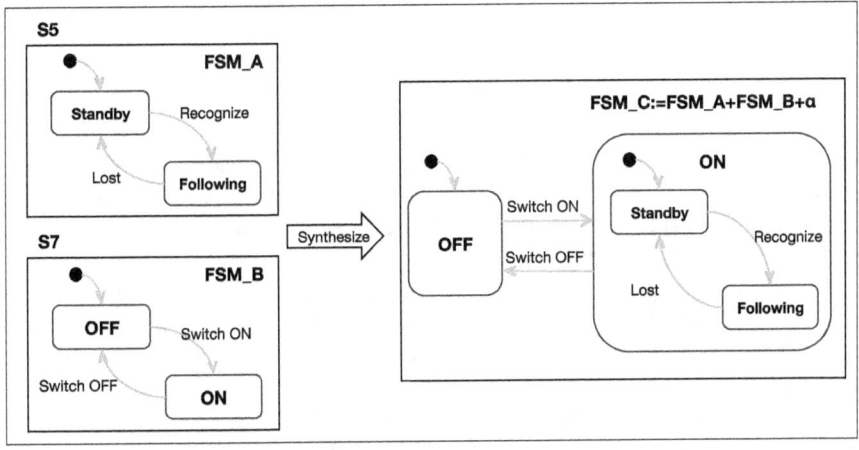

Fig. 2. Item Sketch (Dynamic Model)

It is possible to provide a dynamic representation of an item sketch using a finite state machine model. Consider the next request sentences:

- (S2) When the car ahead is recognized, the system will change to the following mode. If it loses sight of the car ahead, the system moves to "standby" state (see Figure 2 upper left, FSM_A).
- (S3) The system will be ON (mode) by turning on the switch. It will be OFF (mode) by turning off the switch (see Figure 2 lower left and FSM_B).

If the state and the trigger of the state transition are described in the statement in a goal of the goal model, one can easily correlate the requirement to the state transition diagram.

The static and dynamic representation of an item sketch indicates the two facets of the item. For example, the following relationship between the car ahead and the subject car in a static expression (see Figure 1(b)) can be expressed as the following state in FSM_A (see Figure 2). The "switch" in FSM_B will be expressed in the refined static model.

When beginning the concept stage, one must decide the development category, which is the level of abstraction of the item sketch to begin from, that is, if the item is a new one or if it was already analysed (cf. chapter 6 of ISO 26262 part 3). If the item description already exists, and the degree of change is small, it's possible to reuse the item sketch. An analysis is started from the level where the change occurs, with necessary alterations made to the goal model.

The concept stage ends by completing the requirements specification (as for safety, it's the completion of the functional safety requirements specification). At this time, we translate an item sketch into a preliminary system architecture, which will be a basis for abstraction of the hardware and software architecture.

3.3 Refinement of the Goal Model

We use the goal model to consolidate the item function and non-functional requirements, as well as safety requirements. We use the KAOS approach (van Lamsweerde 2009) to describe the goal model. KAOS is the typical goal-oriented requirements analysis method.

In the KAOS approach to refine the goal repeatedly from the top goal mode, one can finally get the required specification. We briefly explain the notation used here. There are several types of node in the goal model: the goal node, the soft goal node, the obstacle node, the solution node and the requirement node. The term goal "is an objective the composite system should meet" (van Lamsweerde, et al. 1998). An object of an item is refined into detailed goals. The goal node represents each goal. The soft goal is a special goal by which the achievement standard of the goal isn't indicated clearly. The obstacle node is the node that disturbs achievement of a goal and indicates the solution to the obstacle using the solution node. The requirement node is the output of the goal model and is equivalent to the leaf of a goal model.

There are two types of refinements. One is the AND-refinement, and another is the OR-refinement. The former (AND-refinement) is the decomposition of an upper goal. The latter (OR-refinement) shows that lower goals are alternatives of the upper goal. Figure 3 shows a part of the goal model of the CACC system.

Next we will show the two ways to combine the models of an item sketch. One is based on refinement, and the other is based on synthesis. The model combination is used to get the in-depth picture of an item sketch; one might find the different candidates hazards in this process.

In the refinement, the upper goal is divided into more than one goal. When seeing a goal model as a diagram, this is equivalent to adding a branch off for the lower parts. How does the item sketch change when the goal changes? Consider the next new sentence:

- (S4) (The subject car can) recognize the forward car by LIDAR, which is one of the active-type devices that measure distance by illuminating a target with a laser.

This is a refinement of S1, and it shows that the item uses LIDAR as an active-type device. This item sketch is shown in Figure 1(d). The basic structure is maintained in the two relationships (that is, (a) and (b) vs. (c) and (d)).

Synthesis is to combine goals that are not directly refinement related. An example in the dynamic representation of an item sketch is shown in Figure 2. One can get several sentences from the goal model (see Figure 3):

- (S5) (The subject car can) follow the identified car.
- (S6) The user can choose the behaviour of the CACC.
- (S7) The switch toggles the ON/OFF mode.

S5 corresponds to the upper left portion of Figure 2 (FSM_A), and S7 corresponds to the lower left portion of Figure 2 (FSM_B). Using these two FSMs, we can synthesize and then get FSM_C. Here, S6 and S7 request us to be able to choose the behaviour in ON mode. So, FSM_A becomes the inner-state of ON state of FSM_B. It will be different from the previous "refinement;" in the case of "synthesis" the meaning is considered, and it is combined.

The obstacle node is one of the nodes used in the goal model. It indicates that it is an obstacle of goal achievement. For example, the goal "high acceleration performance" may have an obstacle node because someone has an unpleasant feeling in the abrupt acceleration.

We use this obstacle node for distinguishing a hazard. Hazards disturb goal achievement. In the next section, we will describe how to find the obstacle node using the guideword.

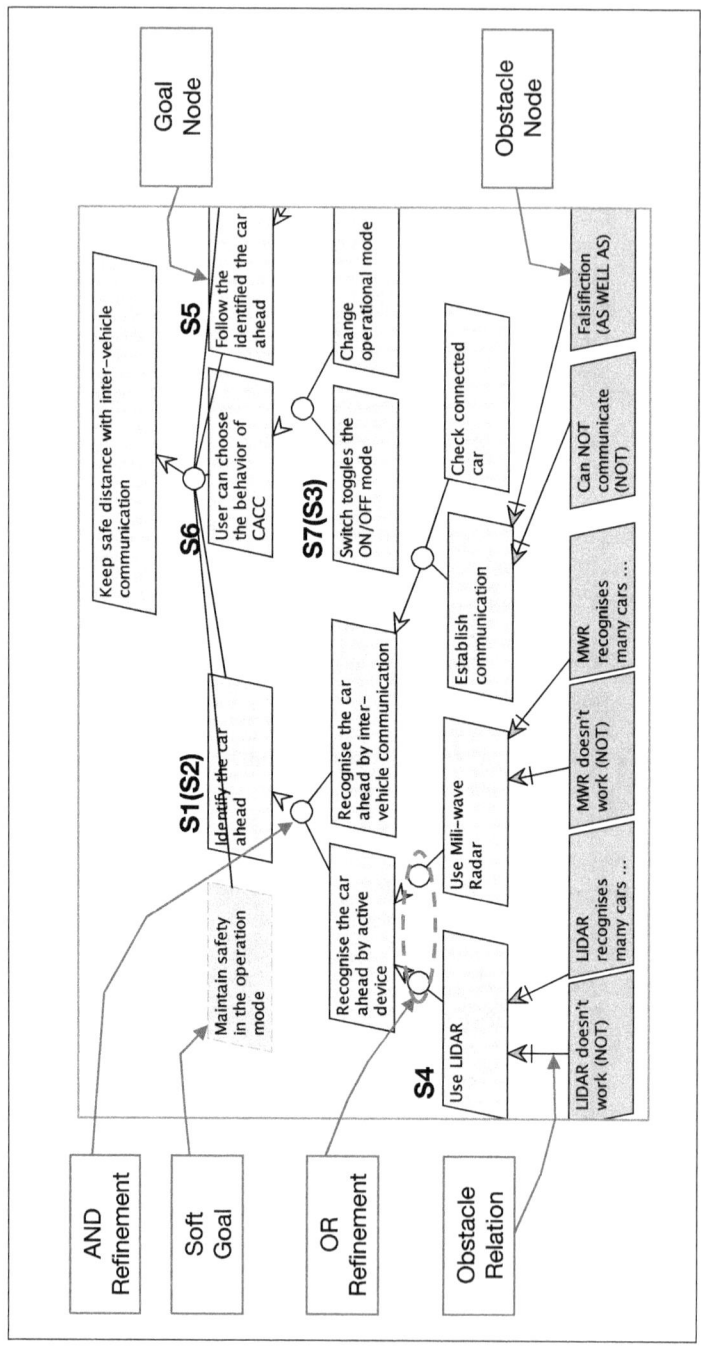

Fig. 3. Goal Model (from sub-goal to obstacle node)

3.4 Applying the Guidewords

We apply the guidewords of HAZOP to the description of each goal. The guidewords are categorized in two types: time relating (see Table 1) and space relating (see Table 2).

Table 1. HAZOP guidewords of the space category

Guideword	Meaning
NO or NOT	Negation
MORE or LESS	Increases or decreases
AS WELL AS	Qualitative increase
PART OF	Qualitative decrease
REVERSE	Opposite
OTHER THAN	Substitution

Table 2. HAZOP guidewords of the time category

Guideword	Meaning
EARLY or LATE	Related to the clock time
BEFORE or AFTER	Relating to order or sequence

With guidewords, one can easily do the what-if analysis systematically (Nolan 2011). This approach doesn't depend on the checklist that is usually used in the early development phase and can find the hazards in more comprehensive way.

Here we show some examples by using the sentence S4 (in the last sentence one applies two guidewords at a time):

- (S4-NOT) (The subject car can) NOT recognize the car ahead by LIDAR.
- (S4-MORE) (The subject car) recognizes the many cars ahead by LIDAR at a time.
- (S4-ASWELLAS) (The subject car) wrongly recognizes the car ahead by LIDAR.
- (S4-LATE) (The subject car can) recognize the car ahead by LIDAR, but notification is too LATE.
- (S4-NOT-LATE) (The subject car can) NOT recognize the car ahead by LIDAR, but the notification is too LATE.

One can get a new sentence (S-*) by applying guidewords to goal description (S). An asterisk means one or more applied guidewords. That is, one can get more than one S-* from one description sentence of a goal.

Using the sentences (S-*) generated by guidewords one can identify the hazard. The following description was already written:

- Item sketch
- The sentence to which a guideword was applied to goal description (S-*)

The main idea is that one can interpret the meaning of the guideword-applied sentence (S-*) on the item sketch. That is, one does the thought experiment on the item sketch by operations by changing the attribute value or model description.

3.5 Hazards

A simple hazard is found by the abnormal behaviour of elements described in the item sketch. It looks like failure mode and effect analysis (FMEA) except for the difficulty defining the failure mode. For example, by S4-NOT: "(The subject car can) NOT recognize the forward car by LIDAR," one can find a hazard candidate from the elements of the item sketch (that is, LIDAR in Figure 1(d)).

The more complicated hazards will be found by manipulating the information on the item sketching (for example, multiplicity). Viewing the car ahead from a subject car, the multiplicity of the car ahead is zero or one (Figure 1(b)). This means the car ahead is one at most if there are several cars approaching, and they run with identical speed. In this case, the system might think there is only one car suggested by the multiplicity, or just is not able to identify the car ahead. This is called, "operation on the item sketch."

This situation can be interpreted in another way. If the system might incorrectly regard two cars as one car, it is relating to another guideword-inducted sentence S4-ASWELLAS (The subject car wrongly recognizes the car ahead by LIDAR). Consider the different multiplicity of a static item sketch. What if the multiplicity of the preceding car is two? What will happen next? One cannot extract hazards immediately, but "operation on the item sketch" and guideword-inducted sentences (S-*) help to find hazards.

A hazard is expressed as an obstacle node in the goal model. We give the next attribute to this obstacle node.

- A situation: relating SSM and the relevant part in it
- Consequences: vehicle level
- Consequences: people level
- The avoidance means: What kind of measures can a driver take to avoid harm after an obstacle is detected?

Finally, in order to create the hazard list we gather the obstacle nodes and arrange them. If we can use the tool (see 4. (6)), the list is automatically generated.

3.6 SSM

In this section, we briefly explain a situation-scenario matrix (SSM) to define an environmental condition. Thinking about a vehicle such as an automobile, it is important to cover the environment in which it could exist. The scenario consists of the combination of elements in the situation, and one can write different scenarios depending on various situations (generally, the element belongs to a category). Each scenario helps identify hazards of the system. In the CACC, the category of the element of situation is like "subject car," "target car," "perimeter (for example, pedestrian)," "road type," "road condition," "regulation," "environment," "radio wave condition," or "driver." For example, the category "subject car" has elements like "speed," "acceleration," "jerk," "engine state," and so on.

When considering control of a built-in system, it's often modelled using the controller and the control targets (plants). But one also must consider the environment, especially when analysing a moving object like the automobile. Usually, the automobile design environment is considered, for example, such as when the CACC system controls a car at constant speed, but there is a slope on a road or friction drag changes by weather condition. The main aim of using an SSM is that one also considers the operational situation in identifying hazards.

3.7 Risk Assessment

In the risk assessment, we give the ASIL value to the hazardous event that is found in the hazard analysis. First, we classify the hazardous event on severity, the probability of exposure or controllability, and give ASIL using the table (Table 4 in the ISO 26262 part 3).

It is impossible that we reflect reality completely on selecting the value in exposure/controllability/severity category. When we are on the highway, and the velocity is 80km/h, we assume that the ACC system broke down. How we exactly know this exposure/controllability/severity value? We have many questions: what does the percentage of a malfunction occur, how long does it take people, the driver and the surrounding people, to avoid the hazard, and to what degree is the driver or pedestrian is injured?

The important thing is to clarify the process and the rationale for selecting the situation, avoidance behaviour of a driver and the extent of damage in an assumed accident. This is similar to the discussion of the safety case.

Exposure probability is the rate of encountering an operational situation of a hazard. The sum of the situation in the SSM is used for this calculation. If there is a hazard in a situation, we calculate the ratio of the situation from the whole of the scenario. The total probability of exposure is calculated by checking all SSMs. Then we select the appropriate class from the Table 2 of ISO 26262 part 3.

The controllability is the ability of people to avoid a specified harm or damage. The person here is not limited to the driver, people in the passenger and other vehicles, including pedestrians.

In the goal model, a hazard candidate is expressed as an obstacle node. The countermeasure, how people can deal with it, is described in the connecting note (Figure 4). The probability of avoidance of harm depends on the situation described in the SSM. When driving at low speed and the distance from the following vehicle is sufficient, avoiding read-end collision is easy. On the other hand, when the distance between the front and rear of the car is short on the highway, it is not easy to avoid a collision. As a chain reaction, the same thing applies to a subsequent driver. To analyze the situation and driver we define the driver model (Ito 2015a). After controllability analysis, we choose the class from Table 3 of ISO 26262 standard.

Severity can be thought of as an extension of controllability analysis. If we do not get enough avoidance means, we suffer harm, that is, we have an accident. The extent of damage caused by the accident is dependent on the situation. As with Controllability, information on the status of such speed and road surface conditions that constitute a situation of hazard events is important. These are also described in the SSM. Of course, we cannot calculate this mechanically. To calculate the damage, it is necessary to have more detailed information. For example, the difference between types of cars or the impact angle between them affects the extent of the damage. By limiting the situation defined in the SSM, we assume a worst-case condition, and we obtain severity from the table 1 of ISO 26262 part3.

We obtain the ASIL to apply each class of exposure/controllability/severity to Table 4 of ISO 26262 part 3.

Then we define the function safety goal for each hazardous event that we assign the ASIL. The first description of the safety goals can be obtained from the negation of the hazard description. For example, in the case of "(S4-LATE) (The subject car can) recognize the car ahead by LIDAR, but notification is too LATE", we get the first safety goal; "When system recognizes the car ahead by LIDAR, it communicates the information without delay."

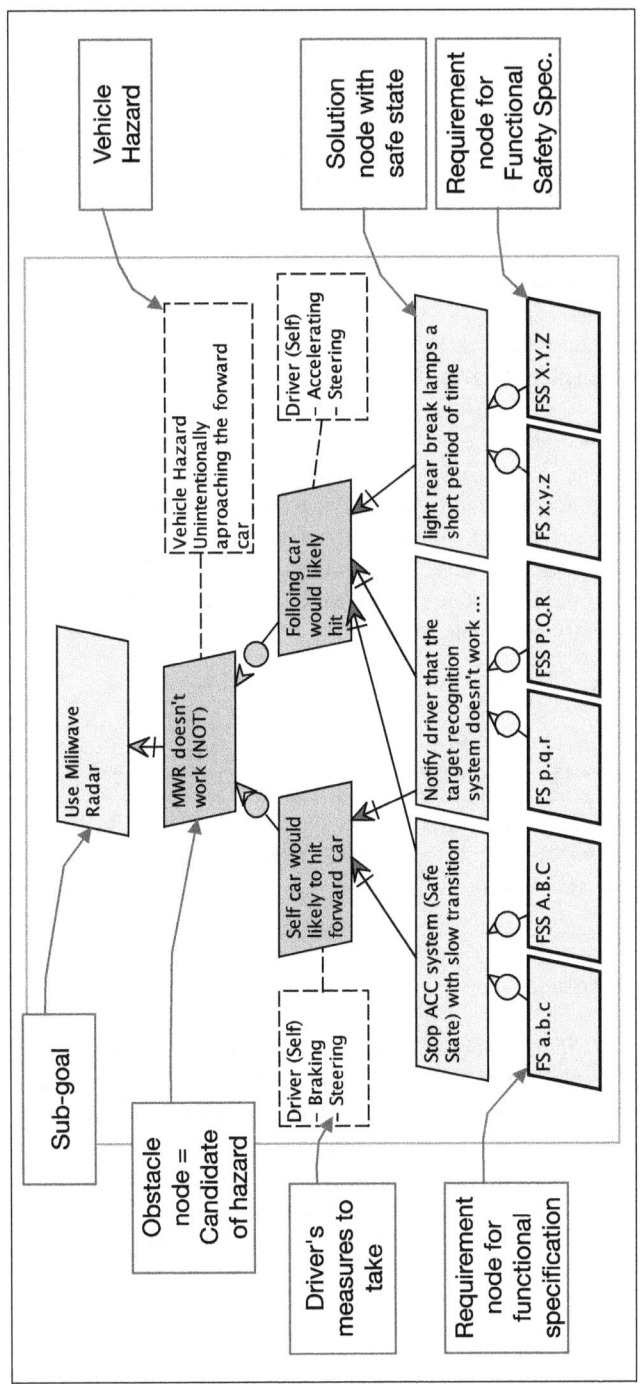

Fig. 4. Goal model (from sub-goal to requirement node)

3.8 Functional safety concept

In this phase, we describe the functional safety specification. A goal defined in the previous subphase and the functional safety specifications described in this subphase are not a one-to-one relationship, so we review the correspondence as needed. The functional safety specification has the contents showing below (3-8.4.2.3):

a) Operating modes
b) Fault tolerant time interval (FTTI)
c) Safe states
d) Emergency operation interval (EOI)
e) Functional redundancies (e.g. fault tolerance)

We have already considered the operating mode and safe states in the previous subphase, and we elaborate them in this subphase. FTTI is the time interval that passes from the occurrence of a fault until the presence of a possible hazard. EOI is the time from when the occurred fault to entering the safe state. Especially in the new item, it is hard to give those two time-intervals. Finally, the functional redundancy is added in the item sketch. Item sketch plus functional redundancy is the base of the preliminary system architecture.

4. Discussion

We already had been describing some ADAS systems using our CARDION approach. We found several issues relevant to ISO 26262. We think that those issues might be typical in the new system like ADAS.

(1) Software component

The framework of functional safety corresponds to the failure of the elements in the system. So, the embedded software is, even as it is large and complex, one element of a system. In other words, its decomposition depends on the hardware, such as CPU and memory partitioning. However, even in hardware boundary, there should be a level that can ensure the independence of the software itself. For example, it is possible to indicate that the component C1 doesn't affect directly the component C2. In particular, it seems reasonable to think at the item level.

(2) Vehicle hazard

The item is the abstraction of the system in the vehicle-level. So, we also have to consider that hazard in the same vehicle-level. However, the standard does not explicitly describe it. In guideline J2980 (SAE 2015), it says that vehicle hazard must be considered after finding the malfunctioning behaviour of a function. For example, "the recognition of forward car is too late" is a hazard of a function and "unintentionally approach the forward car" is a vehicle hazard.

In CARDION, we give the vehicle hazard for each hazard as an attribute. This vehicle hazard is important when we make the risk assessment.

(3) Importance of hazard identification

Software never fails gradually. In other words, there is no deterioration over time. The cause is whether the specification is wrong, or its implementation is incorrect. We think the abnormality of the ACC by the above-mentioned diffuse reflection. This failure probably comes from the error of specification. The software itself is operating based on the specification. The designer might not consider the influence of irregular reflection from the opposite lane. If so, it is not possible to know that it is the false recognition, and also it cannot issue a warning.

(4) Exposure/Controllability/Severity

If we describe the situation in the scenario in detail, classification of severity and controllability becomes easy for us. On the other hand, exposure will be difficult to determine the value (and the number of it is a very small value). Consequently, it is difficult to calculate correctly the risk (i.e., it is difficult to set the appropriate ASIL).

In CARDION, by calculating using a plurality of scenarios (SSMs), it can avoid this type of difficulty. Also, we prepare the scenario set by the culture and characteristics of the destination. However in the selection of the scenario set, there is some arbitrariness. It is important to clarify how we scenario set and calculate risk. It is not realistic to prepare all the real scenarios. The position for the following safety case is also the same.

(5) Safety Case

The standard defines the safety case as follows: "argument that the safety requirements for an item (1.69) are complete and satisfied by evidence" (3-1.106). In the concept phase, functional safety requirements are intended to ensure the safety of an item. The safety case assures it with evidence by indicating that the safety measure keeps us in a safe state. The safety case has an important role, but the standard has no detailed description, and the positioning of safety case in the concept phase is not clear.

In CARDION, we think about safety case in the following manner. After hazard identification and risk allocation, we consider the functional safety concept. In goal model, hazard corresponds to the "obstacle" node. The functional safety concept is corresponding to the "solution" node. We express the final functional safety specifications as "specification" node. After that, we assume that the relationship between the "obstacle" node and "specification" node (via the resolved node) is the claim of the safety case. And we give the reasonable evidence and argument to them.

We describe the safety case for individual hazards rather than describing it for the entire item. Of course, it is possible to describe the safety case covering all the hazards for an item. But the separate description of the safety case is easier to handle.

(6) Tool

In CARDION, we use multiple notations. For the static and dynamic item sketch, we use the notation of UML or CATALYSIS. For goal decomposition of the item, we use the goal model representation of KAOS. For the safety case, we make use of the Goal Structuring Notation (GSN) (Kelly 1999). So, it is desirable to manage the description of these notations uniformly.

We have developed a tool; in the tool, the various descriptions are coupled by a link. For example, it is possible to show the relationship between the functional safety specifications and hazards, and by clicking the link we can know the argument and evidence between them as a safety case.

5. Conclusion

In this paper, we explained our CARDION approach. We can use it in the concept phase of functional safety activity defined by the ISO 26262 standard.

First we define an item with item sketch and refine its goal by the goal model of KAOS approach. The guidewords of HAZOP are helpful to find out the candidate hazards by applying to the description of the (sub-) goal. As it is a top-down approach with the operation of a natural language sentence, we can find the hazards comprehensively. SSM is a scenario that shows the time-series change in the situation. We create multiple scenarios under the assumption of usage of the car and situations. We use this SSM in various activities, for example, in hazard identification or risk assessment. As we use the goal model for refinement, we can create both functional requirements and functional safety requirements simultaneously (see Figure 4.). This is another benefit of our approach.

Moreover, using substantially the same framework we can treat threats for security issues (Ito 2015b), that have become problems in recent years.

References

CEI/IEC (2001) Hazard and operability studies (HAZOP studies) - Application guide, CEI/IEC 61882:2001 IEC

CNN (2015) http://money.cnn.com/2015/09/30/autos/ford-recall-f-150/

D'Souza D, Wills A (1998) Objects, Components, and Frameworks with UML: The Catalysis Approach: Addison-Wesley Professional

Fenelon P, Hebbron B (1994) Applying HAZOP to software engineering models

ISO (2011) ISO 26262 Road vehicles - Functional safety ISO

Ito M (2015a) Controllability in ISO 26262 and driver model in Systems, Software and Services Process Improvement, Springer

Ito M (2015b) Finding Threats with Hazards in the Concept Phase of Product Development. in Systems, Software and Services Process Improvement. Springer

Kelly T (1999) Arguing safety - a systematic approach to managing safety cases. University of York department of computer science-publications

van Lamsweerde A (2009) Requirements engineering: from system goals to UML models to software specifications. John Wiley & Sons Ltd.

van Lamsweerde A, Darimont R, and Letier E (1998) Managing conflicts in goal-driven requirements engineering. Software Engineering, IEEE Transactions on, vol. 24, pp. 908-926

Naus G, Vugts R, Ploeg J, van de Molengraft R, and Steinbuch M (2009) Cooperative adaptive cruise control. IEEE automotive engineering symposium Eindhoven

Nolan D (2011) Safety and security review for the process industries: application of HAZOP, PHA and What-If and SVA reviews, 3rd ed. Oxford: Elsevier

Redmill F, Chudleigh M, Catmur J (1999) System Safety: HAZOP and Software HAZOP: John Wiley & Sons, Inc.

SAE International (2003) Adaptive Cruise Control (ACC) Operating Characteristics and User Interface (J2399)," SAE International

SAE International (2015) J2980: Considerations for ISO 26262 ASIL Hazard Classification. SAE International

Thalen J (2006) ADAS for the Car of the Future http://essay.utwente.nl/58373/ Accessed 1 December 2015

.

Next generation of driver assist systems

Ireri Ibarra, David Ward

HORIBA MIRA Ltd [1]

Nuneaton, UK

Abstract This *paper presents an overview of the challenges faced by engineers and companies in the automotive industry with regards to driver assist systems. Starting with an overview of different assist features made possible by the use of electronics in the last 30 years, the paper explores the most innovative systems that also account for characteristics of the environment surrounding the vehicle, such as objects in the vicinity, road characteristics, traffic signs, etc. and how connectivity is both an enabler and a source of concern for these features. In particular the relationship between safety and cyber security in the context of vehicle systems is discussed. New considerations on both subjects are required given that connectivity inside and outside the vehicle is becoming the norm and will have a great impact on future mobility services. Remotely controlled functions will be used as a case study to present a number of design drivers for these systems.*

1 Introduction

Early driver assist systems consisted of a single or a small number of electronic control units (ECUs) with the corresponding sensors and actuators to deliver the required functionality. For example, the original cruise control required the driver to manually bring the vehicle to a desired speed and press a button to set the cruise

---

[1] Watling Street, Nuneaton, England CV10 0TU. +44 24 7635 5415, ireri.ibarra@horiba-mira.com

control to the current speed. It largely consisted of a single ECU, which controlled throttle demand, originally through a separate actuator. This version is equivalent to SAE Level 1 of automation in Figure 1 *Basic Cruise Control*, as per definitions found in SAE (2014).

Once "by wire" throttle control was introduced, cruise control was integrated as a feature into the engine management system. An upgraded cruise control was introduced some time later, and is known as adaptive cruise control. Adaptive cruise control allows for the host vehicle to keep a set distance from the vehicle in front. Depending on the implementation, the feature may involve the use of radar, lidar or even GPS data; regardless of the actual implementation, what is significant is that the delivery of this functionality requires the interchange of information between several ECUs which are connected through more than a single dedicated network. This version is also equivalent to Level 1 of automation Figure 1 *Adaptive Cruise Control* with the difference that it involves internal communications between ECUs.

In addition to the control of speed, the sensors and systems used for cruise control may also be used to implement additional features, for example providing the driver with a forward collision warning; when both the host and the head vehicle are too close to each other, within the pre-set headway or braking distance.

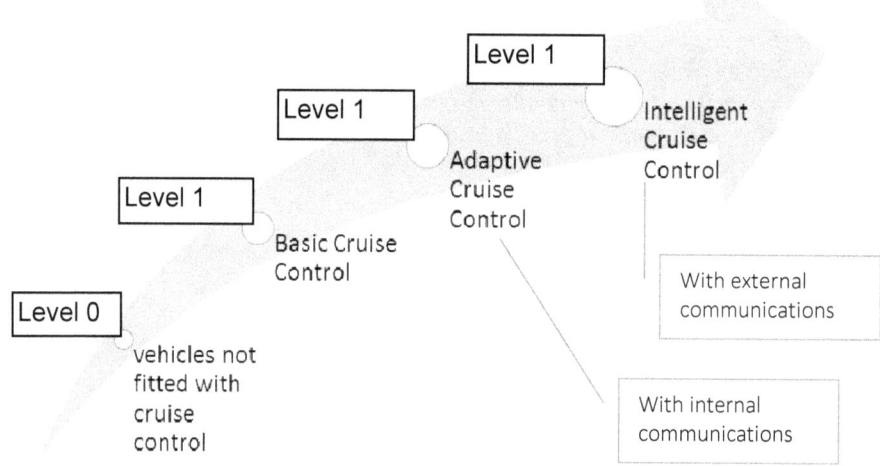

Fig.1. Examples of Cruise Control systems, automation levels and types of communications

1.1 Examples of lateral and longitudinal control of road vehicles

For lateral control, the functionality provided by lane keep assist (LKA) is shown in Figure 2, where the system senses the position of the vehicle in relation to lane markings, if the vehicle deviates from the centre the system initially informs the driver that input torque at the steering wheel is required (1). If the driver fails to provide input torque, the system will apply an overlay toque (2).

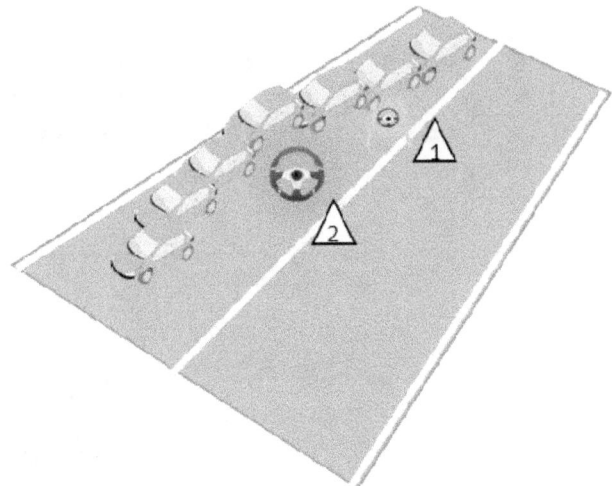

Fig. 2. Lane Keep Assist

For longitudinal control, the functionality provided by Adaptive Cruise Control (ACC) is shown in Figure 3, once adaptive cruise control is set an activated, the host vehicle (orange) is capable to detect vehicles ahead and adapt its speed accordingly. If the vehicle ahead (blue) is travelling at a lower speed than the host vehicle, the host vehicle is able to decrease speed to keep a safe distance. Subsequently if the vehicle ahead moves away or increases speed, the host vehicle can resume travelling at the speed that was initially set.

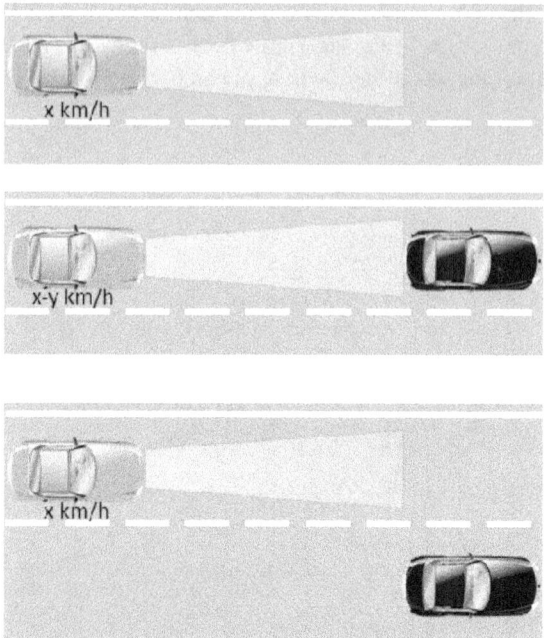

Fig. 3. Adaptive Cruise Control

As a consequence of the introduction of more and more sophisticated features, complex electronic architectures in modern passenger cars now involve over a hundred ECUs and various communications networks suitable for different types of data, which may have different characteristics regarding robustness and availability.

In the near future, the implementation of an adaptive cruise control feature may also involve communication between the host and the vehicle ahead; so that if the vehicle ahead reduces speed dramatically, the expedited transmission of such information may be able to help to reduce the risk of collision even further. Another application involving communication external to the host vehicle, would be adaptive cruise control that also accounts for speed limits on the actual road the vehicle is travelling on; this becomes especially useful if those speed limits are variable. This version, depending on actual capability may also be equivalent to Level 1 of automation in Figure 1 Intelligent Cruise Control[1], but it includes external communications.

Previously the vehicle was deemed a closed environment, however external connectivity brings the possibility of external nefarious actors interfering not only the functionality of the vehicle, but the safety of its occupants and other road users.

---

[1] Note that the terms used to indicate the different versions of a Cruise Control System are representative of their capability and used only as examples, individual vehicle manufacturers may have proprietary commercial names for these.

Vehicle manufacturers have concentrated on improving safety since; from mitigating the consequences of a crash to introducing systems that help with the driving task and remove the variance introduced by inconsistent driver behaviour.

Furthermore significant effort has also gone in ensuring that vehicle system failures do not become hazardous; by managing functional safety risks from concept to implementation and even considering additional aspects such as the manufacturing of a vehicle, its servicing and disposal.

Given that there is a possibility to harm people with the use or the misuse of a vehicle, it is important to explore the relationship between safety and cyber security in the context of vehicle systems. New considerations on both subjects are required given that connectivity inside and outside the vehicle is becoming the norm and will have a great impact on future mobility services.

Remote functions will be used as a case study to present a number of design drivers for these systems.

Note that additionally to longitudinal control, a number of lateral control functions, such as LKA are becoming widely available in some vehicle segments; in order to move up to the next level of automation, Level 2, control of both lateral and longitudinal acceleration is required. Level 3 is possible if the driver can remain out of the loop for a given amount of time, a function to assist with autopilot on a motorway is an example. In this case, consideration as to human behaviour and reaction times take central stage.

2 Human behaviour and road safety

It is believed that human behaviour is the root cause of many road traffic accidents, either:

a. unsafe behaviour where drivers simply do not adhere to road traffic regulations
b. genuinely challenging driving situations where the driver is unable to judge what action is safe to take

It is however, very difficult in practise to understand how human behaviour directly relates to road traffic accidents, there has been a lack of road accident data available that has been acquired consistently, to understand and determine how best to address errors due to human behaviour. In an attempt to do so the IEEE created the first standard for motor vehicle event data recorders (MVEDRs) IEEE 1616 IEEE (2005); this type of system, although relatively new in the road transportation area has been in service since the late 1950's in aircraft and railways.

Although MVDERs still do not provide a full answer and should always be used together with other data sources as reported in DaSilva P. M. (2008), this study nevertheless provides an insight on how the data collected by MVDERs can be used to aid in the development of vehicle safety systems. Some of the data collected and used involved avoidance actions taken by the drivers such as whether they pressed the brakes, steered, accelerated, or took no avoidance action at all.

Another study commissioned by NHTSA, on a collision avoidance NHTSA (2005), the system under trials consisted of an integrated Adaptive Cruise Control (ACC) with Forward Collision Warning (FCW), which was fitted to 11 vehicles to use over a number of months. It is worth mentioning that amongst the data gathered, video of the driving situation and driver face was also included. This study reports the following conclusions:

> "ACC was found to be benign from a traffic safety perspective. Both ACC and FCW reduced the occurrence of short (e.g., <1 sec) headways, with the ACC reductions being substantially more marked and robust across driving conditions. While incidents were found during manual driving in which the FCW may have contributed to a timely driver response to an emerging rear-end crash conflict, the frequency or magnitude of such conflicts were unaffected by FCW presence." NHTSA (2005).

This indirectly supports the approach that driver assistance systems may be able to contribute to the reduction of road traffic accidents.

As a consequence, premium vehicle brands started offering similar systems and now they have permeated to most vehicle segments, from family hatch backs to high end executive saloons.

Another example, is found in the European Commission Policy Orientations EC (2010), which supports the promotion of technologies to increase road safety, where the following systems are included:

Intelligent Speed Adaptation (ISA) a system that compares vehicle speed with local speed limit on a road and modifies the vehicle speed according to such limit.

Advanced Emergency Braking Systems (AEBS) these detect vehicles in front of the host vehicle and can apply the brakes automatically if the vehicle in front is not within a safe distance.

Lane Departure Warning Systems (LDWS) these warn the driver when the vehicles drifts out of the lane it I s travelling in, when it is not performing an overtaking or turning manoeuvre.

Pedestrian Detection Systems combined with **Automatic Emergency Braking (PDS / EBR)** similar behaviour to AEBS but it is based on the detection of smaller objects, such as pedestrians and bicycles.

Blind Spot Detection for Trucks (BSD-T) trucks inherently have blind spots given the geometry of the vehicle, hence these systems detect vehicles side and rear that are not easy to see with the use of mirrors only.

Tyre Pressure Monitoring Systems (TPMS) these monitor pneumatic pressure on tyres, so that tyres do not get overstressed and burst.

These systems have the capability to warn the driver audibly, through a warning lamp shown on the dashboard or sometimes through vibration or tactile means.

All these systems seek to address the variance on driver capability and experience, which is point b) at the start of this section.

There are however systems that can be used to assist enforcement such as Alcohol Interlocks, and Speed Limiters for Light Commercial Vehicles, which aim to address point a) at the start of this section.

3 Connectivity is both an enabler and a source of concern

Connectivity refers to external communications to the vehicle, where GPS data, traffic information, etc. are used as enablers of some driver assist functions.

Currently, most in-vehicle systems have been operating under a controlled access policy, where there is either no available physical connection, or where points of access are available via proprietary solutions and have controls which restrict access, this indirectly has an impact on the likelihood of a successful cyber-attack.

Some of the driver assist systems referred to previously, control lateral and longitudinal driving tasks with various degrees of authority. In some cases these systems have the ability to effectively take control over from the driver.

Technology demonstrators of automated driving have already been trialled by a number of vehicle OEMs, automotive suppliers and research and development establishments, and there is now great public awareness of such technologies.

However these demonstrators, although very capable of performing driving tasks under various environmental conditions, are not ready for series production and the mass market.

One of the main enablers of the highest levels of automation is connectivity, which means that data sources external to the host vehicle are required or could be used to inform the driver, influence the driving task or send automatic alerts in case of a collision. However, unless there is a dedicated process to understand and manage the inherent risk of vehicle connectivity, and technological solutions to prevent and mitigate intentional manipulation of these connectivity points; the main concern is creating safety issues. These safety issues, are not the result of system failure or operational risks, but the result of a cyber security vulnerability, which if exploited successfully has the potential to impact more than a single vehicle on the road. If a vulnerability has been inadvertently designed in, and can be exploited within a short period of time, providing that access to the affected vehicles is also straightforward; the potential for a scenario where multiple vehicles can be compromised simultaneously grows beyond acceptable risk for the general public.

Most current vehicle communications architectures do not support identification or authentication of the data sources used in the control of various vehicle functions; there is an implicit assumption that the data in the network is to be shared and trusted.

4 Relationship between safety and cyber security in the context of vehicle systems

The relationship between safety and cyber security in the context of road vehicles is even more important than in other areas such as power distribution or communications, given the dynamic characteristics of a road vehicle, the fact that the physical actuators that distribute power to the road wheels are controlled by sophisticated software and of course the number of vehicles on the road. These different arguments have been presented by the authors in Ibarra I. and Ward D. D. (2012) a-c, a brief summary and some more points are elaborated in this paper as follows:

4.1 Threat Analysis and Risk Assessment

The original MISRA Guidelines MISRA (1994) drew upon work in the cyber security domain in order to create a process for safety hazard analysis and risk assessment Jesty P.H. and Ward D.D. (2007). This process then was further refined in EVITA Ruddle A. et al. (2009) with the use of attack trees for classifying risks from cyber security threats. Recently HORIBA MIRA has pioneered the use of a proprietary process which is based on the original EVITA approach but is improved in the areas of severity classification and is adapted to be used within a production-intent development lifecycle, such as the one proposed in ISO 26262 ISO (2011).

The HORIBA MIRA FS^2 process is intended to manage both safety and cyber security risks in an integrated approach so that requirements are produced on the basis of analyses and fed into the implementation of the system itself and the solutions used to mitigate risks.

4.2 Development framework for safety related systems in passenger vehicles

Given the inherent constraints of the automotive product lifecycle, where time to market is key to the automotive business model, introducing a new development lifecycle exclusively for cyber security purposes will be impracticable. We believe that ISO 26262 as the industry standard for the development of safety–related vehicle systems provides an appropriate framework to systematically approach risk management of cyber security concerns. A summary of the perceived advantages is found in Figure 4.

ISO 26262   An automotive standard for functional safety

Automotive specific requirements for hazard analysis and risk assessment

Structured systems-led approach to design, including hierarchy of safety requirements

"Goal based" approach to (at least some of) the "methods" for achieving systematic safety integrity

Baseline process used by many companies in the automotive domain, providing commonality

Fig. 4. ISO 26262 advantages

In practise, aligning to the process found in ISO 26262 for safety related systems has been found useful for cyber security risk management because engineers are becoming more and more familiar with the steps and objectives found in the standard, although these are exclusively for functional safety, they can be applied to cyber-security risks with some modifications. Additionally most automotive companies have incorporated or are working towards including the requirements of ISO 26262 into their own product development process; the standard is already considered best practice in terms of identifying and managing risk and these activities also tend to be part of the wider quality management system.

However, ISO 26262 is intended to provide a framework to aid consistency in the supply chain, and not to be used as the sole tool to address either functional safety or cyber security issues.

4.2 Risk management

In our experience, we have found that vehicle functions that involve connectivity, especially those using consumer electronics to access information or provide certain control of the vehicle have to be tackled in such a way that risks from both hazards arising as a result of a malfunction of the electronic systems, and from cyber security attacks are considered in all phases of the product development lifecycle.

Fig. 5. Simplified top level risk management process from ISO 26262

At the start of the safety development lifecycle according to ISO 26262 for a safety related system, the hazards and their risks are assessed and a set of top level safety requirements called safety goals are specified, as shown in Figure 5. The intention is to use these safety goals to derive more detailed requirements that can be incorporated to the rest of the development lifecycle of the product

Fig. 6. Additional scheme for cyber security threats

The authors have been working on a parallel scheme similar to the one for hazards but for cyber security threats instead, as shown in Figure 6. This scheme is presented in Ward D.D. et al (2013) and in practice the following main differences have been noted:

A hazard based approach lends itself naturally to a top-down, system-led process, which relies on overall system behaviour in a given environment; the actual implementation details of the functionality provided by the system are not required at this point of the lifecycle.

In contrast, to produce cyber security goals although, they also are top level requirements, a much finer level of detail on how this functionality is implemented is needed. At least the main interfaces between the vehicle and external sources have to be considered, as cyber security goals are highly dependent on the actual implementation of the functionality of the system.

In order to derive cyber security goals, the authors rely on the use of attack trees, as shown in Figure 7 together with a hierarchical system model that contains enough detail to construct a tree where the top event is a successful attack. One level down from the top event there normally are a number of attack objectives, which indicate how the system vulnerabilities may be exploited. It has been found that cyber security goals fit better at the level of attack objectives as they align to the system architecture, hence the importance of a hierarchical system model:

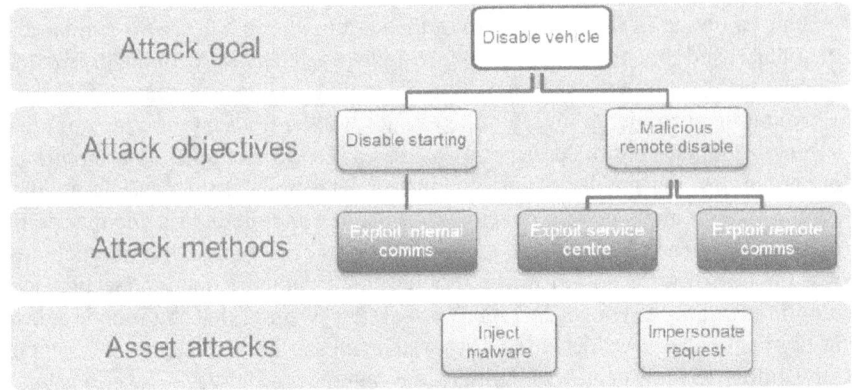

Fig. 7. Example of an attack three

Another point to note is that safety goals are often worded in terms of exposing individuals to harm, in contrast cyber security goals go beyond harm to incorporate other types of loss. EVITA presented severity assessed for different domains (safety, privacy, financial and operational), hence the way we develop cyber security goals is also in these four different domains and the goals are therefore worded in terms of avoiding harm as well as impact on financial loss, reduced operation or privacy. For cyber security goals it is much more reasonable to word then in terms of event or behaviour that must not happen, e.g. avoiding unwanted access or manipulation of the system that could lead to exploiting a vulnerability Firesmith D. (2003); if the vulnerability is found to have an impact on safety, this is fed into the safety analysis. Additionally this helps to ensure all the mitigation measures are consistent, especially those that will not always be implemented solely onto in-vehicle systems, but distributed across cloud services, mobile devices and so on.

As noted in the section about connectivity, data across an in-vehicle network is mostly regarded as trusted and may be used by more than one system; cyber security goals are the top level requirements, however once they are refined, the derived requirements may also benefit from including details as to when data may strictly needed to be trusted and what measures to take if it is not, at least at the level of external interfaces to the vehicle.

Case study: Remote control of vehicle functions

As with in-vehicle systems it is not surprising that those functions that can remotely control the dynamic characteristics of a vehicle have the highest rated cyber security goals related to the safety domain.

In our experience the main aspect of obtaining successful results in terms of cyber security goals, is heavily influenced on how the 'trust' boundary is set. This includes a notion of the architecture of the system and to a degree is similar to the boundary used for safety analysis, but includes the interface with the external data sources and excludes the support services behind those data sources.

A simple example, to highlight key issues will be used, in the shape of a basic vehicle function controlled remotely via a COTS device, such as a mobile telephone controlling the lock/ unlock function; the boundary (orange dotted line) should include the software developed to use the mobile device for this purpose and the connection between the mobile phone and the vehicle but not include the rest of the software hosted in the mobile phone, nor the data connection to the mobile network, as shown in Figure 8, bearing in mind that the mobile network itself should also be subjected to a separate analysis. These boundaries do not usually follow the physical systems that may interact to deliver a function, but the design responsibilities and scope of development; hence a number of trust boundaries will have to be established in order to be able to focus the analysis, whose results will be then used to place requirements on the relevant component systems used to deliver the function.

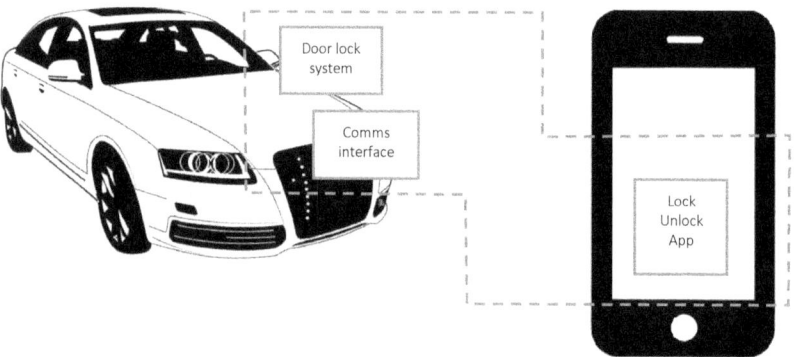

Fig. 8. Remote locking function

Impact on future mobility services

In summary, while driver assist functions are a challenging area to work in, they carry significant promise in terms of contributing to reducing the occurrence of road traffic accidents.

It is fundamental to achieve a high degree of confidence on their safety properties as well as securing them against intentional nefarious attacks, as a first step, without overlooking the fact that functionality updates and changes require to be strictly controlled in order not to introduce emergent behaviour as a result of exploiting a vulnerability.

Connectivity is a key enabler for these technologies but the new vehicle landscape calls for a different approach to manage technology risk.

There are certainly parallel activities and principles that can be drawn from a more mature area such as functional safety, to tackle the challenges posed by addressing cyber security for road vehicles. An integrated approach that systematically applies the basic concepts of risk management, but incorporates both safety and cyber security requirements is already yielding benefits.

For future mobility services where either more assist systems will be incorporated onto road vehicles or high levels of automation may be offered on different types of vehicles, safety and cyber security considerations are fundamental to their adoption.

Acknowledgements

The authors would like to thank HORIBA MIRA and the Functional Safety team at MIRA for their collaboration in this publication.

References

DaSilva P. M. (2008) VRTC Analysis of Event Data Recorder Data for Vehicle Safety Improvement, DOT HS 810 935, Cambridge, MA, US.

EC (2010) Towards a European road safety area: policy orientations on road safety 2011-2020.

Firesmith D. (2003) Engineering Security Requirements, Journal of Object Technology.

Ibarra I. and Ward D. D. (2012)a Applying ISO 26262 and functional safety principles beyond passenger cars, (oral presentation), IZB Technology forum, Wolfsburg, Germany.

Ibarra I. and Ward D. D. (2012)b Beyond cross-cutting concerns for resilient intelligent transportation systems, (oral presentation), ITS World Congress Session - SIS10 - Cybersecurity and the impacts on the Intelligent Transportation Systems, Vienna, Austria.

Ibarra I. and Ward D. D. (2012)c How functional safety contributes to vehicle resilience, Breakout Session (oral presentation), TRB Vehicle automation workshop, California US.

IEEE (2005) Standard for Motor Vehicle Event Data Recorders (MVEDRs), IEEE Std 1616-2004 .

ISO (2011) ISO 26262, Road Vehicles – Functional Safety.

Jesty P.H. and Ward D.D. (2007) Towards a Unified Approach to Safety and Security in Automotive Systems, pg. 21-34, The Safety of Systems Proceedings of the Fifteenth Safety-Critical Systems Symposium, Bristol, UK.

MISRA (1994) Development guidelines for vehicle based software.

NHTSA (2005) Automotive Collision Avoidance System Field Operational Test (ACAS FOT) Final Program Report, DOT HS 809 886

Ruddle A. et al. (2009) EVITA, Deliverable D2.3: Security requirements for automotive onboard networks based on dark-side scenarios.

SAE (2014) J3016 Taxonomy and Definitions for Terms Related to On-Road Motor Vehicle Automated Driving Systems, US.

Ward D.D. et al (2013) Threat analysis and risk assessment in automotive cyber security, SAE World congress, Detroit US.

The challenges facing an autonomous car's risk assessment

M.G. Spencer & N.B. Durston

Osprey Consulting Services Limited

Farnham, UK

Abstract *This paper explores some of the potential challenges facing the intro-duction of autonomous cars, especially in the absence of a clear definition of what an autonomous system is. The automotive industry is currently transitioning to position itself to introduce autonomous cars to the global market. This paper dis-cusses autonomy from the standpoints of the aerospace and automotive industries, focusing on Unmanned Aerial Systems and autonomous cars.*

1 Introduction

In a recent report the United Kingdom Department for Transport (DfT 2015) claim the introduction of autonomous cars onto UK roads will provide significant economic, environmental and social benefits. The (DfT 2015) presents an argu-ment for increased productivity stating that the average English driver spends 235 hours driving annually and it is claimed that this time can be applied in more use-ful ways.

Connected vehicle technologies will enable autonomous cars to communicate with their environment, such as roadside infrastructure and other road vehicles. The (DfT 2015) claim the use of road space will be optimised, therefore reducing congestion and providing consistent journey times. This optimisation is argued to minimise fuel consumption and emissions.

Improvements in mobility are also claimed especially for those unable or disin-clined to drive. The (DfT 2015) states 31%, 14% and 46% of women, men and seventeen to thirty year olds do not hold a full driving licence, respectively. Greater mobility is claimed to intensify social inclusion and enhance quality of life for many.

Most importantly, it is argued that the introduction of autonomous cars and developments in vehicle automation technology will substantially reduce collisions, deaths and injuries. The (DfT 2015) states that over 90% of collisions are attributable to human error. Autonomous vehicle capabilities, such as constant environmental monitoring are claimed to eradicate errors such as driver distraction, inappropriate speed or disobeyed signals or road markings.

Significant public concern exists regarding the responsible use of security and safety of autonomous cars. In order for road users, including drivers, cyclists and pedestrians to accept the notion of vehicles on the road without a human in control, they must have confidence that the autonomous systems in operation are safe and secure. In order to achieve this the automotive industry must prove that there is no inherent danger to the public through a rigorous development process. The aerospace industry has many years of experience in the use of such a process, which has produced results acceptable to the public harmonising the differing requirements of all stakeholders. What exactly is autonomy? This paper will discuss autonomy from the standpoints of the aerospace and automotive industries, focusing on Unmanned Aerial Systems (UAS) and autonomous cars, respectively.

2 Autonomy - "one who gives oneself one's own law"

The feasibility of autonomous aviation was first demonstrated during the 1914 Concours de la Securité en Aéroplane in Paris. (Scheck 2004) states the US aviator Lawrence Sperry and his French mechanic, Emil Cachin entered the competition with a single-engine Curtiss C-2 biplane with a hydroplane fuselage, similar to the one detailed in Fig. 1. This aircraft was equipped with a gyroscopic stabilizer apparatus, designed to improve stability and control. During their first demonstration, Sperry engaged his stabiliser device, held his arms high out of the cockpit, revealing straight and steady flight with the pilot obviously not in control. During their second demonstration, similarly with Sperry's hands not at the controls of the aircraft, Cachin climbed out on the starboard wing and moved about 7 feet away from the fuselage. The aircraft momentarily banked due to the change in centre of gravity, however the gyroscope-equipped stabiliser immediately corrected the attitudinal change. During their third demonstration both the mechanic and pilot left their seats and took to the Curtiss' wings, again revealing straight and steady flight. Unsurprisingly Sperry was awarded first prize in the competition. The succeeding century has witnessed the introduction of many advancements in autonomous technologies following this rudimentary example of autopilot technology.

The UK Civil Aviation Authority (CAA)'s Civil Aviation Policy (CAP) 722 struggles to define autonomy other than using a dictionary definition *"freedom from external control or influence"*. This policy (CAA 2015) state the necessity to demonstrate compliance with safety requirements, defined in the various Certification Specifications means that currently, all UAS are required to perform deterministically. Specifically, their response to any set of inputs must be the result of a pre-designed data evaluation output activation process.

Fig. 1. A Curtiss C-2 biplane with a hydroplane fuselage

As a result, the (CAA 2015) argues no UAS related systems currently meet the definition of autonomous; however (CAA 2015) claims UAS can be categorised either as highly automated or high authority automated systems. Highly automated systems are those that still require inputs from a human operator, for example, confirmation of a proposed action but which can implement the action without further human interaction once the initial input has been provided. Conversely, high authority automated systems are those which can evaluate data, select a course of action and implement that action without the need for human input. High authority automated systems have the capability to take actions and respond through evaluation of a given dataset that represents the current situation including the status of all the relevant systems, geographical data and environmental data. Such systems are required to be deterministic and hence apply consistency in that the system must always respond in the same way to the same set of data. They are usually composed of a number of sub-systems which gather and evaluate data, then respond by selecting an appropriate set of actions and issue commands to related control systems. The (CAA 2015) states that an UAS can have authority over two types of function: general control system functions and navigational commands.

The concept of a high authority automated system demands a range of system authority, ranging from full authority where the systems are capable of operating without human control or oversight to lesser levels of authority where the system is dependent upon some degree of human input. It is within this concept that operational tasks performed by a human operator are delegated to the system.

Autonomous behaviour was identified at an early stage in the Autonomous Systems Technology Related Airborne Evaluation and Assessment (ASTREA) II project (ASTREA II project 2015) as one of the critical technologies that will make civil UAS operations viable. This identification was made without a clear definition of autonomous behaviour pertaining to UAS operations. Autonomy is defined by the (Oxford University Press 2011) as *"the right or condition of self-government"*. The Automatic vs. Autonomous debate has generated many definitions, as emphasised by the national aviation regulator's uncertain provision of a clear definition of autonomy pertaining to unmanned flight. In absence of a clear definition the (ASTREA II project 2015) proposed and focused on specific capabilities and functionality, giving full consideration to equivalence to manned aircraft.

Autonomy with regards to UAS operations is currently not concerned with "unthinking" drones, which demonstrate unpredictable behaviours, and which are fully autonomous. Importantly a human remains in the loop. The (ASTREA II project 2015) states autonomy is a human centric process. This is demonstrated through shared decision-making between operator and the system based on appropriateness, time-criticality and mutual situational awareness. This process allows the operator to perform a supervisory "command" role rather than a hands-on stick-and-throttle "control" role, permitting a greater capacity to plan contingency in managing uncertainty and unplanned external events. This process also enables efficiency gains during mission execution, however ultimate authority always resides with the human operator rather than the system.

Why is autonomy important for unmanned systems? Autonomous systems can replicate and augment the system monitoring and contingency management functions that would otherwise be performed by the human operator. Such systems can supplement overall system situational awareness and permit continued safe operation in the event of system degradation.

Autonomous systems can be more efficient and effective than alternative manned solutions. With regards to transport systems, autonomy can facilitate optimal mission performance with minimum operator input, utilising resources across a network supporting multiple vehicles and sensors. The (ASTREA II project 2015) claims improvements in reliability, maintainability, supportability and sustainability will be realised through autonomous capabilities, such as prognostics and health management, which will provide autonomic logistic support.

The (ASTREA II project 2015) was initiated to develop the technology readiness levels for a range of systems to enable autonomous operation of UAS. The project is comprised of two main sub projects: Separation Assurance and Control and Autonomy and Decision Making.

The Pilot Authority and Control of Tasks (PACT) framework aims to define clear operational relationships between the pilot and the Unmanned Aerial Vehicle (UAV) (the aircraft component of the UAS) utilising military technology. The PACT protocol was designed by Defence Science and Technology Laboratory (DSTL) as part of a proof-of-concept demonstration. This variable autonomy framework, detailed in Table 1, allows different authority to be allocated to specific tasks and responses depending on appropriateness and criticality.

Table 1. Modified PACT Levels of Automation

PACT Level	Computer autonomy	PACT locus of authority	Levels of Human Machine Interface
5b	Computer monitored by pilot	Full	Computer does everything autonomously
5a			Computer chooses action, performs it and informs human
4b	Computer backed up by pilot	Action unless revoked	Computer chooses action and performs it unless human disapproves
4a			Computer chooses action and performs it if human approves
3	Pilot backed up by computer	Advice, & if authorised, action	Computer suggests options and proposes one of them
2	Pilot assisted by computer	Advice	Computer suggests options to human
1	Pilot assisted by computer only when required	Advice only if requested	Computer suggest options and human selects
0	Pilot	None	Whole task done by human except for actual operation

(Taylor et al 2001) state the PACT protocols are based on contractual autonomy. The contract defines the nature of the operational relationship between the pilot and the computer aid during cooperative performance of functions and tasks. The pilot retains authority and executive control, while delegating responsibility for the performance of the tasks to the computer. Table 2 details the PACT contractual levels of authority.

Table 2. PACT Levels of Authority

Automation PACT	UAS Authority	Supervisor Authority
Commanded	None	Full Supervisor Authority
At Call	Advice, only if requested	Request advice
Advisory	Advice	Accept advice
In Support	Advice and if authorised, action	Accept advice and authorise action

Automation PACT	UAS Authority	Supervisor Authority
Direct Support	Action unless revoked	Revoke action
Autonomous	Full UAS authority	Interrupt

In comparison, the automotive industry standard (SAE International's J3016 2014) describes six levels of driving automation ranging from no automation to full automation, which indicate minimum rather than maximum capabilities. A particular vehicle could operate at different levels detailed in Table 3, depending on the driving automation features in operation.

Terms such as dynamic driving task, driving mode and request to intervene are used within the narrative definitions of these autonomous driving levels.

The dynamic driving task includes the operational (physical inputs of control) and tactical (response to driving environment) aspects of the driving task but not the strategic aspect (determining destination, waypoint and route). Driving mode is a type of driving scenario with characteristic dynamic driving task requirements, for example high speed motorway cruising. The request to intervene is notification to a human operator, by the autonomous driving system, prompting initialisation or recommencement of the dynamic driving task.

These definitions are comparable to the aviation domain. For example, the dynamic flying task includes the operational (physical inputs of control) and tactical (response to flying environment) aspects of the flying task but not the strategic aspect (determining destination, waypoint and route). Phase of flight is a type of flying scenario with characteristic dynamic flying task requirements, for example initial climb. The request to intervene is notification to a human operator, by the autonomous flight system, prompting initialisation or recommencement of the dynamic flying task.

Comparable contractual autonomies, including dynamic tasks, scenarios, requests to intervene and levels of authority, have fashioned common technological themes within the design and operation of UAS and autonomous cars. These common technological themes include: sense and avoid; communications security and spectrum; autonomy, decision making and contingency management; and operations and human systems interaction. A major challenge for autonomous system designers and manufacturers is how do they capture and fulfil safety requirements and demonstrate transparency and equivalence to existing manned operations, which possess common technical characteristics, given the differing environments in which they operate.

Table 3. SAE International J3016 Levels of Driving Automation

SAE J0316 Level	Name	Narrative Definition
0	None	The full-time performance by the human driver of all aspects of the dynamic driving task, even when enhanced by warning or intervention systems
1	Driver assistance	The driving mode-specific execution by a driver assistance system of either steering or acceleration/ deceleration using information about the driving environment and with the expectation that the human driver perform all remaining aspects of the dynamic driving task
2	Partial automation	The driving mode-specific execution by one or more driver assistance systems of both steering or acceleration/ deceleration using information about the driving environment and with the expectation that the human driver perform all remaining aspects of the dynamic driving task
3	Conditional automation	The driving mode-specific performance by an automated driving system of all aspects of the dynamic driving task with the expectation that the human driver will respond to a request to intervene
4	High automation	The driving mode-specific performance by an automated driving system of all aspects of the dynamic driving task, even if a human driver does not respond appropriately to a request to intervene
5	Full automation	The full-time performance by an automated driving system of all aspects of the dynamic driving task under all roadway and environmental conditions that can be managed by a human driver

3. Resilience to dynamic air and ground environments

A successful reduction in road traffic accidents and the associated human error rate can be realised if dependable and resilient systems are developed for autonomous cars. (Yeomans 2014) states that 1.3 million fatalities and 50 million injuries result globally and annually from the 93% of road traffic accidents caused by human error. Almost all of these accidents resulted in a collision with another vehicle, object or person. Such successful reduction is dependent on the amount of system authority given to specific tasks and responses to real time environmental situations. Appropriate and critical autonomous system analysis and response is undoubtedly required for any system replicating the human functions of driving a car in such a complex ground environment.

Conversely, aircraft operate in an environment comprised of airspace which is heavily controlled and regulated and are supported by a multitude of systems including: Traffic Collision and Avoidance Systems, Primary Surveillance Radars, Secondary Surveillance Radars and various transponder systems. However vehicles operating on roads generally do not have such external control and support systems. Such road vehicles are greatly reliant on the 'see and avoid' principle undertaken by the driver with no supporting external systems. As witnessed by most drivers, the 'see and avoid' principle is not infallible, increasingly so when other factors are taken into account, including inclement weather conditions and the poor maintenance of, or use of, other vehicle's lights and distractions by modern technology.

The aerospace industry designs and develops products within a strict regulatory framework that ensures interoperability. One such example is spectrum control for radios and data links. Safety-critical aerospace flight control systems are developed around the flight envelope of the aircraft and the rules of the air. It is argued that whilst aircraft, including UAVs, are significantly more complex than cars, they operate in a far simpler environment.

Initially, autonomous cars will be required to operate in a mixed environment. For financial reasons alone, it is an extremely unrealistic supposition that every vehicle currently on the road will be immediately replaced by an autonomous one during the introduction of such vehicles. The current cost of such automotive autonomous technologies presents an affordability challenge to the majority of the driving population. (Edwards 2015) does however forecast incremental decreases in cost following the adoption of such technologies.

Fundamentally, autonomous cars will need to understand the rules of the road, for example, the Highway Code (DfT 2015) and the legal framework, the Road Traffic Act 1988 (Her Majesty's Stationery Office 2015), understanding signage, road conditions, weather conditions and interaction with vehicles both driver operated and autonomous. A strategy for the deployment of autonomous cars needs to be developed in accordance with the levels of autonomy detailed in (SAE International's J3016 2014). One scenario is that autonomous cars may need to be sufficiently sophisticated that they can operate totally independently and require no reliance or support from roadside infrastructure. Such autonomous cars will require sophisticated decision making capability. Alternatively, autonomous cars may require a complete support network from roadside infrastructure deploying a vast array of sensors and situational awareness systems. However it is argued that without a combination of both philosophies, the development of autonomous cars will experience delays and stifle.

The complexity of an autonomous car's operating environment should not be underestimated. (Haddon 1968) famously described road transport as "an ill designed 'man-machine' system needing comprehensive systematic treatment". The actions required and appropriate response made by a driver on a typical journey need to be replicated through a system of systems, utilising multiple technologies including sensors, software and decision making algorithms and artificial intelligence. The number of actions and responses that will need to be replicated by an autonomous car are vast!

One such development which recently entered global operation is the Tesla Motors Autopilot detailed in Fig. 2. (Tesla 2015) states that their Model S is equipped with hardware, such as: a forward radar, a forward-looking camera, 12 long-range ultrasonic sensors (positioned to sense 16 feet around the car in every direction at all speeds), and a high-precision digitally-controlled electric assist braking system to facilitate the incremental introduction of self-driving technology.

Fig. 2. Tesla Motors Autopilot

The Tesla Version 7.0 software release was designed to work in conjunction with the automated driving capabilities and has enabled real-time data feedback from four different feedback modules: camera, radar, ultrasonics, and global positioning system across the global Tesla fleet. The Tesla Motors Autopilot permits a Model S to steer within a lane, change lanes by activation of the turn signal, and manage speed by using active, traffic-aware cruise control. Digital control of motors, brakes, and steering supports collision avoidance from the vehicle's front and sides, as well as preventing the car from deviating from the road. (Tesla 2015) accepts that the introduction of fully autonomous cars is still a number of years from now, however they argue that Autopilot increases driver confidence, increases Tesla driver's safety on the road, and makes motorway driving more enjoyable. (Tesla 2015) importantly states the driver is still responsible for, and ultimately in control of, the car, however the human machine interface provides intuitive access to the information the car is using to inform its driver of its actions. This vehicle would be described as level 2 – Partial Automation by the criteria detailed in SAE International's J3016. The driver still possesses full authority over the vehicle.

Autonomous systems need to be able to monitor the environment they operate in and conduct dynamic risk assessments. With regards to the aviation domain, the level of authority over navigational commands may differ during the mission. This is dependent upon any safety of flight risks to the UAV and the time available for the human operator to intervene effectively. If the UAV is flying in clear airspace without immediately conflicting with obstructions, terrain, or another aircraft the system may be designed such that any flight instructions, for example amendments to the flight plan, are instigated by a human operator. However, if the UAV is faced with an immediate safety of flight risk, such as, a route conflict with an obstruction, terrain or another aircraft in combination with insufficient time for intervention by a human operator, the UAV will need to be able to mitigate that risk. Such mitigations may include functionality provided through the integration and use of full authority automated systems. As a result, such systems are highly dependent on two types of safety-critical data when determining responses to hazardous situations. Firstly, data sourced from on board sensors with data fusion providing an overall real time mission and environmental representation and secondly, high integrity data sets. High integrity data sets including but not limited to UAV performance data, Digital Vertical Obstruction File, Digital Terrain Elevation Data and Notices To Airmen are all controlled through stringent configuration management processes which manage their publication, distribution and update for use.

A key factor in mitigating hazards presented to vehicles when operating in any domain is the immediacy of response afforded to the human operator. Automotive autonomous system hazard identification and management processes must be able to distinguish between various objects which could present a hazard to the vehicle in an extremely short timeframe. For example determining differences between an object such as a parked vehicle conflicting with an autonomous car's route that could present a credible hazard as opposed to something benign as a plastic bag being blown by the wind and floating across the road in front of the autonomous car. Both objects could collide with the autonomous car, however the severity of outcome resulting from the impact is very different in reality. This poses the question, how will such systems be sufficiently sophisticated to differentiate between hazards and respond to them in a given timeframe, without degrading the safety performance benchmarked by manned vehicles?

(Bonnefon et al 2015) introduce a morbidly interesting concept. They state some accidents involving autonomous vehicles will be inevitable, since some situations will require an autonomous car to make a decision which could result in running over a pedestrian on the road or a passer-by on the side; or alternatively choosing whether to run over a group of pedestrians or to sacrifice the occupants in the autonomous car by driving into an obstacle following a decision to take avoiding action. (Bonnefon et al 2015) define three traffic situations involving imminent unavoidable harm. Firstly, the car can stay on course and kill several pedestrians, or swerve and kill one passer-by. Secondly, the car can stay on course and kill one pedestrian, or swerve and kill its occupants. Thirdly, the car can stay on course and kill several pedestrians, or swerve and kill its occupants. (Bonnefon et al 2015) identify an arduous challenge to define the algorithms that are responsible for guiding autonomous systems when confronted with such moral dilemmas. In particular, these algorithms will need to accomplish three potentially incompatible objectives: being consistent, not causing public outrage, and not discouraging buyers of such vehicles. How will such systems make such decisions? This example emphasises the necessity for new (and revisions to) extant legal and regulatory frameworks to address such concerns. However this issue, among others, will present substantial challenges before the introduction of fully autonomous cars onto UK roads.

Depending on the sensors deployed, the differences between day and night may be eliminated, however other weather conditions such as rain, snow, ice and fog will require careful consideration and not just from the subject vehicle but other road users. It is also important to understand that the road infrastructure is dynamic with changes to road layout, often resulting from the planned building or improvement of roads, such as, temporary layouts and road works. Nature also contributes to the dynamism of road infrastructure. The growth of trees and hedges can obscure signage or road markings. General weathering can degrade and eventually remove road markings. International and regional differences of road signage are other factors that will need to be taken into consideration by the software and hardware sub-systems employed in the design of high authority automated systems.

All of these factors will, as a minimum, require the setting of system and software assurance levels, as well as rigorous testing, commensurate to the risk presented. The demonstration of compliance with safety requirements, confirming transparency and equivalence to existing manned systems will undoubtedly increase public confidence that the autonomous systems are safe in operation.

4. Conclusion

This paper has explored some of the potential challenges facing the introduction of autonomous cars with reference to UAS The absence of a clear definition of what autonomy constitutes has presented a major challenge to the introduction of such technologies. In response the aerospace industry, largely driven by initiatives, such as PACT, have developed frameworks that have largely been adopted by industries involved with autonomy as a whole, modified and revised appropriately for suitability of use.

Comparable contractual autonomies, including dynamic tasks, scenarios, requests to intervene and levels of authority, have fashioned common technological themes, which have validated comparisons within the design and operation of UAS and autonomous cars.

A successful reduction in road traffic accidents from the associated human error rate can be realised following the development of dependable and resilient systems which are integrated into autonomous cars. These systems need to be able to monitor the environment in which they operate and conduct dynamic risk assessments. However, whilst aircraft, including UAVs, are significantly more complex than cars, they operate in a far more controlled and simpler environment. Successful reductions in accidents and human error are dependent on the amount of system authority given to specific tasks and responses to real time environmental situations. Appropriate and critical autonomous system analysis and response will be required for any system replicating the vast human functions concerned with driving a car in such a complex ground environment.

A key factor in mitigating hazards presented to vehicles when operating in any domain is the immediacy of response afforded to the human operator. Autonomous cars will typically have less time to respond than UAS. However when things go wrong the severity of an accident in the air is far more severe with regards to speeds travelled and the energies involved in collision.

Another challenge facing autonomous systems is how will they make decisions during situations involving imminent unavoidable harm? The necessity for algorithms responsible for ethical decisions accentuates the requirement for new, and revisions to, extant legal and regulatory frameworks.

The ultimate goal of increasing public confidence that the autonomous systems in operation are safe can only be achieved through demonstration of compliance to safety requirements, confirming transparency and equivalence to existing manned systems.

References

Autonomous Systems Technology Related Airborne Evaluation & Assessment. "Current Projects", ASTREA, http://astraea.aero/current-projects-2, (2015)

J F Bonnefon, A Shari, and I Rahwan "Autonomous Vehicles Need Experimental Ethics: Are We Ready for Utilitarian Cars?" (2015).

Civil Aviation Authority CAP 722 "Unmanned Aircraft System Operations in UK Airspace – Guidance" 6th Edition (2015).

Department for Transport. "The Official Highway Code new edition" (2015).

Department for Transport. "The Pathway to Driverless Cars: A detailed review of regulations for automated vehicle technologies", pp. 15-17, ISBN 978-1-84864-152-5, (2015).

T Edwards. "Connected and automated vehicles: Concepts of V2X communications and cooperative driving", Autonomous Passenger Vehicles Essential Strategies in bringing driverless cars to the market, The Institution of Engineering and Technology, (2015).

W Haddon. "The Changing Approach to the Epidemiology, Prevention and Amelioration of Trauma: The Transition to Approaches Etiologically Rather than Descriptively Based", Am J Public Health, vol 58, pp. 143--1438 (1968).

Her Majesty's Stationery Office. "Road Traffic Act 1988". (2015).

Oxford University Press. Concise Oxford English Dictionary: Main Edition Twelfth Edition (2011).

W Scheck "Lawrence Sperry: Autopilot Inventor and Aviation Innovator" Aviation History November 2004 (2004).

Society of Automotive Engineers (SAE) International Standard J0316 "Taxonomy and Definitions for Terms Related to On-Road Motor Vehicle Automated Driving Systems", pp. 15-17, (2014).

Taylor, R. M., Abdi, S., Dru-Drury, R., and Bonner, M. C., 'Engineering psychology and cognitive ergonomics Vol. 5, Aerospace and transportation systems', Ashgate, Aldershot UK, Ch. 10, pp. 81-88, (2001).

Tesla Motors Team Blog. "Your Autopilot has arrived" http://www.teslamotors.com/en_GB/blog/your-autopilot-has-arrived (2015).

G. Yeomans. "Autonomous Vehicles Handling Over Control: Opportunities and Risks for Insurance", Lloyd's, pp. 4-23, (2014).

Modelling the Data Safety Guidance

Dave Banham[1]

Rolls-Royce PLC

Derby, UK

Abstract *An ontological model can be used to define a rich semantic landscape of terminology for a technical domain. The advantages of doing so are a high degree of self-consistency between the domain's technical language terms, which increases the quality of their natural language usage. This paper explores an initial attempt to construct an ontological model of data safety terminology for the Data Safety Initiative project. In doing so it introduces the systematic approach ontological modelling brings to not only defining but also understanding a domain's terminology.*

1. Introduction

There is growing appreciation of the contribution of data in the causation of accidents that have resulted in harm to people, harm to the environment, the loss of assets and production capabilities, and the loss of reputation that in turn may result in a loss of future business [1], [2].

The ways in which data can be involved in such accidents are many and varied. There are the straightforward cases of incorrect data, such as a corrupted process parameter, and there are the more complex cases of misunderstanding correct data. This may be due to a confusion of the data's units of measurement, as happened to the Mars Climate Orbiter [3]. Worse still, the way in which the data should be used to achieve system safety can fail, for example the digital display of hydrographic mapping data where the display scale controls whether or not small under water features are shown. This becomes hazardous when zoomed out for navigational purposes and then inadvertently charting a course over shallow water features that are invisible on the digital display due to the displayed chart scale. [4]

---

[1] Software Centre of Excellence

On top of this is the growing volume of data that a system may have to work with. Aligned to this, are the problems of data verification and system verification with such large data sets. Data complexity comes to a head when a multiplicity of data sets are required, as is the case with enterprise class systems used by industry and public service agencies such as health care and the police. A simple case in point is the premise that an individual's database record can be accessed correctly with a unique identifier, such as a social security number. Yet there have been many cases of mismatched records being accessed because either the wrong identifier value was used or the identifier was not a unique identifier in the data set [5], [6], [7]. In both cases, a basic property for the system's correct operation is invalid and incorrect behaviour results.

The normal approach to dealing with the safety of complex systems is to simplify, separate the concerns, use appropriate standards, methods, and techniques, and then introduce mitigations commensurate with the risk for dealing with failures of the safety system. A common architectural device is to use safety monitoring where one (simpler and highly assured) system, monitors the behaviour or outputs of a larger, more complex system.

How does this work for data? Can a safety system keep the primary system safe when the primary system is using complex data sets or there is a common, data-related factor, between the two systems? In some cases it can and in others it cannot. For example, the risk that medical records incorrectly detail a person's blood group can be mitigated by a simple and fast blood test to establish the truth before administering a blood transfusion. For other medical factors such as allergies, the only immediate information is the person's medical records, and this can be matter of life or death when a doctor is faced with a patient that is in septic shock and the designated medical intervention is a large dose of intravenous antibiotics. The patient could be allergic to the administered antibiotic and may die from the resulting anaphylactic shock.

Dealing with these data problems requires that we know the purpose, nature, and limitations of the data in a system and can ensure that it is demonstrably used correctly and appropriately to achieve system safety. Yet data has a *life* outside of the safety related system's operational context. There are other systems that create, transport, store, and ultimately destroy data. All of these systems, along with the way the safety related systems uses data, are part of the data lifecycle and all need to successfully collaborate to achieve a safe operational system.

We create safety assurance cases to argue for the safe operation of the system. We use data to describe data. How do we know that data utilised in the safety case, as created, is the same data that was reviewed and accepted? How do we know that the references to evidence in the safety case are not corrupted? How do we know that the data in the supporting evidence is valid and unchanging? Fundamentally, what is the data integrity of the safety case itself? Again, either implicit assumptions creep in to the framework of the safety case or we have an explicit argument as to why such assumptions are valid and can be upheld. In that sense, we have a recursive process for reasoning about the validity of data.

Even the very notion of what "validity of data" means should be questioned; it is certainly possible to have correct data, but for the system to make improper use of it (as in the story of the digital navigation displays of hydrographic data), or for the data to be used without regard to its actual meta-data (as illustrated by the Mars Climate Orbiter story). This raises the question of what are the properties of data that can be used to describe its safe use?

In response to these complex issues, Mike Parsons established the Data Safety Initiative Working Group (DSIWG) in early 2013, supported by the UK's Safety Critical Systems Club.[1] The DSIWG comprises a number of volunteer safety specialists that have produced initial guidance for dealing with "data safety". Two successive revisions of this guidance were published in early 2014 and 2015, respectively [8], [9], along with two associated reports on the group's progress [10], [11].

The DSIWG realised early on that it had a significant validation challenge for the guidance that would need to be addressed if the guidance were to be useful to safety practitioners as well as system architects. Some of these validation concerns are:

- Completeness of subject matter treatment, irrespective of the end application domain;
- Internal consistency across the breadth of the guidance;
- Soundness of the approach;
- Ease of application across multiple domains with their own sector system safety standards;

The DSIWG has a number of strategies for tackling these concerns, the most straightforward of which is the widespread systematic review of the guidance. The real validation comes from showing the practical benefits of the application of the guidance in retrospective case studies and then in pilot studies. However, the challenge that these approaches have is that they have a tendency to be selective in their review or application of the guidance due to domain knowledge bias. This in turn can lead to change requests that create a push and pull on the specifics within the guidance as one domain proposes clarifications in favour of its own position, when the guidance is intended to be domain, sector, and safety standard neutral.

The validation method discussed by this paper is that of modelling and specifically ontological modelling.[2] The benefit of such an approach is that it will establish, at least on a semi-formal basis, the conceptual language for reasoning about data in a system safety context. In doing so, the model will both substantiate the language and provide a demonstration of its completeness and soundness. The ontology will also lend itself to a more systematic review than a review of the written guidance will achieve. Moreover, the use of the language defined by the

---

[1] http://scsc.org.uk/

[2] Ontology: *Noun* A structure of concepts or entities within a domain, organized by relationships; a system model. (https://en.wiktionary.org/wiki/ontology accessed 17 August 2015)

ontology in the data safety guidance document will be consistent with its intended meaning.

2. Ontology Modelling

The purpose of an ontology model for terminology is to both define the meaning of language terms and the relationships between these terms for use with a specific subject domain.

Three types of semantic relationship are used. They are:

1. Classification
2. Composition
3. Relation

Ontological classification relationships are similar to those found in taxonomies. Whereas, a system of compositional relationships are similar to those described by meronomies.

A taxonomy[1] defines a single classification system. For example, a car (automobile) *is a* type of road vehicle. This implies that for the set of things that are classified as "road vehicle" there is a subset of those things that are classified as "car (automobile)". The set of classifications that exist at a specific level in a taxonomy are *disjoint* from one another, which means that a thing can only be a member of exactly one of the sub-sets; the sub-sets are non-overlapping. For example, if it is also the case that truck *is a* road vehicle, then a road vehicle cannot be both a car and a truck; it can be classified as a car or as a truck. A thing may be classified by two or more taxonomies in the ontology, in which case their classifiers are overlapped. For example, a person can be classified by their gender (male or female) and by their role in an organisation (big boss, boss, minion, safety engineer, etc.). A person simultaneously belongs to the set of gender classifiers and to the set of organisational role classifiers. It is also useful to know whether a set of classifiers forms a complete or incomplete set for the container classification. In the example of organisation roles the use of the "etc." implies that the set of classifiers in the list is incomplete and that others are possible.

A meronomy[2] defines a single compositional system of whole-part relationships. For example, a car *has a* wheel and a wheel is *part of* a car.

An ontology contains multiple taxonomic and merological systems that are interlinked with other cross-referencing relationships to create a comprehensive set of semantic relationships between its terms.

---

[1] Taxonomy: *Noun* The science or the technique used to make a classification. (https://en.wiktionary.org/wiki/taxonomy accessed 17 August 2015)

[2] Meronomy: *Noun* A hierarchy that deals with part–whole relationships rather than the discrete sets of a taxonomy. (https://en.wiktionary.org/wiki/meronomy accessed 17 August 2015)

A significant challenge when constructing an ontology is the establishment of the semantic relationships between the terms in the model. Relationships that exist between terms have to be captured by the model, and in doing so those relationships become asserted by the model. The model can deal with optionality of relationship as well as a multiplicity (i.e. the plurality of the term), but it cannot deal so well with capturing the idea of a temporal or transitory relationships. A technique in ontology modelling is the use of abstraction as a means of separating out more general concerns from more specific concerns and then creating relationships between terms that appear at specific levels of abstraction. This allows specific and specialised forms of a more general term to be created that deal with specific circumstances. For example, the general term of *data set* is specialised into an *ordered data set*, an *un-ordered data set*, and a *structured data set*, which neatly allows a *structured data set* to structure (that is aggregate) any type of the abstract *data sets* (see Figure 6).

In summary, three distinct types of semantic relationship are possible between terms:

1. Taxonomical (*is a*)
2. Meronomical (*has a*)
3. Relational (*refers to*)

To capture these relationships a modelling language is used. Since a multiplicity of relationships may exist between a large set of terms, a 2-d graphical modelling language was used to maximise spatial cognition in the model's development. The graphical language requirements are for box and line to represent the terms and the relationships between the terms, respectively. There are three types of relationship to be denoted in the model and the terms need to have the ability to capture their definitions. Moreover, because the model was anticipated to be larger than can sensibly be presented on a single document page, the modelling language needs to have the ability to allow subsets of the model to be diagrammatically expressed. These factors, and the author's background experience, led to the OMG UML [12] class modelling language being used.

3. UML as an Ontology Modelling Language

The OMG UML [12] modelling language is primarily intended to model software based systems and can be thought of as a collection of interrelated modelling languages that individually focus on different aspects of understanding and specifying a software based system. One of these modelling languages is the class modelling language. Class models can be used to model the static structure of a system of object classifiers; that is the relationships between the object classifiers and their individual specifications. UML Class model relationships cover the three identified relationships for ontology modelling and the UML class model element can be used to capture the terms and their definitions in the ontology.

Figure 1 shows the graphical notation subset of UML class diagrams that is used to model the data safety guidance ontology. A UML Class may have more than one distinct relationship with one or more other classes.

The *constraint note* that is attached to the *is a* taxonomical relationship indicates that the classification relationship is *disjoint* (non-overlapped) and *complete* (all the classifiers are shown). Hierarchies can be *overlapped*, as well as *incomplete*. A general principal followed in the model is that the classifications within a single taxonomy are disjoint and where a term belongs to two or more taxonomies, each taxonomy is named and are overlapped with one another.

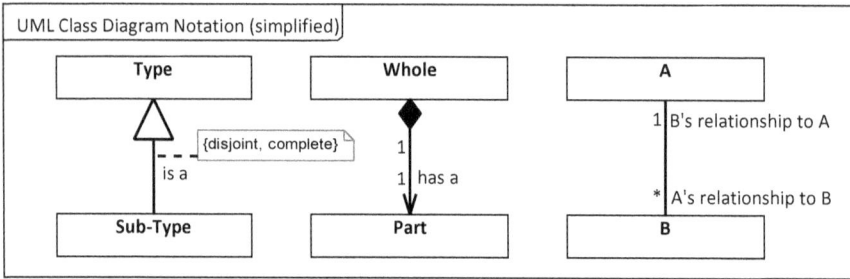

Fig. 1. UML Class Diagram Notation (Simplified)

The multiplicity shown at each end of a *has a* and *refers to* relationship specify the optionality and plurality of the class with respect to far end of the relationship. In Figure 1, the *Whole* has one *Part*, and *A* is related to **B*, where "*" should be read as *may have some*. Table 1 details the multiplicities to be found in the model.

Table 1. UML Class diagram relationship multiplicity designators

Multiplicity designation	Read as
1	One
0..1	May have (i.e. none or one)
1..*	Some (i.e. one or more)
*	May have some (i.e. none, one, or more)

4. Safety Standards Integration

There are many safety standards, such as IEC 61508 [13], SAE ARP 4761 [14], and ISO 26262 [15] to name but a few. There are multiple integrity measures including SILs (with various definitions), DALs, and ASILs; there are qualitative approaches and quantitative approaches; there are objective and proscriptive approaches; terminology varies; the point of view varies; and some sectors, such as civil aerospace, already address elements of data safety, which should not be contradicted by the DSIWG's work. This all makes addressing data safety specifically for each one with a single set of guidance challenging.

One area of commonality, despite all the variance, is that an assurance claim for the system's safety has to be made at some level. The data safety guidance is therefore presented from a risk reduction point of view with an assurance claim framework in mind. The ontology elaborates the relationships that a *data safety assurance claim* has to have, as shown in Figure 2. These relationships are:

1. A *data artefact* about which a claim is being made;
2. The *data safety requirements* for the *data artefact*;
3. The *artefact data integrity* the claim is to achieve.

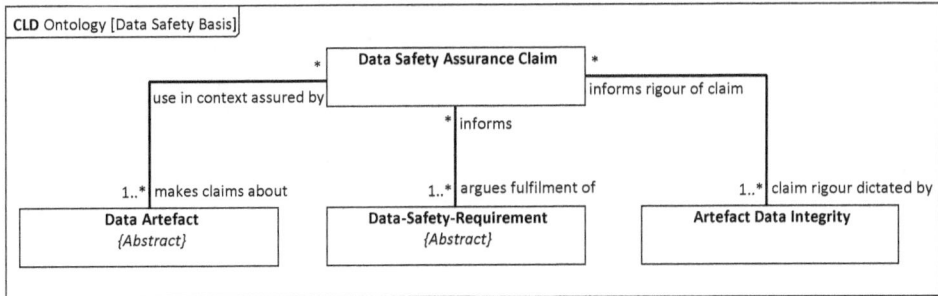

Fig. 2. Fundamental relationships in a data safety assurance claim

This is not too dissimilar to a system safety assurance approach whereby system safety requirements are allocated to aspects of the system that then have to be shown to meet their requirements to a satisfactory level of rigour that is a function of the allocated SIL, DAL, or ASIL. A *data safety assurance claim* is thus a *type of* safety assurance claim that is made for a safety related system. Such claims are usually composed of sub-claims until a level of argument rigour has been achieved commensurate with the system's required integrity level (shown by the meronomical, i.e. diamond headed, association in Figure 3).

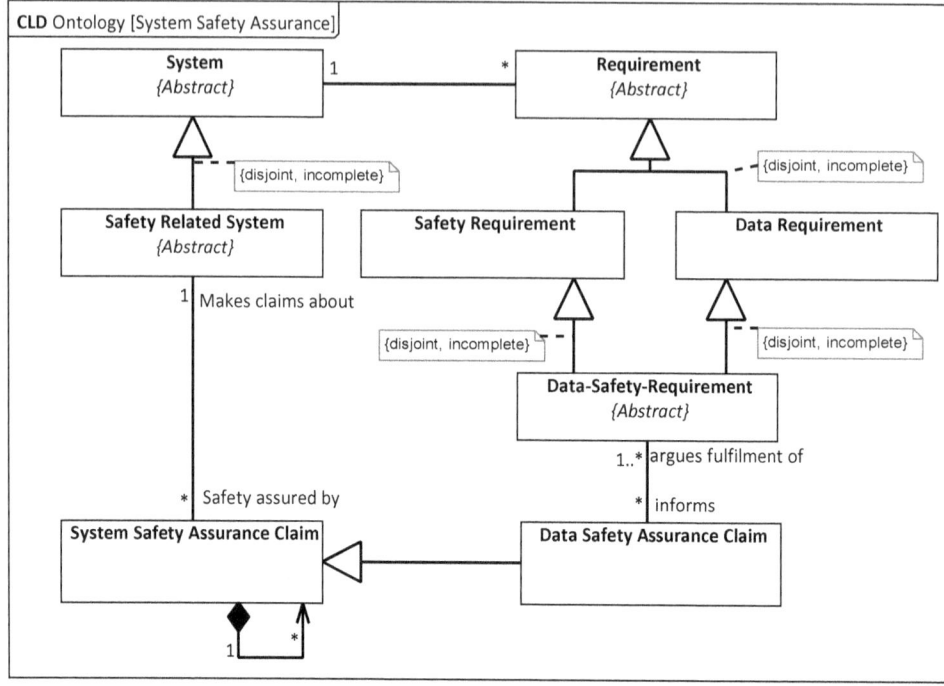

Fig. 3. Data safety assurance in the context of system safety assurance

A common factor in the sector standards is that of a safety requirement driven approach. Hence, the need for data safety requirements is established, which can be further decomposed into five distinct classes (Figure 4):

1. Data Artefact Requirements
2. Data Safety Property Requirements
3. Data Production Requirements
4. Data Use Requirements
5. Data Process Requirements

These requirement definitions provide focus on distinct areas of the assurance problem space and drive the assurance arguments along the following lines of questioning:

a. What data does the data artefact contain?
b. What properties is the data required to exhibit?
c. How is the data produced such that it has the desired properties?
d. How is the data to be used such that its properties are preserved?
e. How is the data to be handled by system processes such that its properties are preserved?

Before we consider what properties of data might be of concern in a safety assurance claim, we first need to consider the context of the claim.

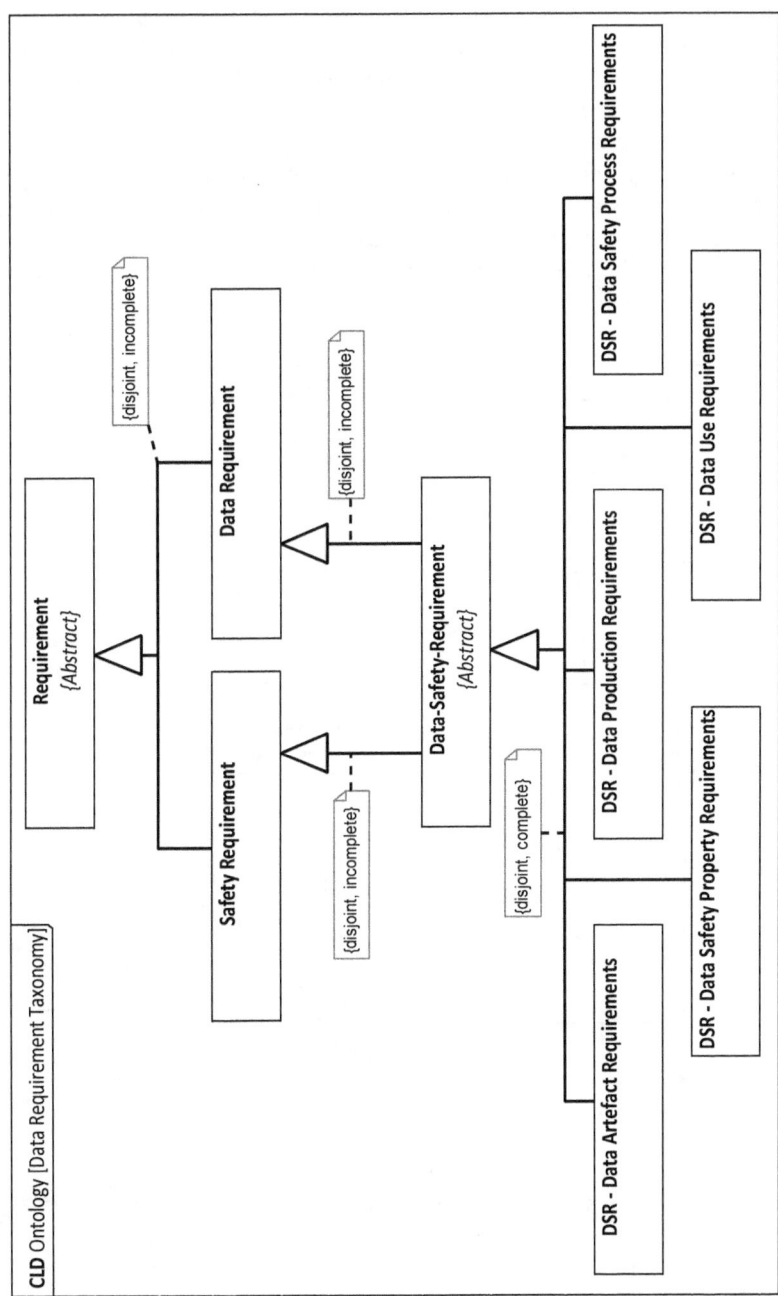

Fig. 4. Taxonomy of Data Requirement terms

Data flows from one system to another. From a data lifecycle point of view this can be seen as a flow from the system that produces the data to the system or systems that use or simply store that data and finally to the system or systems that destroy that data. A *data producer* can produce data mechanistically (e.g. from an algorithm), by measurement of sensory inputs, by transforming input data, or by fusing (aggregating) multiple data inputs. We therefore need a way of dealing with assurance arguments about the systems that handle data rather than just focusing solely on the assurance arguments of the safety-related system alone. The colloquialism "rubbish in, rubbish out" exemplifies the danger of solely focusing on system functions to the detriment of the data flowing between them.

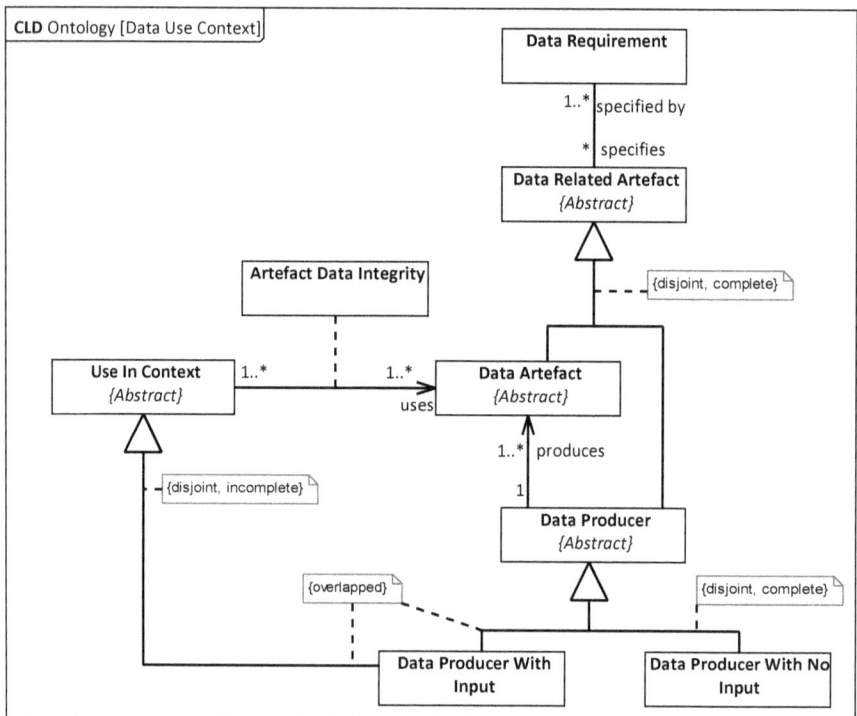

Fig. 5. The use in context of data artefacts dictates their assurance integrity level

The ontology deals with this convoluted problem by using abstractions and circular relationships as shown in Figure 5. A *data artefact* has one or more *use in contexts*. Each of these relational bindings dictates an *artefact data integrity* (which is discussed later on in this paper). A *data artefact* is produced by a *data producer*, which may have the need for input data. This option for different types of *data producer* is handled by the specialisations of *data producer with input* and *data producer with no input*. This allows for their semantic difference to be shown in the model, because the *data producer with input* is also shown as being an overlapped specialisation of a *use in context*, as denoted by its specialisation from the *use in context* term in Figure 5. This creates a semantic loop in the ontology, whereby a *data producer with input* is-a *use in context*, which is associated with *data artefact(s)*, that are produced by *data producer(s)*, which may (or may not) have data inputs. Hence, the assurance argument recurses back up the data production chain until there is no input or there is a relatively simple assurance claim to be made. Such data chains could be quite long and in some cases a never ending story. Such assurance scope expansion can be prevented by also taking into account the assessed *artefact data integrity* and stopping when this falls below the minimum data-integrity threshold.

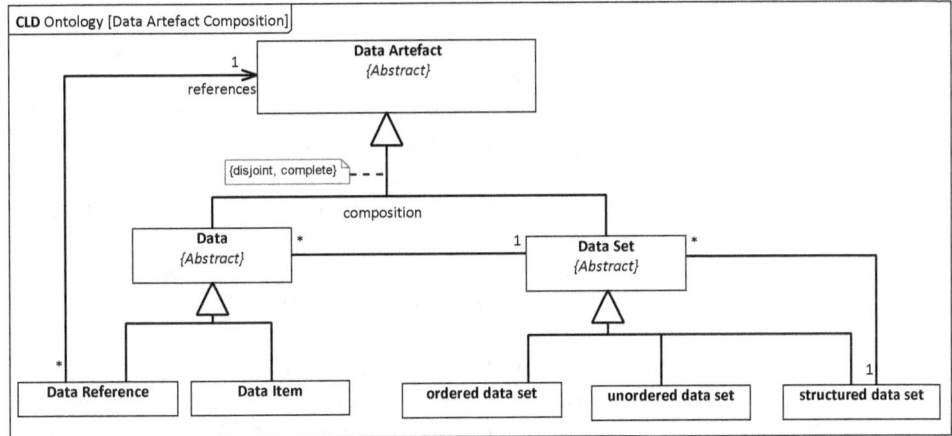

Fig. 6. Data artefact composition

This exact same reasoning can be applied to the data in the safety related system, the data the system consumes, the data the system was tested with, and to the data in test systems that were used to verify the safety related system, as dictated by the system safety assurance claims.

At this stage it is probably useful to describe the nature of a *data artefact*. First, we can describe the composition of a *data artefact* and second we can classify the abstract application or use of data.

As a general statement, we do not necessarily care specifically about the make-up of data. What is important is that the safety assurance properties for a data artefact are appropriately established and demonstrated for that data artefact's specific use. This means that at one end of the spectrum we might have a single piece of critical data to assure and at the other end of the spectrum a large data set that is to be assured as a whole. The model in Figure 6 shows how the ontology currently deals with this problem by defining the terms *data* and *data set* as being derivative terms of *data artefact*. A *data artefact* may therefore be either *data* or a *data set*.

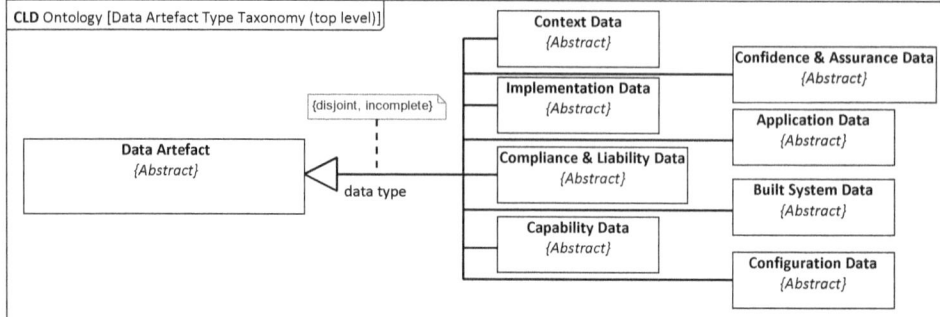

Fig. 7. Data artefact type taxonomy (top level)

Data is defined to be either an actual item of data (*data item*) or a reference to another data artefact (*data reference*). The ontology currently does not refine this any further because the underlying format of the data is currently not a concern. However, it is conceivable that specific assurance treatment for distinct types of data may have benefit, for example by reasoning about numerical properties of arithmetic types, or the semantic and encoding properties of strings of character data.

A *data set* is composed from *data*. There are three types of *data set*: ordered, unordered, and structured. A *structured data set* composes (i.e. may have some) *data sets*.

The DSIWG originally conceived a long list of the ways data can be used in or around safety related systems. There have been a number of attempts to structure this list and Figure 7 shows the top level of a recent attempt at a "data type" breakdown. Each of these subordinate terms has a further list of derived terms, the details of which are contained in the main guidance document [9]. Figure 8 shows the complete data type taxonomy from the ontology model.

From the ontology model's point of view the "data type" taxonomy (as shown in Figure 7) is of limited direct value because it does not connect with anything else in the model. It exists because it was a point of entry for the DSIWG's discussions and data-safety guidance validation. The model might benefit from a refinement of the *use in context* term shown in Figure 5 and, from specific contexts, find the linkage to the specific use of data. The problem the DSIWG had is that it is much harder to define a reasonably definitive set of *use in context* classifier terms than it is to define a list of *data type* classifiers by data usage application. The *data type* taxonomy therefore exists as a bridge to real world data *use in contexts*.

There are more than twenty actual *data type* terms in the current draft of the guidance and the use of an abstract classification hierarchy for them (as shown in Figure 8) has currently been removed, because, as already noted, it did not add value to the guidance's application. The subject is slowly evolving as the work on the guidance document and its ontology mature.

5. Data Integrity

The notion of data integrity was previously introduced in Figure 2. The term *integrity* is used in two different ways by the guidance. The first is as *artefact data integrity*, which has a value property of a *Data Integrity Level* (DIL). The second use is a specific *data safety property*. In some respects this duplicity of meaning is unfortunate and care must be taken to qualify which "integrity" is being referred to.

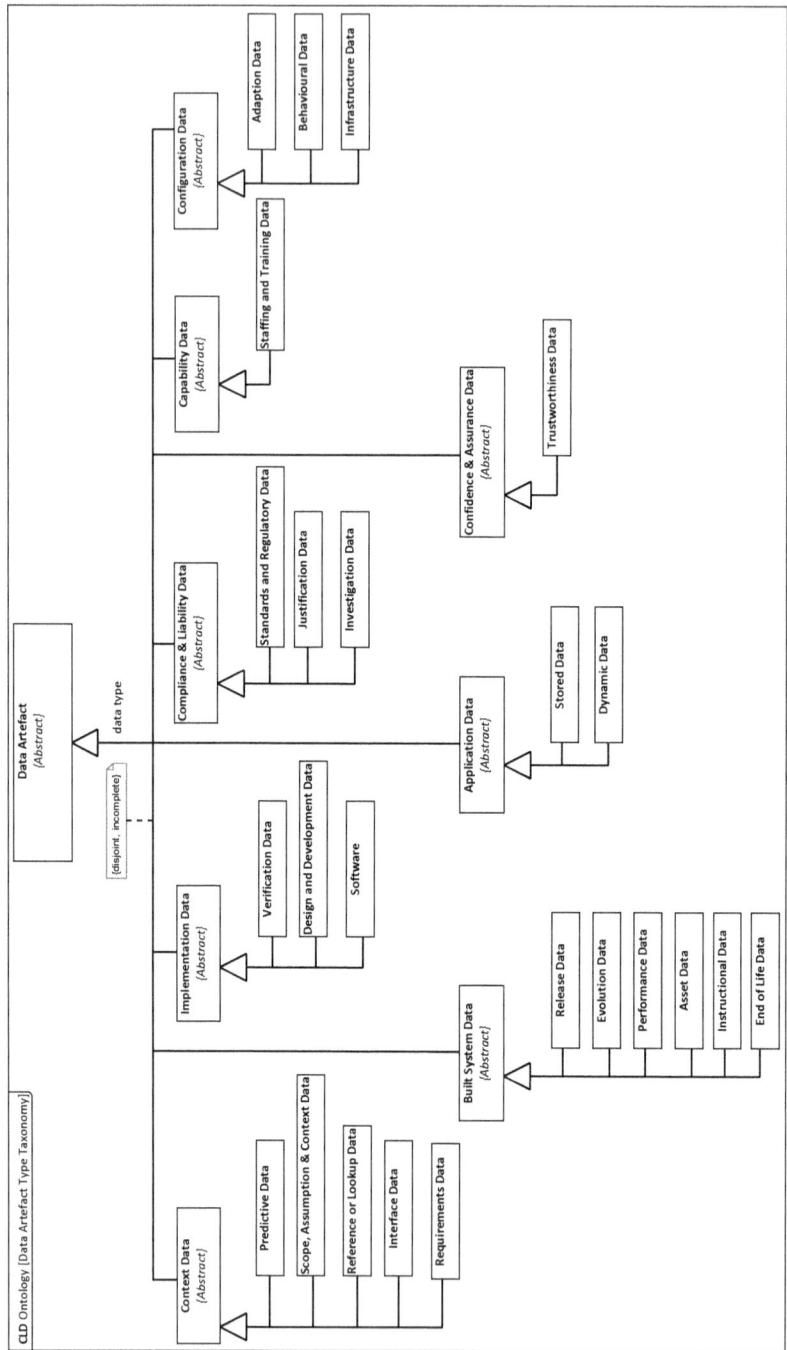

Fig. 8. Data Type Taxonomy

The term *Data Integrity Level* or DIL was coined early in the DSIWG's work and is intended to be a commensurate concept with the functional safety standards' SIL, DAL, and ASIL concepts in that it defines a level of rigour that the assurance claim must demonstrate in the engineering of the associated system. Broadly speaking, the greater the risk of data resulting in, or leading to, harm, the higher the DIL score on a 1 to 4 scale (with a score of 4 demanding the highest level of rigour). This is comparable to the SIL 1 to 4 scale defined by IEC 61508 for safety function integrity [13]. Out of completeness the idea of a zero DIL score can be used to mean that there are no assessed safety implications for the use in context of a data artefact.

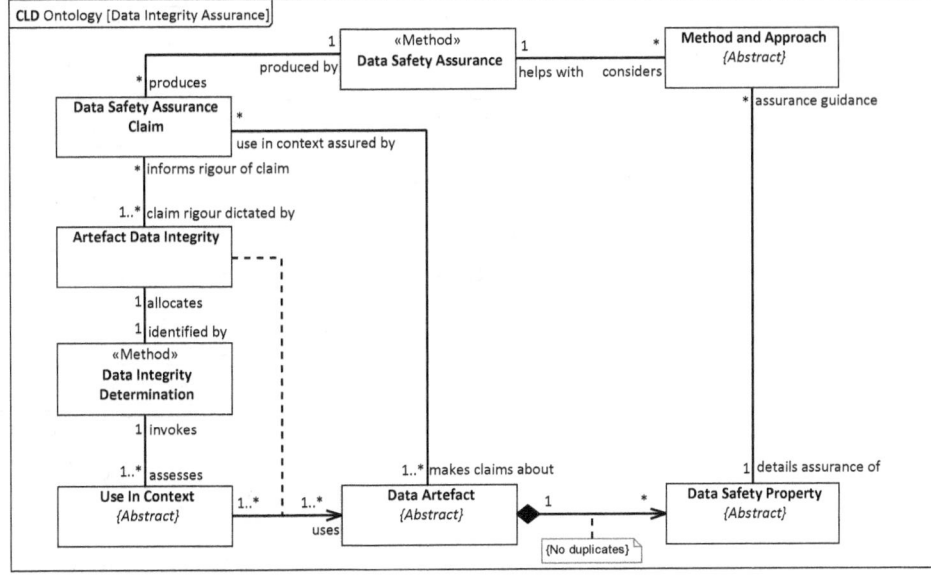

Fig. 9. Data Integrity Assurance

The model in Figure 9 shows the basic relationships between terms that establish the core framework for data safety assurance. It also introduces specialised terms for methods (or process areas) that are denoted by a double angle quoted «Method» annotation.

The intention of this model is that a *use in context* for a *data artefact* is used to determine the level of data integrity required. The ontology model shows that the term *artefact data integrity* is linked (by the broken line in Figure 5 and Figure 9) to the relational association between the *use in context* and *data artefact* terms. This means that we can only refer to an *artefact-data-integrity* when we can contextualise it with a specific *data artefact* and a specific *use in context* for that artefact.

Having established the appropriate *data integrity level*, the *data safety assurance claim* can then make claims about the *data artefact*'s safety properties to a commensurate level of rigour defined by it. The guidance provides method and approach suggestions for each *data safety property* across the range of DILs.

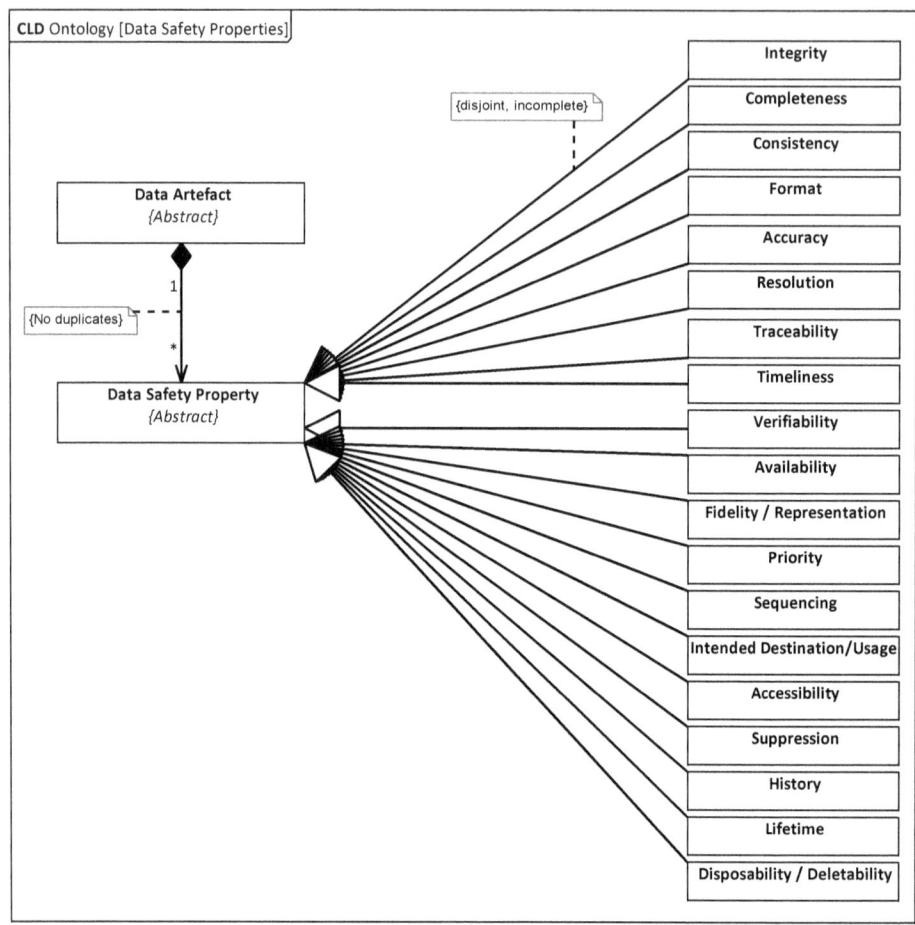

Fig. 10. Data safety properties

Figure 10 shows the current set of data safety properties in the guidance. The list is comprehensive, but not necessarily definitive. It is not intended that all of these properties apply to all data artefacts. In fact, Figure 9 shows the relationship between data artefact and data safety property with a cardinality of zero or more, meaning that there is choice in the matter. But which ones apply?

The DSIWG's solution to this problem is the comprehensive assessment of the data safety properties for a *data artefact* under the circumstances of its use (i.e. *use in context*). Figure 11 shows this part of the ontology. For example, if the *data artefact* lacks the property of "completeness" the question of "how will the system respond?" can be asked. If the *data artefact* in question is a key parameter in a safety related process then the outcome may be a failure of the system to deliver its intended function or worse a malfunction that compromises the overarching system's operation. If the lack of data completeness is levelled against a test data set then the consequence might be interpreted as a lack of test coverage for the system under test and thus undermine an assurance claim of full test coverage.

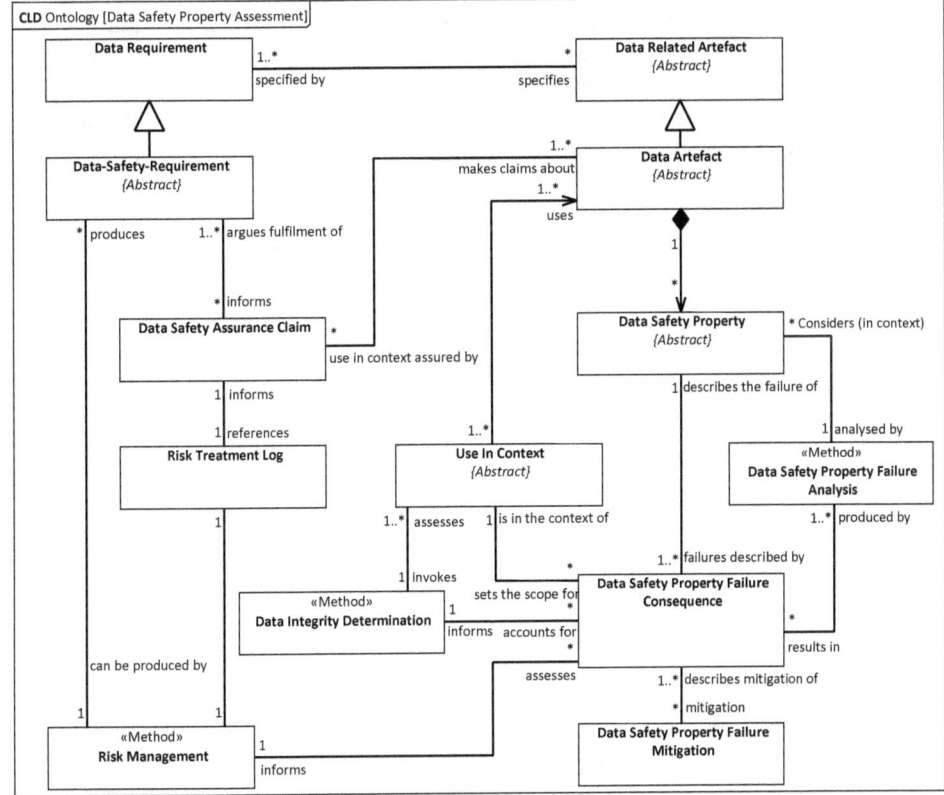

Fig. 11. Data Safety Property Assessment

It is of course legitimate for an assessment of a data safety property to determine there is no impact, or that it does not apply for the data artefact in its usage context. In most applications, it is to be anticipated that the list of properties provided by the guidance can be pruned down to a smaller set.

For those properties that have been assessed as having sensitivity to system safety then a risk treatment approach is advocated. The guidance itself advocates an ISO 31000 risk management approach [16]. The model shown in Figure 11 shows the relationships being formed between the data safety properties and their assessed failure consequences with possible mitigations and the risk management and DIL determination processes. Here the premise is that the greater the risk of an effect on system safety arising from the failure to maintain the safety properties of a data artefact the higher the DIL score that should be assigned. To counter this, mitigations may be identified or the risk transferred and new data safety requirements generated as a result.

The ontology terms concerned with the failure analysis aspect are shown in Figure 12. There are a number of general considerations that each data safety property can be assessed against, which are mostly life cycle concerns, and then there are a specific set of concerns for each one. The guidance [9] currently includes "HazOp" (Hazard and Operability) style guide words for this purpose and this is reflected in the model's incorporation of *HazOp* as a type of method for performing Failure Analysis. This hints that there could be other approaches and the ontology model somewhat loosely includes Fault Tree Analysis (FTA) to illustrate this concept, even though this is not currently included in the guidance.

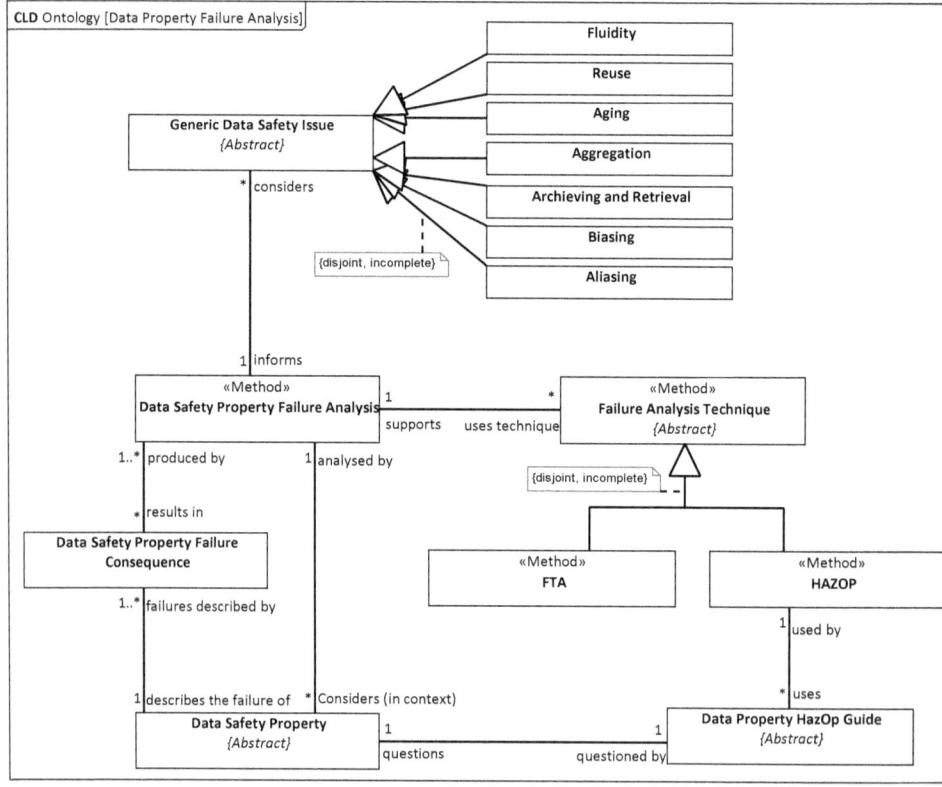

Fig. 12. Data Safety Property Failure Analysis

As shown, the model indicates that there is a one to one correspondence of a *Data Property HazOp Guide* with a *Data Property*. In the fuller ontology model that is not shown in this paper, the concrete specialisation of these two terms and their individual one to one relationships are modelled. This manifests as a table of data safety properties with their HazOp guide words in the guidance document.

6. Conclusions

The production of the ontology has helped the DSIWG understand the problem space it has been working with much more clearly than had been achieved from the written submissions. Much of this has come from the questioning process involved in the formation of terminology and the semantic relationships that these terms have. This in turn has strengthened the semantic definition of the terminology, which in turn will feed into future revisions of the guidance to improve its clarity. It has also provided the framework for the approach the guidance has taken towards its treatment of data safety from an assurance perspective.

In any field of complex terminology the use of an ontology model to both explore the problem space and then to define the domain language to describe it is highly recommended.

7. Further Work

The SCSC Data Safety Initiative project continues to evolve its guidance and the ontology will evolve with it. The ontology work found that the guidance could be more detailed about some of the methods to be used. Conversely, there are parts of the existing guidance [9] that are still poorly served by the ontology, such as the "Organisational Data Risk Assessment" and data lifecycle, especially that of supply chains. The guidance document's current use of "method and approaches" is more sophisticated than the ontology might suggest because it is linked into the safety system's lifecycle stages.

The DSIWG has recently become aware of the European OPENCOSS project[1] that is proposing a standardised meta-model for capturing safety assurance claims (i.e. safety cases). We need to assess this meta-model to see if it reveals gaps in our own, or in theirs. We also need to consider whether there is a benefit to either mapping or aligning our terms with the OPENCOSS meta-model for a safety certification language.

The DSIWG is also aware of the work that Holloway [17] has undertaken with modelling the aerospace critical software standard RTCA DO178C/EUROCAE ED12C [18] using an assurance argument approach to examine its consistency and completeness. The work has the objective of producing an assurance argument, by retrospective scrutiny, of a published standard. In some respect the DSIWG would like to move towards the position of having an assurance argument for its own data-safety guidance, but to have achieved that by design.

The ontology already hints at an upper ontology by making explicit distinction of terms that refer to methods and process as opposed to information. There may

---

[1] http://www.opencoss-project.eu/, accessed 17 August 2015

be some benefit in formalising an upper ontology when more of these meta-terminology concepts become clearer.

Finally, the DSIWG would like to build connections with the safety-standards' working groups for the direct integration of data safety as an explicit system-safety design and assurance concern. In this respect the DSIWG has already made a submission of a draft data-safety annex for possible inclusion in the UK MOD Safety Critical Software procurement standard Def-Stan 00-55 Issue 3 [19].

Acknowledgements

The author would like to acknowledge and thank the following people for taking the time to review drafts of this paper and for their support with the DSIWG On-tology modelling initiative:

- Robert Oates, Rolls-Royce Plc
- Victor Malysz, Rolls-Royce Plc
- Mike Parsons, NATS
- Rob Ashmore, Dstl
- John Spriggs, NATS
- John Bragg, MBDA

The author would also like to acknowledge the DSIWG members for their con-tributions to the field of data safety. A fully list of the contributory members to the guidance can be found towards the end of the guidance document [8], [9].

References

[1]   "How to stop Data causing Harm Seminar," Safety-Critical Systems Club, 2012. [Online]. Available: http://scsc.org.uk/e209. [Accessed 17 August 2015].

[2]   "How to Stop Data Causing Harm II Seminar," Safety-Critical Systems Club, 2015. [Online]. Available: http://scsc.org.uk/e343. [Accessed 17 August 2015].

[3]   "Mars Climate Orbiter Mishap Investigation Board Phase I Report," 10 November 1999. [Online]. Available: http://sunnyday.mit.edu/accidents/MCO_report.pdf. [Accessed 17 August 2015].

[4]   "Report on the investigation of the grounding of Ovit in the Dover Strait on 18 September 2013," September 2014. [Online]. Available: https://assets.digital.cabinet-office.gov.uk/media/547c6f2640f0b60244000007/OvitReport.pdf. [Accessed 10 September 2015].

[5]   "20 Million Americans Have Multiple Social Security Numbers Associated With Their Name," 11 August 2010. [Online]. Available: http://www.idanalytics.com/news-and-events/news/20-million-americans-multiple-social-security-numbers-associated-name/. [Accessed 10 September 2015].

[6]   D. Gehrke-White, "Social Security wrongly labels disabled man as inmate," 28 January

2014. [Online]. Available: http://articles.sun-sentinel.com/2014-01-18/business/fl-disability-withheld-wrong-prison-identity-20140117_1_social-security-office-prison-inmate-prison-officials. [Accessed 10 September 2015].

[7]   L. Cockcroft, "National Insurance mix-up has left dozens of elderly people without their pension," 28 July 2008. [Online]. Available: http://www.telegraph.co.uk/news/uknews/2464627/National-Insurance-mix-up-has-left-dozens-of-elderly-people-without-their-pension.html. [Accessed 10 September 2015].

[8]   "Data Safety by the SCSC Data Safety Initiative Working Group [DSIWG] V1.0," 31 January 2014. [Online]. Available: http://scsc.org.uk/paper_127/Data%20Safety%20(Version%201.0).pdf?pap=954. [Accessed 17 August 2015].

[9]   "Data Safety by the SCSC Data Safety Initiative Working Group [DSIWG] V1.2," 23 January 2015. [Online]. Available: http://scsc.org.uk/paper_128/Data%20Safety%20(Version%201.2).pdf?pap=958. [Accessed 17 August 2015].

[10] M. Parsons and P. Hampton, "Stopping Data Causing Harm," in *Addressing Systems Safety Challenges*, 2014.

[11] P. Hampton and M. Parsons, "The Data Elephant," 2015. [Online]. Available: http://scsc.org.uk/paper_129/Hampton%20-%20The%20Data%20Elephant.pdf?pap=976. [Accessed 17 August 2015].

[12] *Information technology - Object Management Group Unified Modeling Language (OMG UML), Superstructure*, ISO/IEC 19505-2, 2012.

[13] *Functional safety of electrical/electronic/programmable electronic safety-related systems Part 1: General requirements*, IEC 61508-1, 2010.

[14] *Guidelines And Methods For Conducting The Safety Assessment Process On Civil Airborne Systems And Equipment*, SAE ARP4761, 1996.

[15] *Road Vehicles - Functional Safety*, ISO 26262, 2011.

[16] *Risk Management - Principles and Guidelines*, ISO 31000, 2009.

[17] C. M. Holloway, "Explicate '78: Uncovering the Implicit Assurance Case in DO–178C," 2015. [Online]. Available: http://scsc.org.uk/paper_129/Holloway%20-%20Explicate%2078%20Uncovering%20the%20Implicit%20Assurance%20Case%20in%20DO-178C.pdf?pap=969. [Accessed 17 August 2015].

[18] *Software Considerations In Airborne Systems And Equipment Certification*, RTCA/DO-178C and EUROCAE ED-12C, 2012.

[19] *Requirements for Safety of Programmable Elements (PE) in Defence Systems Part 1: Requirements and Guidance Issue 3*, Ministry of Defence DEF.STAN.00-55, 2015.

Glossary of Acronyms

Acronym	Definition
ASIL	Automotive Safety Integrity Level [15]
DAL	Development Assurance Level [14][1]

[1] NB The revised SAE ARP4754A replaces DAL with FDAL (functional development assurance level); SAE ARP4761 has not (currently) been revised to account for this.

DIL	Data Integrity Level
DSIWG	Data Safety Initiative Working Group
IEC	International Electrotechnical Commission[1]
ISO	International Standards Organization[2]
OMG	Object Management Group[3]
SAE ARP	Society of Automotive Engineers Aerospace Recommended Practice[4]
SCSC	Safety Critical Systems Club[5]
SIL	Safety Integrity Level [13]
UML	Unified Modelling Language [12]

[1] www.iec.ch

[2] www.iso.org

[3] www.omg.org

[4] www.sae.org

[5] http://scsc.org.uk/

Safety of Socio-technical Systems from a Perspective of Enterprise Engineering

Xiaocheng Ge[1]

University of York

York, UK

Abstract *It is not new applying a socio-technical approach to analyse the safety of complex systems. Early works from Reason (Reason et al. 1998), Rasmussen (Rasmussen 1997), and Leveson (Leveson 2004) already provided frameworks of socio-technical approach by identifying layers in a system actually involved in the control of safety. However, as systems are more and more complex, the challenge in these socio-technical approaches to system safety is now a problem of modelling. It is widely accepted that architecture is the foundation of good system engineering. Thus the model in a systems theoretic approach of system safety should be embodied in all components (both social and technical) in the system and their relationships to each other and the environment. The key objective is to explore whether safety analysis on a socio-technical system can benefit from model-based approach in which system engineers and safety engineers share a common model. To evaluate and demonstrate our approach, we developed a software tool to help the application of our approach. The case study analyses a tram accident: the derailment at East Croydon in February 2012. The analysis is purely based on the information from the official investigation report (RAIB 2012) so the architecture of entire organisation may not be represented completely; but it is adequate enough for the discussion of a general architecture-based approach to the safety of social-technical systems.*

1 Introduction

Safety is an emergent property of a system, so the safety analysis should take into account all the behaviours of a system as a whole in the context of its environment. Now model-based safety analysis is one of directions in safety engineering.

---

[1] Department of Computer Science

It is an approach in which the system and safety engineers share a common system model created by using a model-based development process. The main advantage of this approach is that the system and safety engineers are able to work on a shared, unambiguous model of the system leading to a tighter integration between the systems and safety engineering processes. The shared model ensures that safety analysis results are relevant and up-to-date as the system architecture evolves, and allows safety assessment early in the system design process.

Modern safety-critical systems, e.g., a train control system, are more than just about engineering their technical subsystems (usually computer-based systems). As the concept of a system has already extended beyond the boundary from a technical system to include its social (operational) environment such as managerial activities and procedures, they are seen as socio-technical systems. The term "socio-technical" here emphasises the interactions among the "social" and "technical" components in the system since these interactions usually create the decisive conditions for successful (or failure) behaviours of the entire system. The idea of adapting socio-technical approaches to safety engineering is not new, and it had been identified that risk management in a socio-technical system is a problem of modelling (Rasmussen 1997), which means that management of safety risks should be based on an adequate abstraction of the actual system that systematically captures the fundamental information of various types of interactions in the system so that engineers from every layer in the system's hierarchy can obtain the benefits and avoid the obstacles of complexity in the system.

Enterprise architecture seems to be a mature option for conducting system-level analysis, design, planning and management, and several well-defined frameworks of how to create and use enterprise architecture, such as MODAF (MOD Architecture Framework) (MODAF), have been proposed. Therefore, the safety of a socio-technical system is not actual a problem of how to construct adequate and comprehensive models for system engineers and safety engineers, instead is a problem that the links between system architecture and safety analysis model are indistinct as the system becomes more complex. In other words, the system architecture conducted by following an enterprise architecture framework becomes too complicated so that the safety analysis will lose the benefits what a model-based safety analysis provides.

The aim of our research is to explore a possible approach to bridge the processes of enterprise architecture and safety assessment of a socio-technical system so that safety analysis can be relevant and up-to-date as the system architecture evolves, and gain the benefits of allowing safety assessment early in the system design process.

In this paper, we will first discuss what the problem of a modern safety critical system as a socio-technical system is, and then introduce an approach to map between the elements in the system enterprise architecture and in the model of safety analysis. We will also demonstrate how the analysis can be supported by a software tool and application of our approach in the case study.

The rest of paper is structured as follows. In section 2, we will briefly review the approaches of enterprise architecture framework and TRAK. In section 3, we argue what the problem of the safety of a socio-technical system is and in section 4, we introduce the model driven approach to connect system development with enterprise architecture and safety assessment. Section 5 is a demonstration of how our approach is applied in a case study and the discussion on the safety analysis of socio-technical systems.

2 Enterprise architecture framework and TRAK

Interactions will dramatically increase the complexity of a system when elements in its social environment are taken into account, then what is needed for systems engineering is a new way of modelling – a systematic modelling approach that capture the fundamental relationships of information to complexity so that engineers in the "*enterprise*" can obtain the benefits and avoid the obstacles of this change. By enterprise we mean an entity comprised of interdependent resources (e.g., people, processes, organisations, and technology) that interact with each other (e.g., coordinate functions, share information, and allocate resources) and their environment to achieve goals.   An enterprise is always a purposeful collection of activities of interest that exhibits multiple states across time, and may exhibit emergent behaviour that can be identified or suspected beforehand, often at different scales (Troche et al. 2011).

The architecture of a system is the structure of components, their relationships, and the principles and guidelines governing their design evolution over time (IEEE Std 610.12). More precisely, architecture is a model (but not necessarily a single model) that details what a system does (*activities*), the relationships among these activities, how a system performs those activities (*processes*), the functions of these tools that are used to perform the activities in total or in part (*system functions*), the actual tools themselves that perform those system functions (*subsystems*), the skills and experience of those performing the activities (*roles*), all related through a documented, repeatable, and agreed-upon structure (*taxonomy*), and rendered through an equally approved documented and agreed-upon rule set (*framework*). Each of these activities, processes, system functions, sub-systems, and roles is described in terms of **attributes**.

There are a number of important theories and practices (Rebovich and White 2011) proposed to enterprise systems. One of them is that enterprise architecture (EA) can be used as a primary tool in support of enterprise-level assessment because a good systems engineering starts from an adequate architecture. There are a number of EA frameworks, and TRAK is one of them. TRAK is a general enterprise architecture framework whose original intent was to develop a railway-specific architecture framework in adapting MODAF. After any defence or domain-specific content was removed from MODAF, TRAK provides a domain-free meta-model and viewpoints that were only based on representing complex systems.

In TRAK, there are 22 architecture viewpoints which are grouped into 5 architecture perspectives. Each viewpoint belongs to a single perspective and specifies a single view (type). And each viewpoint specifies what sets of types of architectural description element and relationships (tuples) can appear. These architectural description element types and relationships are specified by the TRAK meta-model (shown in Figure 1).

In this meta-model, the 5 architecture perspectives are coloured:

1. Enterprise Perspective (EVp). A perspective covers the enduring capabilities that are needed as part of the entire enterprise. These are high level requirements that everything else contributes to and form part of the long term strategic objectives that need to be managed.
2. Management Perspective (MVp). A perspective provides views that describe the architectural task and those relationships that are common across other perspectives. It provides ways of defining the scope and findings of the architectural task – structuring the approach and modelling. The management perspective provides ways of describing the normative standards that apply. It contains views that provide supporting information to aid the portability and understanding of the model(s).
3. Concept Perspective (CVp). The perspective covers the logical view of what is needed in response to the capabilities required by the enterprise in the EVp. It covers the logical connections of nodes, for example a service control centre, to other nodes with no recognition of how this might be realised either by organisation or technology. It also implies no particular part of a life cycle, which covers everything from concept to disposal.
4. Procurement Perspective (PrVp). A perspective provides a top level view of the solution to the enterprise capability needs outlined in the EVp and developed in the CVp. It provides a way of showing how projects deliver the solutions described in the SVp to provide capability. It also provides a way of showing time dependency between projects and is an essential for investigating capability gaps.
5. Solution Perspective (SVp). A perspective provides views about the solution whether proposed or realised. It covers the part of systems (whether social or technical), their exchanges and protocols. The SVp describes how organisations and equipment are organised and governed. It also describes how the logical requirements outlined in the CVp are realised and shows how the solution(s) realise capability needed by the enterprise and described in the EVp.

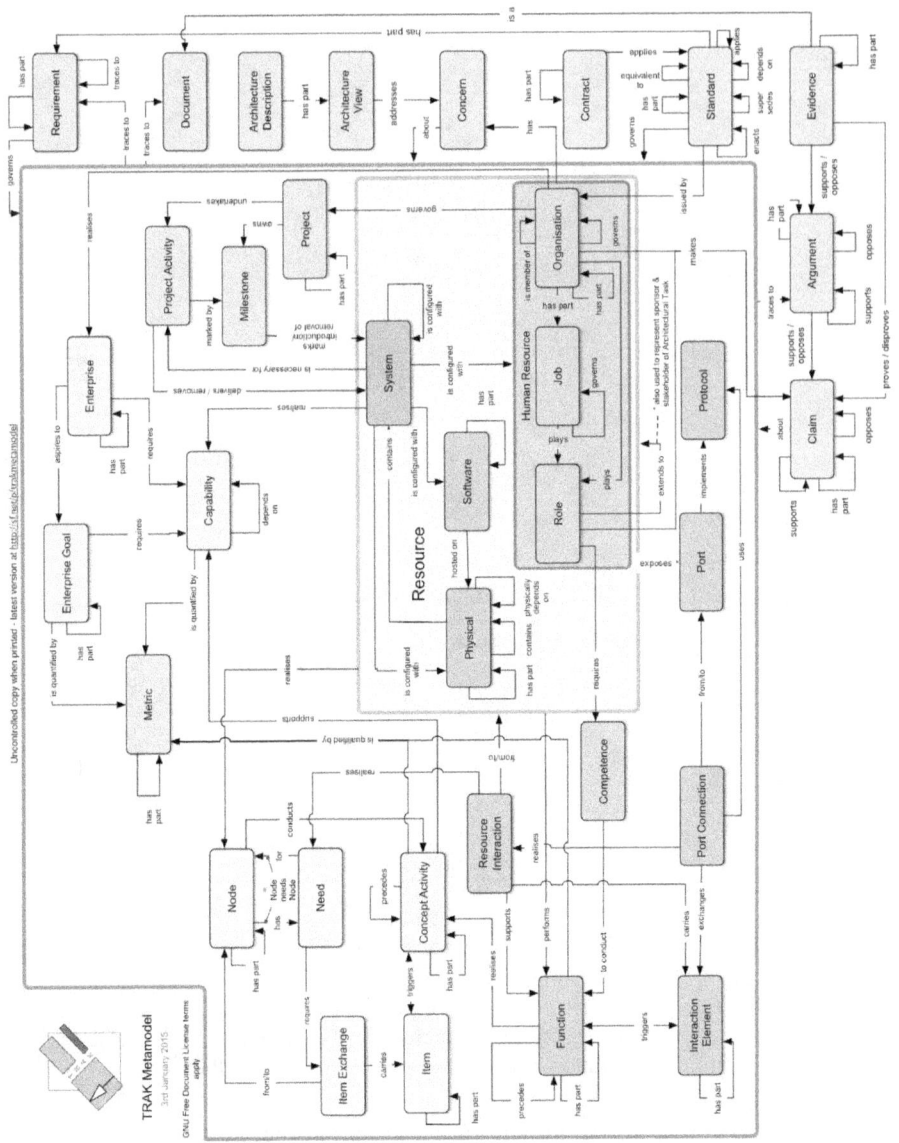

Fig. 1. TRAK meta-model (TRAK)

For example, Figure 2 is a viewpoint of signalling control from solution perspective. Not only the elements related to signalling control have been identified, but also their types, e.g., <<node>>, <<system>>, <<role>> etc, the relationships between various elements, e.g., <<plays>>, <<isConfiguredWith>>, etc. are identified. Based on the model, the structure of system (the part of signalling control) can be understood correctly, and it also provides a foundation for the safety analysis.

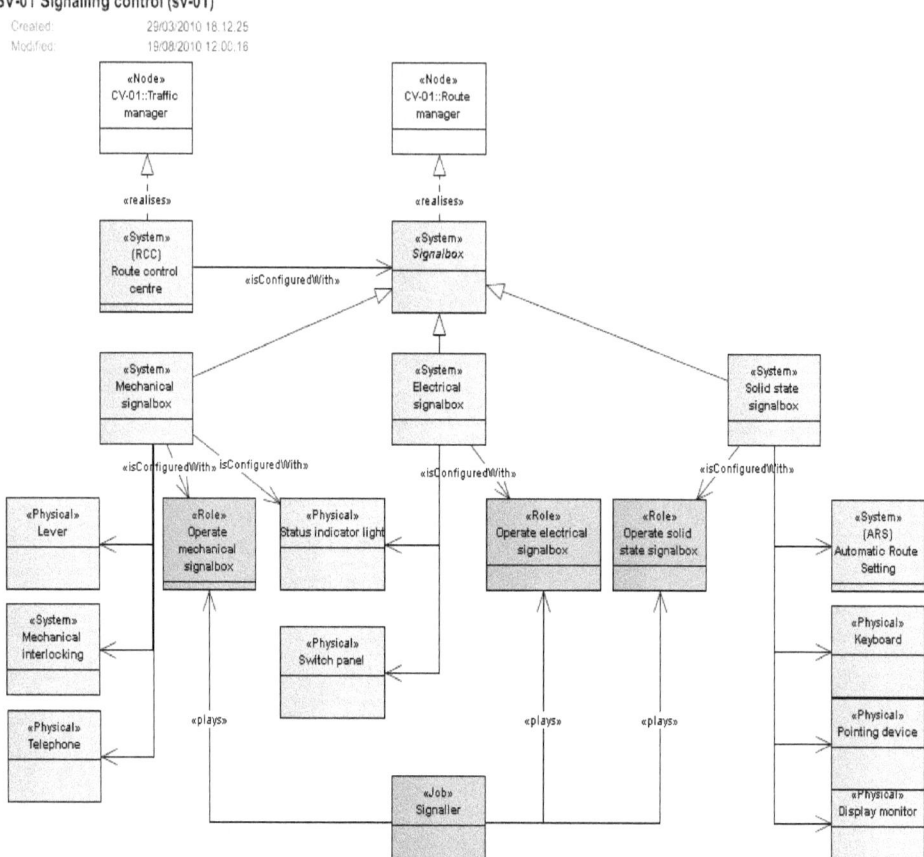

Fig. 2. An example of TRAK viewpoints - SV-01 Signalling Control (RFA)

A modern safety-critical system is comprised of a large amount of interconnected entities from various systems (including social and technical systems) and
most importantly it exhibits various emergent properties as they evolve over time
(for example in Figure 2).

There are several socio-technical approaches of analysing the safety of system in a wider context, for example, Rasmussen adopts a system oriented approach based on a hierarchical socio-technical framework for the modeling of the contextual factors involved in organisational, management and operational structures that create the preconditions for accidents (Rasmussen 1997); Leveson (Leveson 2004) proposes a model of accident causation called STAMP (System-Theoretic Accident Model and Processes) that focuses on the control structure in the complex socio-technical system and considers the inadequate controls among the technical, human and organizational factors in the system; and Hollnagel (Hollnagel 2012) proposed an approach based on a qualitative accident model that describes how the functions of system components may "resonate" and create hazards that can run out of control and lead to an accident. All of these techniques are useful to identify some problems in the system. However none of them utilises the enterprise architecture, instead focuses on a special-purpose and single-view model which is not in the enterprise architecture. In the next section, we firstly argue that the safety of a socio-technical system is a "soft" problem, which means the safety engineering in a socio-technical system is about improving the safety that is "problematical" (Checkland and Scholes 1990) rather than a problem which can be clearly identified.

3 Safety of a socio-technical system

If the ultimate goal of safety engineering is to avoid accidents, then the safety of a socio-technical system (STS) is a "soft" problem, which is hard to define adequately. Safety engineering of a STS is to improve the situations which are "problematical", rather than to solve the problem completely. Challenges to the safety of a STS mainly come from its complexity. It is complex not just because it involves many technical components with wide-apart technologies, but because it additional involves social elements, which are naturally different with technical components. As Redmill (Redmill 2002) pointed out, technical systems fail in relatively predictable ways for certain reasons (e.g., manufacturing defects, component wear-out, improper maintenance, etc.), but humans fail for a host of reasons, some of which cannot be predicted or even foreseen (e.g., sabotage, psychological breakdown, whim, etc.). Thus safety becomes a dynamic characteristic of socio-technical system because the interactions between social and technical subsystems often create the decisive conditions for successful (or unsuccessful) behaviours of the entire system. The interactions in a STS are comprised partly of linear "cause and effect" relationships, which are normally specified and designed for example in various operational procedure, and partly from "non-linear", complex, even unpredictable relationships, which can lead to emergent properties. An inevitable consequence of considering social components as well as technical components when designing and analysing a system is that these social components do not necessarily behave like the technical, people are not machines, paradoxically, with growing complexity and dependences even the technical components can start to exhibit non-linear behaviours.

It is perhaps intuitively obvious that growing complexity becomes a problem for the assessment and assurance of system safety. Although hazard and risk assessment had been undertaken for the systems through rigorous systems engineering approaches, the hazard was "missed" apparently at least in part. There is some explicit evidence to support this view, for example the tram derailment at Croydon East in 2012 was both involved a serial of technical failures and human errors.

In general, safety engineering is about minimising safety risks, but not eliminating them. Given the challenges brought by the complexity of a socio-technical system, the objective of safety analysis is to identify the weakest link in terms of the safety of the overall system, rather than to provide a complete solution to assure the system is safety. The best way of doing it is to keep the traceability of safety analysis with the system architecture and then consider the safety controls in the context of entire system. Thus it needs a mechanism to synchronise the safety analysis with the enterprise architecture. It is easier for a relatively simple system when applying a model-based approach of safety analysis, but it is much harder for the case of a socio-technical system because the architecture of the system consists of multiple viewpoints (models). In next section, we introduced a model-drive approach to map the safety analysis results with the system enterprise architecture; and demonstrate a software tool which can help the safety analysis and visualise the mappings with system architecture.

4 Model-driven approach and tool support

4.1 Meta-model of fault tree

Fault tree analysis is a mature and commonly used safety analysis technique. It is a deductive failure analysis in which a top event is analysed by using Boolean logic to combine a series of lower-level events. In a model-driven approach, it is important to have a meta-model of a fault tree. We built one (see Figure 2) for the purpose of our approach and implemented in the software tool.

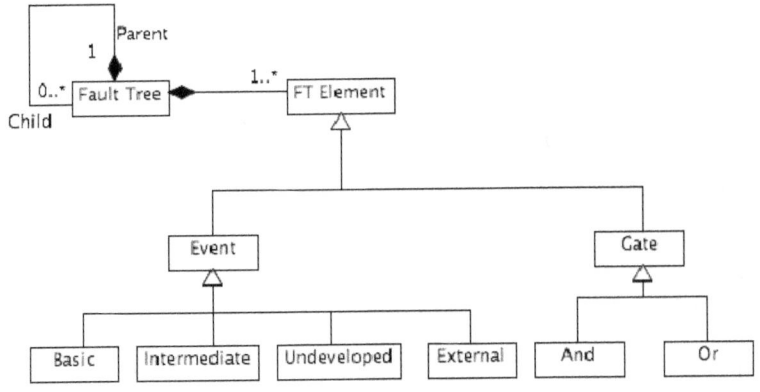

Fig. 3. Meta-model of fault tree

As seen in Figure 3, a Fault Tree model contains another fault tree or elements named FT Element, which can be either Event or Gate. There are 4 special types of events: Basic, intermediate, undeveloped, external events; and 2 types of gate: and gate and or gate. Each unique event in a fault tree is usually a description of the system in a curtain state (mode), for example an event is "the tram is derailed". Instead using a sentence of natural language to describe the event, we proposed to express the event in a format of *(system element).mode* in order to highlight the relationships of an event in a fault tree and the element in the system architecture.

$$event = system\ element.mode \qquad (1)$$

In order to keep a record of the traceability between system architecture and fault tree, the information to be documented includes element's type, name, viewpoint, perspective, mode, and event in the fault tree. In practice, these can be logged in a table, for example in Table 1.

Table 1 Example of Element-Event table

Name	Type	Viewpoint	Perspective	Mode/Event	Indexed in FT
Points motor	Physical	SV-01 points	Solution	Out of control	E2
...					

Thus the meta-model of the fault tree can be extended as shown in Figure 4.

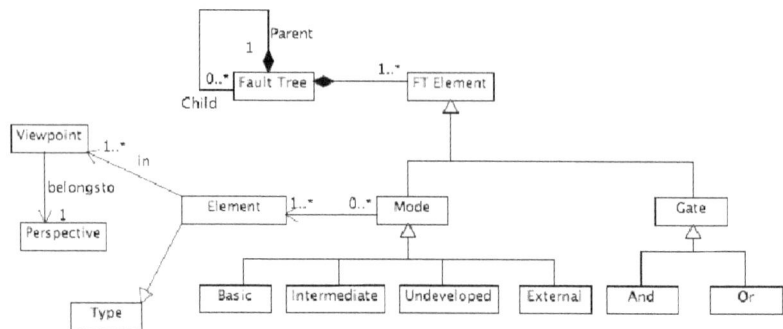

Fig. 4. Extended meta-model for enterprise system

4.2 Failure propagation analysis

The intention of allowing safety assessment early in the system design process is to build safety into the system architecture. In a fault tree the events always belong to different components and in the case of complex socio-technical system these components may be described in different views. In other words, the failure propagation path identified in the fault tree may cross multiple views. So it is difficult to have a good understand of the failure dependencies among components in different views. In order to help the analysis, we developed a software tool to visualise the failure propagation path through the enterprise system. Figure 5 is a screenshot of the tool.

4.3 Tool support

The purpose of software tool is to visualise the failure propagation path in the fault tree. After construct a fault tree, the safety engineer can flip-flop an event to examine the failure propagation and by referencing to the element-event table, the element can be quickly located in the enterprise architecture. Thus the system and safety engineers can quickly allocate the related components in different views and analyse the dependencies in terms of safety failure. With the help of the software tool, it seems that we have found a solution to the problem that the traceability between system architecture and safety analysis is indistinct. In next section, we firstly demonstrated the approach of safety analysis on a socio-technical system and the use of our tool, then had a discussion on the general safety analysis of socio-technical systems.

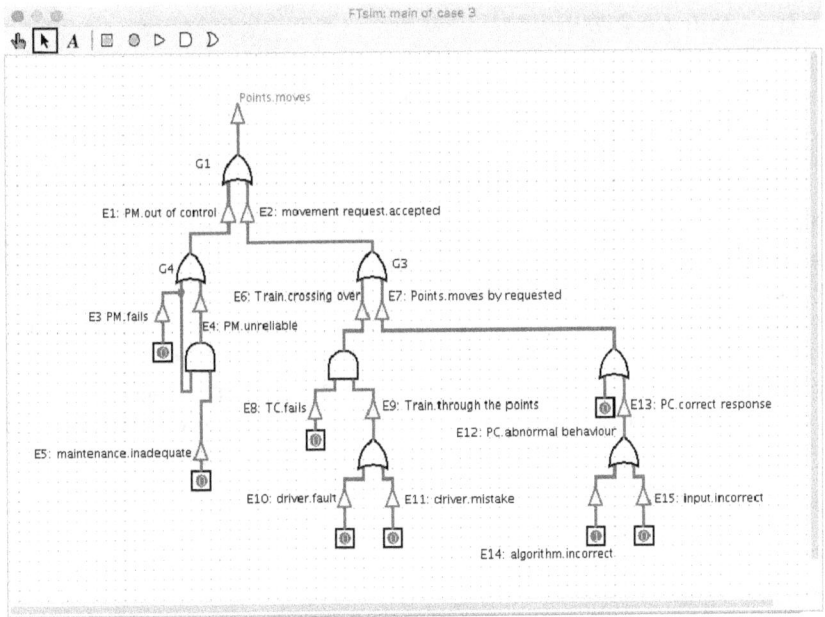

Fig. 5. Screenshot of software tool

5 Case study

System safety analysis should be a process that takes into account both social and technical factors that influence the functionality and usage of computer-based systems. The rationale for adopting socio-technical approaches to systems safety analysis is that systems sometimes meet their technical 'requirements' but accidents still occur. The root of this problem is that technical-centric approaches to systems do not properly consider the complex relationships between people who enact operational processes to deliver services of the overall system and the system or systems that support these processes. In the case study, we deliberately selected a tram accident: a tram derailment at Croydon East in 2012. In the accident report not only the failure of technical systems but also errors of humans/social systems are identified. In the paper, there is no space to illustrate the complete enterprise architecture and the whole fault tree analysis. We show a part of the fault tree analyses which is adequate to demonstrate the application of our approach.

5.2 Croydon tram derailment

The case study is a simplified operational scenario of a tram system. In the system, the tram is driven by a human driver. Under the supervision of a signalling system, the driver stops the train at one of platforms at a station for a couple of minutes and then drives the tram from the station.

At the station, two running tramlines divide to serve three platforms. Platform 1 serves eastbound trams, platform 3 westbound trams, and platform 2 serves trams travelling in either direction depending on the operating mode set by the tram control room. In this case study, we focused on the scenario that station is set to operate in mode A, which means that the signaling system would route a tram to platform 2 if platform 3 has been already occupied when it approached. Figure 6 demonstrates the schematic diagram of the station. In the following sections, the component systems in this case study are analysed in detail.

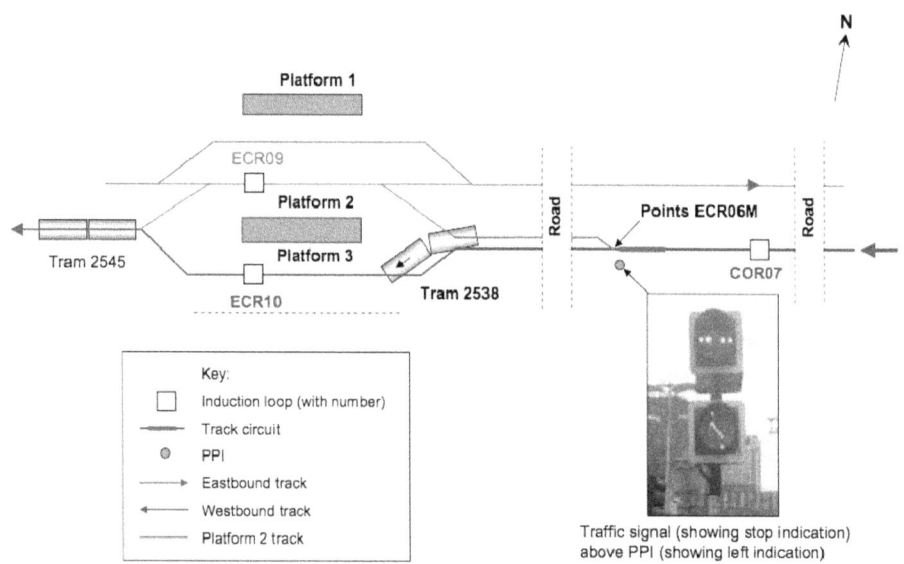

Fig. 6. Schematic diagram of station (derived from (RAIB 2012))

During the operation, an approaching tram is detected by a track circuit (COR07), which triggers the signalling system to control the point (point ECR06M) to direct the route of tram to either platform A (platform 3 in the case study) or platform B (platform 2 in the case study). Normally, the drive will stop the tram in front of PPI (Points Position Indicator) and wait for the signal system showing the route information on PPI. Meanwhile the tram operation company also allows that the drive can drive through the point (point ECR06M in our case study) without waiting for the signal at PPI if the drive can see the front tram is leaving the platform. When a tram is stopped at a platform, track circuit at each platform (ECR09 at platform 2 and ECR10 at platform 3) will send a signal to indicate the platform is occupied.

On Friday 17th February 2012, Tram 2538 was approaching the stop in the west-bound direction. After observing that the PPI for points ECR06M was displaying the indication to platform 3, the driver applied power to take the tram over the points. The leading bogie of Tram 2538 was directed towards platform 3, but the points moved as the bogie was passing over. As a consequence of the points moving, the centre and trailing bogies derailed. The accident has been investigated and the findings were published in (RAIB 2012). In this accident report, the direct cause of this accident has been identified that the track circuit inside of the points ECR06M did not respond to Tram 2538, which is the reliability problem of the TC (Track Circuit).

The fault tree from the analysis is shown in Figure 7. In this case study, the components PM (points motor), points, tram, TC (track circuits), and interlocking were modelled in the Procurement Perspective and the movement request and input data were modelled in the Solution Perspective. And by triggering the events in the software tool, we analysed different failure propagation paths and considered the safety protections in the system architecture.

The case study is small but fits for our purpose to demonstrate how our approach is applied to analyse a real-world case study. Based on this case study, we discussed the safety analysis of socio-technical systems in depth in next section.

5.2 Discussion

Safety engineering in general is about managing safety risks. This can be done by implementing barriers to prevent certain hazardous events from occurring. A control can be any measure taken that acts against some undesirable safety events (triggers), in order to maintain a desired state. In general the system engineering needs a systems thinking that the safety controls may come from different layers of system, i.e, engineering activities, maintenance activities, and operations activities. All these activities can be modelled in enterprise architecture. For example, the engineering activities could be in the viewpoints in solution perspective; maintenance activities are in the viewpoints in the management perspective and procurement perspective; and the operations activities are in the views from concept and solution perspective. However, as the system architecture becomes more and more complex, in which the necessary information has to be captured in multiple views, it is difficult to see the path of how a trigger event evolves to a serious loss through the system. In the case study, the driver's role and responsibility were described as resources in the concept perspective, it is easier to see the importance of these in the fault tree, but it becomes difficult when considering the implementation of safety controls in the enterprise architecture because tram's activities were modelling in the views of procurement perspective. The software tool was designed to visualise the path of failure propagation which can help such analysis.

6 Conclusion

In the paper, we argued that on one hand enterprise architecture is a well-defined practice to provide a relatively comprehensive platform for system engineers to conduct architecture-based analysis, design and planning, and on the other hand it makes the safety analysis lose the benefits that model-based approach can bring to. In this paper, we proposed a model-driven approach to bridge the system development process and safety assessment process by identifying the links in their meta-models. Last but not least, we also developed a software tool to visualize the mappings between safety analysis results and enterprise architecture and the failure propagation paths.

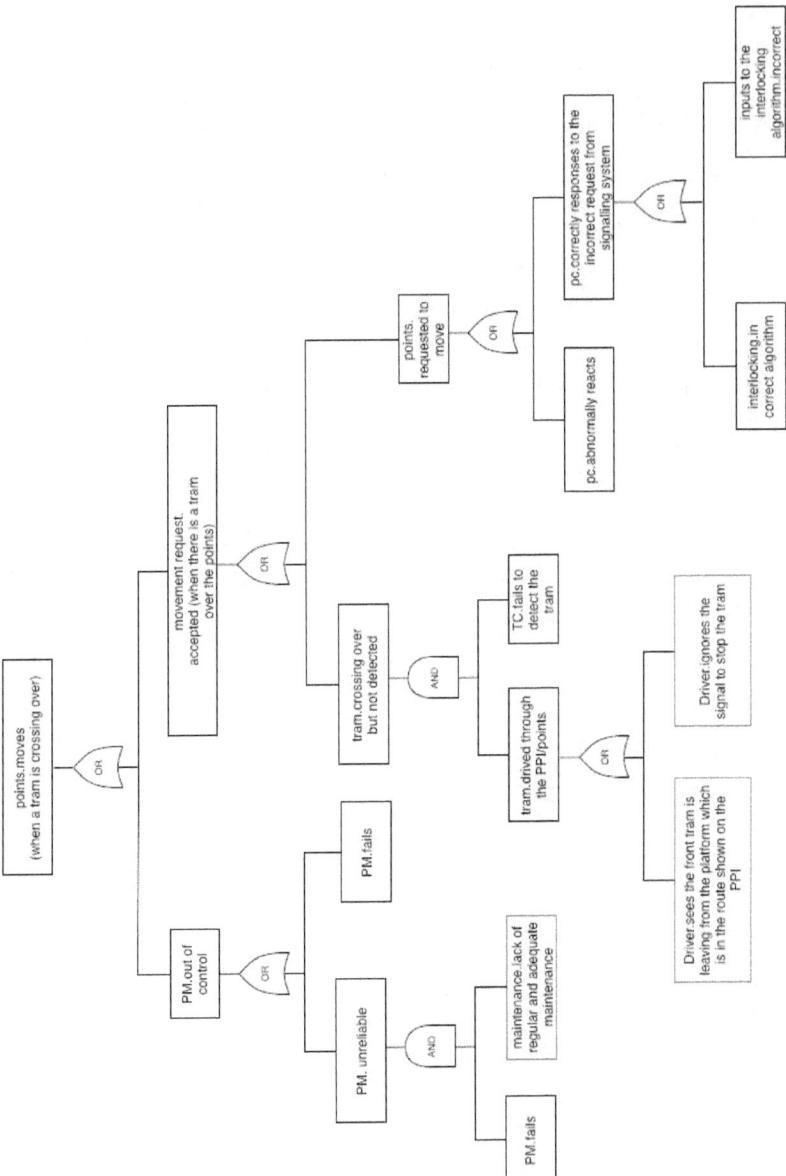

Fig. 7. Fault tree analysis of tram's derailment

References

Reason J, Parker D. Lawton R (1998) Organizational controls and safety: The varieties of rule-related behavior, J Occupational and Organizational Psychology, 71, 289-304

Rasmussen J (1997) Risk management in a dynamic society: a modelling problem, Safety Science, 27, No. 2/3, 183-213

Leveson N. (2004) Engineering a safer world: systems thinking applied to safety. MIT Press

RAIB (2012) Rail accident report: derailment of a tram at East Croydon 17 February 2012.

MODAF   (2012)   The   MOD   Architecture   Framework   (MODAF),
https://www.gov.uk/guidance/mod-architecture-framework

RFA,   Railway   Functional   Architecture,
http://www.futurerailway.org/research/Pages/EA%20HTML/index.htm

Rebovich G, White B, (2011) Enterprise systems engineering: advances in the theory and practice. Edited, CRC Press.

Checkland P, Scholes J (1990) Soft systems methodology in action. John Wiley & Sons.

IEEE Standard 610.12 (1990)

TRAK, TRAK Enterprise Architecture Framework, http://trak.sourceforge.net/

Redmill F (2002) Human factors in risk analysis. Engineering Management Journal, Auguest, 171-176.

Troche C, DeRosa J, Bigelow J, et al (2011) Architectures for enterprise systems engineering. In Book of Enterprise systems engineering, edited by Rebovich and White, Ch 6, CRC Press

Hollnagel E (2012) FRAM: the functional resonance analysis method: modeling complex socio-technical systems. Ashgate

Formal Data Validation in the Railways

Thierry Lecomte, Erwan Mottin

ClearSy

Aix en Provence, France

Abstract *Safety-critical systems and software require particular care when their parameters have to be verified and validated, as any mistake may lead to a catastrophic scenario during their operating use. A recent technique, called formal data validation, enables an improvement in the level of confidence of the verification/validation process by associating a formal data model to the parameters, and by formally checking that these parameters fit within the model. This paper reports on the development and use of such tools for industrial railway applications.*

1 Introduction

Historically, the B Method (Abrial 96) was introduced in the late 80's to design safety-critical software correctly. With B, a software model is built before its implementation; the model is proved to be consistent with static and dynamic properties; the implementation is proved to refine the model. The approach is entirely aimed at modelling and proving software behaviour. Promoted and supported by RATP (Régie Autonome des Transports Parisiens who are the operator of bus and metro public transport in Paris), the B method and Atelier B, the tool implementing it, have been successfully applied to the industry of transportation. Today, Alstom Transport Information Solutions and Siemens Transportation Systems are the main actors in the development of B safety-critical software. They share a product-based strategy and reuse as much as possible existing B models to develop future metros.

In the mid '90s Event-B (Abrial 2005) enlarged the scope of B to analyse, study and specify not only software, but also whole systems. Event-B has been influenced by the work done earlier on Action Systems (Back 1991) by the Finnish School. Event-B is the synthesis between B and Action System. It extends the usage of B to systems that might contain software but also hardware and pieces of equipment. In that respect, one of the outcome of Event-B is the proved definition of systems architectures and, more generally, the proved development of, so called, "system studies" (Sabatier et al 2006, Sabatier et al 2008, Hoffmann et al 2007, Lecomte et al 2007, Lecomte 2008), which are performed before the specification and design of the software. This enlargement allows one to perform failure studies right from the beginning in a large system development. Event-B has been used to perform system level safety studies in the Railways (Sabatier 2012), allowing to formally verify part of the whole system specification, hence contributing to improve the overall level of confidence of the railways system being built.

The verification of behaviour, based on Event-B system specification or B software specification, is quite easily achievable by semi-automated proof. However the verification of parameters (that tune the system or the software) against properties may turn out to be a nightmare in case of large data sets. For the Meteor metro (line 14, Paris), software and data were kept together in a B project (Behm & al 1999). Demonstrating data correctness regarding expected properties was really difficult because of required iterations over large sets of variables and constants (and their domains). Indeed the Atelier B main theorem prover is not designed for this activity (Milonnet 99) that requires more a model checker or constraint solver rather than a theorem prover. Later on, software and data started to be developed and validated within two different processes, in order to avoid a new compilation if the data are modified but not the software. Data validation started to be entirely human, leading to painful, error-prone, long-term activities (usually more than six months to manually check 100,000 items of data against 1,000 rules).

In this article, we present a formal approach, based on the B/Event-B mathematical language and the ProB model checker and constraint solver, designed and experimented on several projects for the validation of safety-critical railways data.

2 Data validation and formalism

Verifying railways systems covers many aspects and requires a large number of cross-verifications, performed by a wide range of actors including the designer of the system, the company in charge of its exploitation, the certification body, etc. Even if complete automation is not possible, any automatic verification is welcome as it helps to improve the overall level of confidence. Indeed a railway system is a collection of highly dependent sub-system specifications and these dependencies need to be checked. They may be based on railways signalling rules (that are specific to every country or even each company in a single country), on rolling stock features (constant or variable train size or configuration) and operating conditions.

By data validation, we mean the validation of the parameters (i.e. constants) that determine a specific behaviour of a software/system over a wide range of possible sets of values. Microsoft Excel defines data validation in terms of type checking: a cell may contain a date, an integer, a string or a floating point number. In our case, the data to validate are not only scalar but also represent more complex structures like graphs. A metro line is seen as a graph, made of connected tracks with distributed signals and switches implementing signalling rules. Graphs are encoded through a large number of tables.

In figure 1, as an example we consider a set of track circuits {t1, t2, t3, t4, t5} defining a simple line L1 (no switches). In the raw data entering the validation process, this set is encoded as a table TrackCircuit where TrackCircuit[0]=t1, TrackCircuit[1]=t2, etc. The function Next models the connectivity between track circuits: as a partial function, it associates zero or one track circuit to any track circuit of the set TrackCircuit. Next(t1)=t2 means that the successor of t1 is t2. The first track circuit of the line L1 may be defined as the track circuit that can't be reached through the Next function (it should not belong to the range of Next). In our case, it is t1. In case the track circuits are looped (i.e. Next(t5)=t1), it is not possible to define FirstTrackCircuit as it belongs to the empty set. If we consider the abscissa of the beginning of each track circuit (function KpAbs), we can also assert that the abscissa of successive track circuits are ordered, that is for any track circuit tt, KpAbs(Next(tt)) > KpAbs(tt). The KpAbs function may be provided as an input table or has to be calculated by using the length of each track circuit, considering that the FirstTrackCircuit has a null abscissa and by summing the length of the successive tracks circuits. In this case, the verification is composed of several, sequential steps; the verification is complete only if all steps are successful.

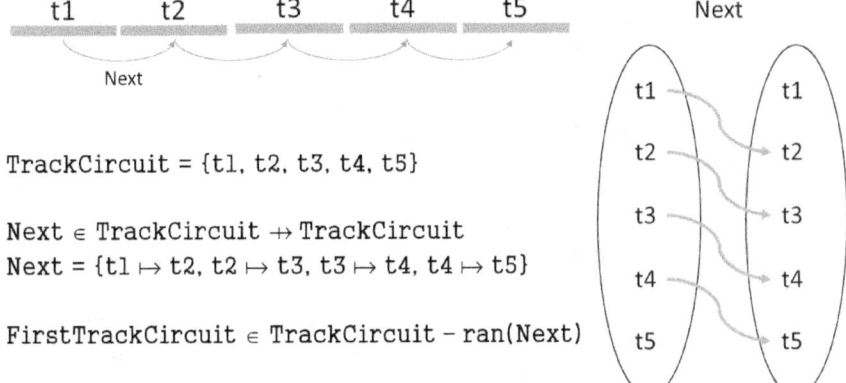

TrackCircuit = {t1, t2, t3, t4, t5}

Next ∈ TrackCircuit ⇸ TrackCircuit
Next = {t1 ↦ t2, t2 ↦ t3, t3 ↦ t4, t4 ↦ t5}

FirstTrackCircuit ∈ TrackCircuit – ran(Next)

KpAbs ∈ TrackCircuit → ℕ
∀x·(x∈ TrackCircuit ∧ x ∈ dom(Next) ⇒ KpAbs (Next(x)) > KpAbs(x))

Fig. 1. Modelling track circuits and related properties like connectivity (the function Next) and abscissa (Kilometre point or Kp). Next function set-based representation is on the right.

We could also express more complex properties like "a path exists (made of successive track circuits) between any two different track circuits" i.e. all track circuits are connected to the signal network. The notation used, set theory, allows for writing properties elegantly and concisely that fits well with the graph-based track topology. A railway network is made of a collection of sub-networks where different signalling rules apply and for which this formalism appears adequate: properties are specified to apply only on subsets that are identified through a mathematical predicate.

The modelling part of this approach consists in translating natural language sentences into mathematical predicates expressing typing information and constraints over several mathematical entities. Some intermediate constants need to be computed prior to complete the verification; if these intermediate constants are used in several verification rules, their definition is localized in a specific file containing the so-called "intermediate constructs": they are computed once and their value is reused when necessary, in order to save computation time. Even if the natural language is most of the time subject to interpretation / inconsistent / wrong / incomplete, the formalization activity allows to save the knowledge of expert engineers in a persistent, sound way. The verification of the tracks data, usually counted as 50,000 to 100,000 Excel cells (see figure 2), against these signalling rules constituting the data model requires a validation engineer to spend several months into error-prone manual / human checks.

Name	ID	IP	Type	UpLink	DownLink	Length	GPS 1
Route_tx_001	243		R	Route_tx_005	Route_vx_002	345	
Route_vx_002	128		R	Route_vx_002	EndLine_000	128	
Switch_w_003	256	192.16.4.55	S	Route_vx_128	Route_tx_006	23	
Relay_s_004	12	192.16.4.10	Y				N 50.85 963
Route_tx_005	3		R	Route_tx_006	Route_vx_128	291	
Relay_s_001	55	192.16.4.125	Y				
Route_tx_006	22		R	EndLine_001	Route_vx_002	110	
Route_vx_128	127		R	Route_tx_006	Route_vx_002	145	
Switch_w_009	242	192.16.4.10	S	Route_vx_128	Route_tx_005	34	
EndLine_000	0		E		Route_vx_002	1	
EndLine_001	1		E	Route_vx_002		1	
Signal_xs_002	32	192.16.4.12	G	Route_vx_128		22	
Signal_xs_003	33	192.16.4.13	G	Route_tx_006		51	
Balise_b_001	301		B	Route_vx_128		0	N 50.85 933
Balise_b_002	302		B	Route_tx_005		0	N 50.86 123

Fig. 2. Raw input data containing respectively equipment ID, IP address, associated routes, length and GPS position. Data may be missing, incorrect, IP addresses duplicated, etc.

This activity is really difficult and demanding as data is evolving during the whole period (CAD data is replaced by real plant data, topology is modified after *in situ* testing, etc.) and several interleaved iterations are required. The challenge here is to design a tool or diverse tools able to automate these verifications while improving the level of confidence of the verification (the tool has to be generic and not specifically designed for this activity).

2 Tools are everything

As seen on figure 3, our experience is linked to multiple tools developed and applied to several cases during both research and industrial experiments. The tools, the formalism and the method have been improved over the last 12 years to reach a stable state and are applicable to industry-strength systems.

In France, "Atelier de Qualification Logiciel" RATP laboratory is in charge of qualifying railways applications before their installation. A specific tool, initially developed for validating Paris metro line 14 data, representing more than 300,000 lines of C++ code, was too difficult to maintain and to adapt to other lines and hence was not reused for other lines. RATP initiated the development of a generic tool to verify trackside data for the metro line 1 in Paris that was being automated. Initially tested on Paris metro line 13 configuration data, the tool has been able to check 400 definitions and 125 rules in 5 minutes.

The approach that has been kept during 12 years consists of formalizing properties with B mathematical language (set theory, first order logic), and in generating a B machine containing both the properties (the data model) and the data to verify. The compliancy is then checked by a generic tool. For this first application, the PredicateB predicate evaluator, developed by ClearSy, was able to parse data (XML, csv or text-based formats), load rules and verify that data complies with rules. The PredicateB tool is a symbolic calculator able to manipulate B mathematical language predicates in order to animate a B formal model: constants and variables initial values are calculated, then operations are executed depending on their guards (enabling conditions) and their substitutions (variable modifications). Symbolic values are scalars, sets, functions, etc. However PredicateB has limited capabilities for non-deterministic computations ("find an element such as ...."): it is not able to find all possible values for any non-deterministic substitution or to find all counter-examples. Moreover the way the errors are displayed may lead to difficult analysis when the faulty predicate is complex as it requires injection of the data into the predicate.

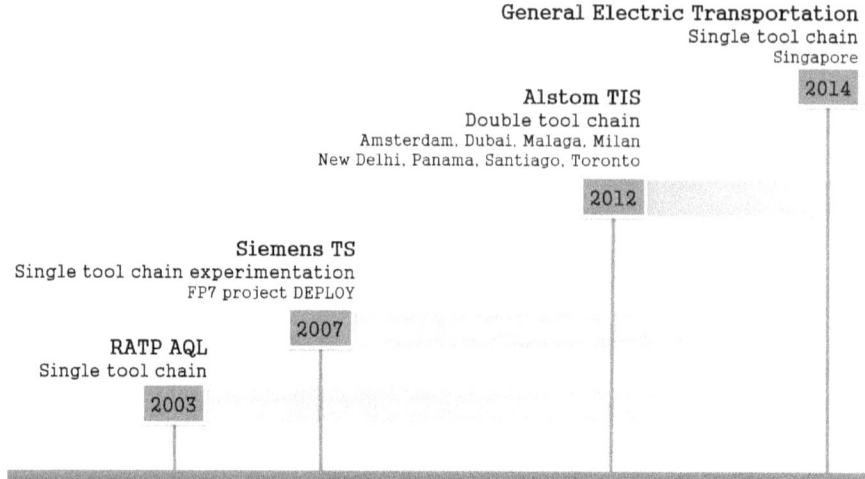

Fig. 3. Formal Data Validation History reported in this article. Formal data validation technology is today used by all companies mentioned.

During the DEPLOY project, the University of Düsseldorf and Siemens Transportation Systems have elaborated a new approach, based on the ProB model checker. The ProB model_checker embeds several well performing heuristics for reducing search space (symmetry detection for example), is able to better handle non-deterministic substitution and to provide a more complete set of counter examples. The major outcome of this decision was a dramatic reduction of the validation duration from about six months of human verification to some minutes of computation (if we set aside the time to formalize verification rules) while being able to take into account all properties. Data provided by Siemens contained a number of added, identified errors but after the verification, undetected errors were uncovered by the tool. The discovery of this error that remained unnoticed by validation engineers while the target metro was in active use clearly demonstrated the added value of this approach and its ability to reproduce results in minutes. Data was extracted from ADA source code and properties came from B models used for the software development. In the case of the San Juan metro (Central America), 79 files with a total of 23,000 lines of B were parsed to extract 226 properties and 147 assertions. The verification took 1017 seconds and led to the discovery of 4 false formulas (one was not expected by Siemens).

ProB was then experimented with great success on several projects: Roissy Charles de Gaule airport shuttle, Barcelona line 9, San Paulo line 4, Paris line 1 and Algiers line 1. On that occasion, ProB was slightly improved in order to deal with large scale problems and well validated in order to ease its acceptance by a certification body. However analyzing false properties remained difficult as it requires browsing the complete model valued with the complete set of data.

Alstom Transport Information Solutions decided to experiment with a new approach by reusing successful features of previous experiments. A new tool was designed and implemented, still based on ProB. The verification rules are expressed using the B mathematical language and structured as B operations. Instead of having to deal with large, quantified predicates, a verification rule is decomposed in small steps that allow displaying accurate error messages helping to determine the source of the error. Figure 4 shows a rule searching for all signals associated with an interlocking territory. The clause WHERE allows filtering of data: it should be a signal, with an ID, and a geographical position (geopoint) included into the interlocking zone. The clause VERIFY specifies the conditions expected for all filtered signals. In case the predicates of this clause are not verified, an error message is displayed for each signal found.

```
FOR
        sig
WHERE
        sig : sys_sud_er::Signal &
        sig : dom(sys_sud_er::Signal__dptId) &
        sig : dom(ic::sys_sud_er::signal_geopoint) &
        ic::sys_sud_er::signal_geopoint(sig) : ic::sys_sud_er::zone_GPZone
(sys_sud_er::IXL_Core__singleZone(ixl))
THEN
        VERIFY
                sys_sud_er::Signal__dptId(sig) : ran(sys_sud_er::IXL_Core__signal(ixl))
        MESSAGE
                «The signal %1 belongs to IXL_Core %2 territory but is not referenced
                among its signals.»
                ARG sys_sud_er::Signal__name(sig) TYPE STRING
                ARG sys_sud_er::IXL_Core__name(ixl) TYPE STRING
        ENDVERIFY
ENDFOR
```

Fig. 4. Example of verification rule. Signals belonging to an interlocking territory are searched (clause WHERE); such signals have to be linked to this interlocking (clause VERIFY). If not, an error message is displayed for each faulty signal found (clause MESSAGE).

ProB is the central tool for the verification. It has been modified in order to produce a file containing all counter examples detected and slightly improved to better support some B keywords. The resulting tool has been experimented with success on several ongoing developments (Mexico, Toronto, Sao Paulo, and Panama) to verify up to 50,000 Excel cells with up to 200 rules. A first round allowed defining required concepts, intermediate constructs (predicates used by several rules) and formalizing a set of generic rules that are shared by all projects. During the next stages, specific project rules and data files were added. A complete verification is performed in about 10 minutes, including the verification report. The process is completely automatic and can be replayed without any human intervention when data values are modified.

For certification purposes (the overall process should be SIL4 compliant), a diverse tool, PredicateB++ (a newer version of the first tool developed in 2003), has been added to the toolchain in order to provide a confirmation of the ProB decision: it reuses the values computed by ProB (especially coming from non-deterministic substitutions) and performs symbolic calculation. The PredicateB++ weak point is hence solved and joint positive decisions lead to final agreement on the rule verification. This tool has been used for the verification of several metro lines data (not only the data for the automatic pilot, as it was the case in 2003). A methodology and a process were defined to handle up to 2000 different rules. Some of these are generic (800), checked for consistency once and then reused from project to project. Some others are specific to lines or stations, and require specialization of existing rules. Defining entirely new rules is now exceptional.

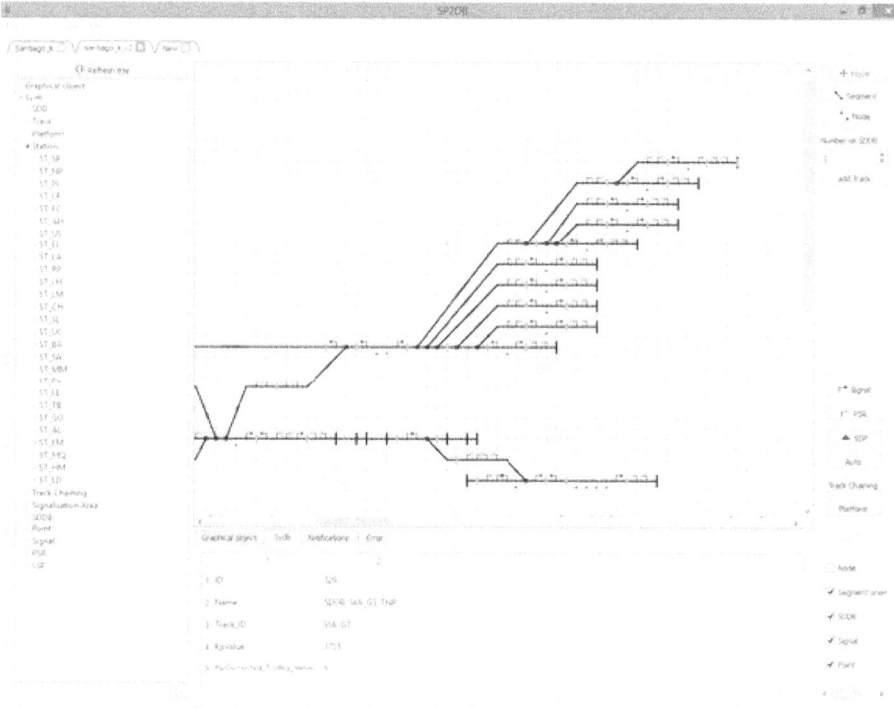

Fig. 5. A tool aimed for specifying graphically non-trivial test cases that validate the formal data model.

Most of the rules fit in one page, but some rules are really large, up to 10 pages, as they embed several small steps or they contain a lot of implicit information. To validate these particular rules, a specific process was devised: rules have to be cross-read and tested by an independent engineer. If reading activity is quite straightforward, testing rules require to fully understand the objective of the rule. Then the identification of the different kind of situations where the rule may be not verified by data enables designing a non-trivial network configuration by inventing from scratch raw input data (could be csv, json, XML, RailML, etc.). In this case, the risk of mistake is high to define a bad configuration where one could wrongly validate a rule by not checking an error condition adequately. A specific tool was the designed to fulfil this need (see figure 5): it allows the tester to design graphically a specific network configuration and to generate track data in the same format as it is provided to the verification tool. For complex rules, many different network configurations may be designed, in order to reduce the probability to negligible of having a software error covering exactly a rule design error.

2 Conclusion and perspectives

Formal data validation appears to be of paramount importance in safety-critical systems. Formal modelling adds semantics that allow the definition of more consistent verification rules, by removing ambiguities and most errors. These rules encode the knowledge and know-how of a company in a domain: knowledge is incrementally built-up, made permanent and more easily applicable to new systems. The rules can be read and checked independently, so data model verification may be easily distributed.

Using generic symbolic verification tools increases the level of confidence of the verification results, as the tool is not specifically built for a project/system and makes a better profit on a considerable return of experience from a large user base. Diverse tools, using different technologies and developed by different teams, can be used for highest safety critical levels. The overall verification time is then lowered from several months to few minutes or hours, that may be replayed at will.

However some verification rules imply large quantities of implicit knowledge that have to be made explicit in the rules. In that case the predicates in the verification rules could be lengthy, difficult to read and to validate. So manual testing of verification rules is required, checking that for any category of error a rule is designed to detect, the tool should trigger an error message. Interactive test configuration tools have been designed to build test scenarios and reduce the risk of human mistake.

Finally introducing formal modelling and symbolic tools has improved the level of confidence of the overall verification process but at the cost of requiring higher qualifications (not only signalling engineers) for formal modelling and verification. Similarly time to complete the verification has been shortened by several orders of magnitude on real size systems. Even if a human is still required for testing the validity of the modelling, formal data validation is being used by most major railways players and it is definitely a good signal to the formal methods community.

Acknowledgments   Work reported in this article has been partly funded by European Union (FP7 projects Rodin and DEPLOY), Alstom and General Electric companies.

References

Abrial, J.R. (1996) , The B-book: Assigning programs to meanings, Cambridge University Press

Abrial J.R (2005) Rigorous Open Development Environment for Complex Systems: event B language

Back R.J. (1991) Stepwise Refinement of Action Systems, Structured Programming #12 p17-30 (Springer verlag ed)Alexander C, Ishikawa S, Silverstein M et al (1977) A pattern language: towns, buildings, construction. Oxford University Press

Behm P, Benoit P, Faivre A, Meynadier J.M (1999) Météor: A Successful Application of B in a Large Project

Hoffmann S (2007) The B Method for the Construction of Micro-Kernel Based Systems, ZB 2007

Lecomte T (2007), Formal Methods in Safety Critical Railway Systems, SBMF 2007

Lecomte T (2008), Safe and Reliable Metro Platform Screen Doors Control/Command Systems, FM 2008

Lecomte T, Burdy L, Leuschel M (2012) Formally Checking Data Sets in the Railways, S-Event-B 2012: Workshop on the experience of and advances in developing dependable systems in Event-B, in conjunction with ICFEM 2012 - Kyoto, Japan, November 13, 2012

Lecomte T (2015) Formal Virtual Modelling and Data Verification for Supervision Systems, FM 2015: Formal Methods, Volume 9109 of the series LNCS pp 597-600

Leuschel M, Falampin J, Fabian F, Plagge D (2009) Automated Property Verification for Large Scale B Models. In: Proceedings FM 2009. Springer-Verlag.

Leuschel M (2012) ProB, ProR and Data Validation with B, FM'2012, Industray Day.

Milonnet C. (1999) B Validation Book; Internal document ref (Matra Transport International)

Sabatier D (2006) Use of the Formal B Method for a SIL3 System Landing Door Commands for line 13 of the Paris subway, Lambda Mu 15

Sabatier D (2008) FDIR Strategy Validation with the B method, DASIA 2008

Sabatier D (2012) Formal proofs for the NYCT line 7 (Flushing) modernization project, DE-PLOY Industry Day, Fontainebleau

Model-Based Risk as a Path to
Safer Medical Devices

Carmen Castaño[1], Robert Schmitt[1,2]

[1]Fraunhofer Institute for Production Technology IPT

[2]Laboratory for Machine Tools and Production Engineering (WZL) RWTH Aachen University

Abstract *All major trends in MedTech have one thing in common: medical devices are going to be more complex than ever - and so are the networks they belong to. A necessary measure to cope with this and other emerging challenges is a satisfactory risk management process. Today, manufacturers address safety hazards with a multitude of techniques, all of which are document-based approaches. This paper presents research on how applying model-based risk management could eliminate disadvantages that are endemic to existing methods, like uncertainty of coverage, incompatibility of professional mind-sets or typical bias-by-design flaws. Risk management, based on a structured, computerised model of both the physical product and its lifecycle, has the potential to improve processing in all stages. We explain how our concepts allow for comprehensive risk identification, interconnected expert judgements and standardisable classification for better risk evaluation; they also help enforcing risk treatment by reducing process cost.*

1 Introduction

1.1 New Challenges

Nowadays, most companies are forced to shorten product development cycles as they are coping with fast technological changes and competitive time-to-market. In addition, globalisation of the marketplace significantly increases the number of competitors and the complexity of the product life cycle as more companies and partners have to interact in the value chain (Oehmen et al. 2010). An adequate Risk Management (RM) is one of the most crucial tools to face all these changes. RM in MedTech is fundamental to guarantee device usability, safety, and regulatory compliance (Ganeshkumar 2002).

Manufacturers of medical devices traditionally have been small or medium enterprises (SME) operating in national or regional markets. While their venture capital is more or less the same, innovation cycles shorten and products become more complex, as do the services and the networks they work in. To achieve rapid turnaround, specialized medical devices are introduced into new market fragments (Pammolli et al. 2005, MedTech Europe 2014) that do not have the same backgrounds as the traditional markets. Many developing and emerging nations do have the need for specialized medical care and would like to acquire modern medical devices. However, manufacturers struggle with lower technology levels and missing a supply chain to support their products with high quality consumables and spare parts (WHO 2010, Chan and Larsen 2010). Also, globalized production implies taking responsibility for RM of your component suppliers as well. One can imagine many other examples that show the need for a consistent and cost-effective RM for the whole product life cycle and alongside the complete process chain. Nevertheless, everyday work life in said companies reveals a different picture.

1.2 Barriers for Comprehensive and Pervasive RM

A survey among 180 manufacturing companies, conducted by Fraunhofer Institute for Production Technology (IPT) (Zentis et al. 2011), showed that the majority of those companies who think of their own products as innovative and complex, state that they implemented an RM process and value the capability of avoiding failure in early stages of product development. The very same panel of companies commentated that they cope with challenges such as insufficient clarity and precision of RM methods, costly risk assessment methods, inaccurate methods to determine primary risks, and inadequate methods of risk analysis to determine risk causes.

Interestingly, with 59% the financial security of the respective companies comes in first in the list of motivations to run risk analysis, even before preventing product failure and compliance to laws, standards and guidelines. At the same time, 62% admit placing risk analysis procedures mainly after failure has occurred. Now, when this very same companies argue to not go through with risk treatment because they fear it to bring low return on investment (ROI), it shows that many top managers have not understood how RM affects production processes or at least that RM reports and recommendations do not sufficiently pervade their decision making. If a risk is already identified and evaluated, inhibiting its treatment may not be financially profitable, but will likely provoke unnecessary flaws down the line in product life cycle. Additionally, it will often bring the documentation of, and communication about, the particular risk to a halt. Decision makers seemingly tend to disregard the iterative nature of RM alongside the process chain and thus see investments in RM as lost when they still might pay out later on (Zentis et al. 2011).

These behavioural patterns are not well understood yet and, looked at from outside, often contradictory. We need to learn more about the obstructing and promoting factors, the underlying circumstances and prejudices. That is why we are currently preparing an empirical social study directed at the motives and mindsets of the RM players in MedTech companies. We wish to question those performing RM as well as the leaders controlling them in semi-structured expert interviews.

2. Deficits of Document-based RM

In this section will be described seven deficient areas in RM that were identified with the help of the studies mentioned and an extensive literature research links directly to the document-based approach. Naturally, those are areas of opportunities for model-based risk management (MBR).

2.1 Missing Comprehensiveness

As Risk Identification is the first of a number of consecutive steps, it must be implemented in a comprehensive fashion. While the execution cannot be infallible, methods and processes must be designed comprehensively (Grubisic et al. 2011, ISO 31000 2009). As IEC/ISO 31010 illustrates in a coherent manner, document-based methods cannot master the concurrence of complexity and comprehensiveness. Methods like FMEA that show good procedural control are only realisable for product structures of very limited scale. The product life cycle of a medical device is more often than not a highly-hierarchical entity with multi-level depend-encies and plenty of manipulators; complexity is implied. The resulting workload will inevitably destroy the comprehensiveness. However, failure to identify a risk in MedTech is fatal. A risk not identified is a risk not evaluated and therefore is a risk not treated. On top of that, all document-based approaches drop in compre-hensiveness step by step due to transcription and copy errors, bureaucratic loops or simply files getting lost (IEC/ ISO 31010 2009, comp. Delligatti 2013).

RM that does not deal with the complexity of the analysed system and its like-wise complex interactions leads to residual risk. This risk will likely appear in a later RM iteration in the product life cycle, but odds are it resurfaces as failure. In MedTech, that means people being harmed too often (Radermacher et al. 2004).

2.2 Uncertainty of Coverage

Usually, organisations do not possess a significant database of their legacy prod-ucts which would allow them to build the probability distribution function or de-termine statistically sound risk probability (Škec et al. 2013). While uncertainties concerned with a single risk and organization can be faced with sensitivity analy-sis, solid documentation and risk communication (IEC/ ISO 31010 2009), the un-certainty of coverage cannot be dealt conveniently with these methods. The doc-uments informing the RM panel about product breakdown structures of prior ver-sions, application scenarios, functionality criteria or regulatory work come in all sorts of styles and formats, typically not accompanied by any quantitative estimate or calculation of error. The workload that would have to be shouldered to quanti-tatively equalise all demands, suitably discuss and agree on acceptable coverage levels and how to calculate their probabilities, operate the accounting and finally interpret, document and communicate the results, would be enormous and in most cases unbearable. In fact, QA departments regularly struggle with the straightfor-ward task of finding and executing measures for linear processes that can be agreed on from shop floor to top-tier of the organization.

2.3 No formalization of Risk Identification as Single Step

On the majority, organisations implementing general RM processes (as against applying RM by team or project) accomplish a reasonable overall level of stand-ardisation. But taking a closer look at the singular steps often reveals them being executed in different manners according to their perceived importance and accura-cy. It is perfectly human - though false - to assume that those steps dealing with exact values would need to be more formalised than the rest. Hence, the risk iden-tification: the recognition of critical characteristics, is regarded as a primary stage of the risk analysis rather than a to-be-concluded step of a consecutive process.

This bears consequences: substandard documentation quality, vague procedural instructions and lose ends in form of undiscovered critical characteristics.

2.4 Incompatibility of Results: Multitude of Techniques

Numerousness RM techniques are available on the market which all have their own procedures and dissimilar features concerning complexity, need for expertise, etc. (Grubisic et al. 2011). All these techniques are based on human observation, judgment and creativity. The selection of techniques follows those skills and, hence, depends on the skills recognised by the selector and, of course, his/her own skills. This chosen set has an effect on the hazard identification as every technique has its own strengths and weaknesses. As no available method for risk identification can find all the hazards, a combination of the techniques is more likely to increase the chance of success (Redmill 2002). The selector may chose a combination of techniques that makes the best use of the panel's skill set, one that minimises structural information loss in the transcription of the documentation between techniques or try to find reasonable compromise. For example, in early RM iterations, when there is little known about a new product or a new version brings a major feature change, it is a powerful approach to let a Preliminary Hazard Analysis (PHA) in risk identification be followed by FMEA. Later on, PHA usually will not unravel sufficient detail. Although it still fits the panel's skill set, it should be exchanged for another technique. The selector will have to put up with one of the shortcomings or actually exchange the techniques after a fixed number of iterations and accept even more workload and incompatibility (IEC/ ISO 31010 2009, Škec et al. 2013, p.2).

In many cases though, selection of techniques is not an arbitrary process. Suppliers, authorities, operators, investors and many more bring their own choice of techniques or are themselves bound to standards and guidelines that favour different techniques than the ones considered by the selector. The amount of RM standards around is not beneficial to the selection process (Hall 2011, p.173). So as not to be misunderstood, we praise the line of thought behind most of these standards and guidelines, but the considerable differences in form and ingredients of RM processes they promote proliferates more incompatibility, which is surely contrary to the intended guidance.

2.5 Human Factor

Bias by design. Asking experts for their opinion is often the most promising way of adding new information to your RM process. Their contribution does not only depend on their expertise, but in large parts also on their social skills, creativity, professional background and readiness to judge. These factors are amplified when brought together (which is why an expert panel is worth more than the sum of its "parts"); fueling them will enhance coverage and outcome volume, but also the emergence of any bias held by the experts. For this matter, a biased expert should not be regarded as a liability though. Just as the condition carries negative implications like prejudice against other disciplines so it may feature perseverance at documenting controversial findings and also a certain sophistication in tackling problems. As long as the impact of the bias is clarified, it does not need to deteriorate the results. Again, this is where document-based RM fails. HAZOP, for example, strongly relies on the expertise of the product developers who involved in the RM process show to restraint pointing out flaws in and proposing changes to their own assemblies. The HAZOP structure, though, does not provide a form to record such behaviour and on this aspect proves quite vulnerable to bias.

Mindsets and value systems. Throughout the whole process, good RM actively entails multidisciplinary groups such as design engineers, clinicians, service personnel, and quality and regulatory personnel (Rakitin 2006). One important reason for it is the demand that product life cycle assemblies should not be treated as isolated technical systems as this could lead to missing recognition of the critical characteristics within the interactions (Schmitt and Zentis 2011). Yet, document-based RM approaches are very limited in addressing different professional mindsets. Very much the same way that forms which are used in one profession on a daily basis can perplex outsiders, mindsets might make the difference whether RM documents trigger whole trains of thought - or not. On that matter, coalitions do not always form along traditional departmental lines, because risk culture is influenced also by our value system. Engineers and pharmacologists usually work within feature-driven mindsets, whereas both surgeons and IT security specialists would by any means prefer to avoid failure. Much of a mindset is tacit and not communicated explicitly. This hinders capturing the reasoning behind decisions in interdisciplinary groups, especially in heavily-structured document-based techniques such as FMEA or Hazard Analysis and Critical Control Points (HACCP).

Environmental influence. Additionally, all consecutive steps of risk analysis - identifying consequences, evaluating their severity and determining likelihood of occurrence - are widely influenced not only by the choice of experts joining the panel, but even by environmental factors affecting the panel . Although as an experiment unlikely to be practical, one can easily imagine that the same panel with the same level of expertise, scope of knowledge, mindsets etc. analysing the very same device described by an identical documentation, just given a different time or place, would not reproduce the results of the risk analysis process (Redmill 2002).

2.6 Incompatible work environments

RM alongside the process chain implies confronting stakeholders with RM tasks in their own work environments. Although it is common practice for engineers and medics to collaborate in MedTech RM, they tend to know too little about each other's daily routine. Whereas panel meetings are low on resources, implementing RM tasks conceived in a different work environment can be anything between discomforting and impractical. Missing software resources, regulatory restrictions, unknown routines or the absence of trained personnel can render simple requests impossible.

2.7 Poor Risk Treatment

In comparison to the steps in risk assessment, risk treatment has hardly been re-searched. A review of the existing literature showed its place in current research not to reflect its importance at all. Similarly, Oehmen et al. found that there is no scientific discussion on evaluating alternative treatment options. While their con-clusions are about RM in general, our findings support them for the field of medi-cal devices. Thus, the indifference towards improving risk treatment processes among scientists seems to level with those of practitioners. Both barriers: the fear of lowering ROI and the disinterest in evaluating alternatives, are so high that many decision makers consider the benefits of risk treatment as very low (com-pare section 1.2). Aversion to accept challenges such as high initial investments (time, money, personnel) and tedious tracking of treatment measures is evident and seems to grow with company size (Zentis et al. 2011, Zentis and Schmitt 2011).

3. About Model-Based Risk

Model-based systems engineering (MBSE) is the formalized application of modeling to support system requirements, design, analysis, verification and validation activities beginning in the conceptual design phase and continuing throughout development and later life cycle phases (INCOSE-TP-2004-004-02 2007).

In that sense MBR is a construct that provides a structured model, independent of the type or classification of medical devices, supporting the risk management process by formalization and giving systematic guidelines to all the stakeholders during the whole product life cycle. As all RM documents would be generated in real time from the underlying model, MBR takes control over them away from stakeholders and panel members. This step drastically reduces human error and red tape, but also creates the need for a software layer between database, model and tools on the one side and human beings and their decision processes on the other, for all stages of a RM iteration. Figure 1 shows a rough scheme of the information flow in an iterative RM approach with such a software layer. As the process chain is continuous, a servicing point (a feasible point for recurring install of the MBR), conveys a virtual halt to start the iterative RM process at a certain status quo. From there on through all stages of the process, all input is strictly separated from the actual changes in model and database by the measures of the software layer tools. And while we acknowledge the potential for conflict and refusal to cooperate because of the perceived loss of control, we estimate that the advantages of ubiquitous access, an environment-sensitive display (e.g. an API for the tools you are already using at your workplace) and the bias-reducing uncoupling of content and contributor outweigh them by far. Experience from transformation in other engineering areas and also the INCOSE SE Visions support that estimate (INCOSE 2007, comp. Murray 2012).

RM alongside process chains means picturing RM stages as a sequence of events in an iterative process, which penetrates the continuous product lifecycle at what we have named servicing points. At each of these points in time, all risk treatment actions of the former iteration have to be finished, so all changes are fed to the model and its old version now becomes part of the database. As for the implementation of our MBR approach, we are analysing the potential of various UML derivatives. At the time of this article going to press, we are envisioning the launch of a SysML fork, as SysML seems to be most adequate in specification and documentation, but lacks some functionalities we need to consider for the human-machine-interfaces. It would also allow us to connect our MBR tools to IPT's in-house RM technique iFEM (innovative function-effect modelling). These techniques currently do not hold an underlying modelling language, but were planned with a possible SysML implementation in mind (Schmitt and Zentis 2011)

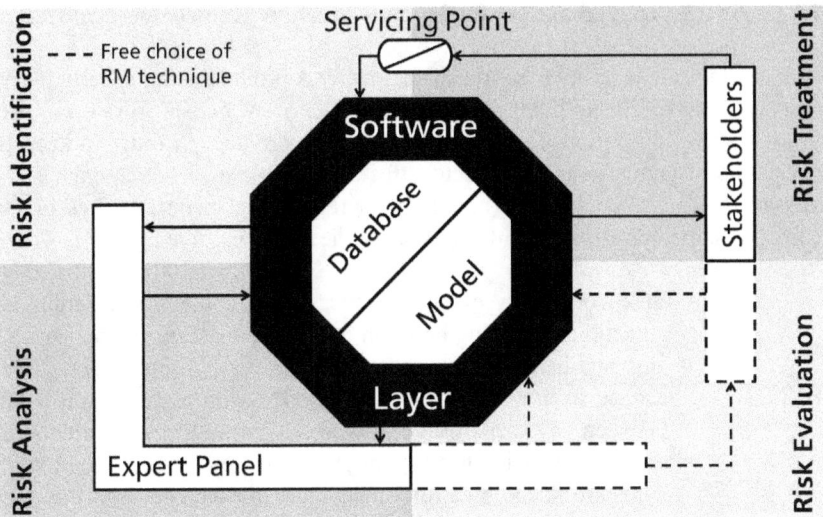

Fig. 1. MBR Approach with Software Assistance

4. Possible Improvements through MBR

With MBR, the deficits mentioned earlier (ch. 2) of a document-based approach can be addressed or at least be dramatically improved. The areas of opportunity span through all stages of an RM process and are tackled with software tools.

4.1 Comprehensive Risk Identification

An organization already using MBSE is able (with some support) to provide a finite model of the whole product life cycle, which is the main requirement to start a MBR process. Others can transform their document-based information about the product life cycle into a valid model through data inquiry, e.g. parsing builds from CAD/CAE or human-machine interface (HMI) as in wizards.

We are establishing building rules for such databases including all relevant standards and existing best practice. As we aim to develop tools for a broad application for any kind of medical device, it is crucial to take all points of view of all stakeholders and all guidelines they follow into account. At the same time, all queries must be engineered in a way that excludes unnecessary data sets as early as possible, while guaranteeing to ask every question necessary to complete the model.

4.1.1 Nomenclature and Syntax for Human-Machine Knowledge Transfer

Every discipline has its own approach to specify a product using specific terminology well known in its field. As a consequence, every person from each discipline will have a different description of the same product. In order to keep the descriptions of known critical characteristics comparable by a search engine, it is necessary to find a common terminology to reduce the inconsistencies of language. Our approach to this mainly consists of the following five parts:

1. A finite vocabulary of interactions. In this context, interactions are all actions occurring between one or more active components and any number of components influenced by this action. Users choose from a list of describing verbs with as few intersections between any two words as possible. This reduces the probability of list items being used synonymously. Nevertheless, any two list items are linked by a relevance value to encompass substitutional use when computing search queries. Suitable list items may come from a broad variety of literature, including international standards on manufacturing and machine tools, guidelines for implementation and maintenance of medical devices (e.g. implant check lists, user manuals for medical disposables), practical training and service material, manufacturer's instructions, health and safety plans and many more. For instance, in the standard DIN EN ISO 4135, estimates of the number of verbs used in modern English vary widely, but are almost always considered five digits. A feasible verb list should probably have some hundred items, but surely no more than a few thousand. So here, further literature review is necessary to decide on a short list that then can be interconnected with different long lists.

2. A set of adpositions clearing relation, location, direction, orientation etc. This set needs to describe each one-to-one correspondence between all physically and logically existing instances of components of the product life cycle in a clear way that states how the instance pointed affects the reference (One component can have many logical instances, even pointing to each other) In comparison to the interactions vocabulary, this is a rather manageable task. Human language has a very limited amount of adpositions1.

3. A finite list of possible component types within MedTech products. The standard ISO 15225 gives the requirement for the development of the Global Medical Device Nomenclature (GMDN), which is a generic way of supplying information to identify medical devices. The GMDN Agency, the organization responsible for development, control and distribution, is supporting us with their proprietary databases required for the licensing and registering of medical devices under GMDN. Their generic terms pro-

vide us with the "drawers of our cabinet", while our building rules decide where to place which drawer in order to receive short sorting times, a very important threshold when aiming for accuracy in human beings Plus, medical devices, whose predecessors have been classified and tagged in GMDN, will be much easier to sort in.

4. A hierarchic classification of MedTech products by function and application. GMDN's collective terms cover, among others, sorting by medical condition, application background or special features and by that allow us to classify assemblies in hierarchies from general to specific. Decision trees based on those hierarchies would reduce inquiry by cutting off branches irrelevant to the device. But even more important, the risk identification queries could contain straight dependencies that can retrieve cross-section information undiscovered when using only generic terms. For instance, peristaltic pumps can shear blood cells and hereby provoke clotting, which is one limiting design factor pumps for bodily fluids are subjected to. Classification under the collective term for the medical condition "renal failure" would make this a critical characteristic that needs to show up in risk identification. If, instead, the pump used for the application background "production of blood serum", where the blood is coagulated on purpose, it may be irrelevant.

5. A classification of possible application and maintenance cases. As we advocate RM alongside process chain, the whole product life cycle needs to be classified, hierarchically organised and fed to the product breakdown structure. GMDN's collective terms can help to realise this task for application and maintenance. Design and production phases are usually well-documented through CAD/CAM which we envision to be integrated using software interfaces and parsers. At present, we do not have a suggestion on dealing with the phases from obsolesence to disposal. Today, those phases are not the focus of RM. But MedTech scenarios that may put them in focus like the use of nuclear materials or information sources for green markets in this case e.g. radiation protection rules or environmental compliance forms.

---

[1]While German and English feature several hundred, the about twenty to twenty-five in Spanish already exceed the needs of our syntax.

4.1.2 Identification of Critical Characteristics

After establishing a finite syntax to capture modelling information, modelling fragments may be received from CAD/CAM, guidelines, field data, whitelists, RM documentation from legacy products, or a wizard querying the user directly etc. and will end up in the database for known critical characteristics. In the next step, a highly customisable search engine will compute the identification of critical characteristics which lead to known hazards that then can be grouped on multiple levels, e.g. according to the potential sources of harm, and prepared for graphic display (Fig. 2). Our approach to model interactions of two elements as a new element within the breakdown structure allows us to treat them the same way than actual components. The identification tool will deliver comprehensive results which only depend on data quality and not on processing, sampling known critical characteristics for discussion in the expert panel and - just as important - alerting the panel of all lose ends (each node intersecting at least two components where no record was found). Results obtained in computerised risk identification are reproducible and comparable, added information can be traced in input/output tests. A change of information can be done in the model and automatically will spread the change to all the points related to it. In terms of cost, investments can be reduced as all documents are real time outputs of the underlying model and substantial parts of the risk identification/comparison process are switched from man- hours to more cost-effective compute-time, the better the data collected, the more so. Besides, compute-time is much easier to estimate than panel sessions, which reduces process delays and time pressure on the experts.

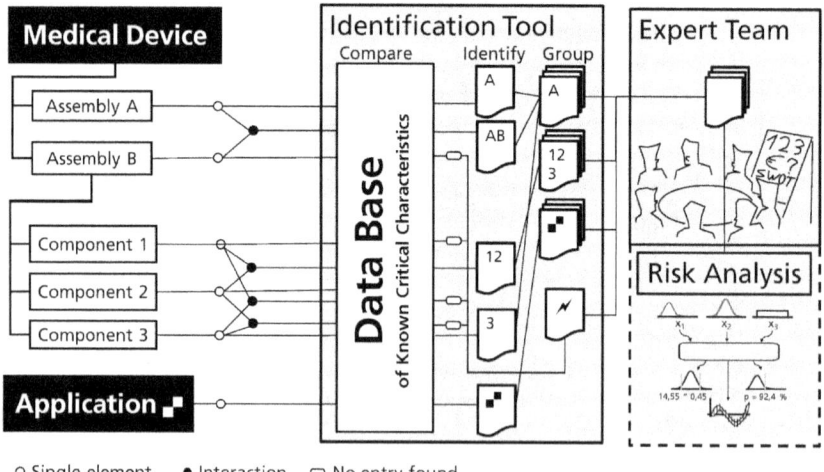

Fig. 2. Formalised Identification of Critical Characteristics

4.2 Statistical Control of Coverage

Database calculations allow for comprehensive coverage of known critical characteristics. Statistical tools can be applied revealing level of coverage, coverage probabilities and numeric error. Even though the database will grow steadily, there will be minimal process overhead and the time consumption will stay predictable. Computerised RM data makes it possible to statistically compare a RM project to similar projects giving RM stakeholders an idea which areas need more development as well as which areas are saturated, improving the impression of spread and profundity of coverage. An organisation computerising the search for known critical characteristics will gain more statistical grip with each product rolled out and each RM iteration finished.

4.3. Risk Identification as a Single Formalized Step

At this point, it is advisable to explain why we have chosen the structure of the more general ISO 31000 over DIN EN ISO 14971 even though it actually specifies RM application to medical devices. By no means is it meant to reject the guidelines found in each step, but rather highlight the importance of a self-consistent risk identification, as ISO 31000 does. If RM alongside the process chain is understood as iterative, the values of contained, consecutive RM steps become clear. Only a risk identification, whose inputs and outputs stay comparable when repeated, makes changes in risk measurable between iterations or set alternatives. Moreover, the question of how comprehensive risk identification has been managed should be answered while concluding risk identification and when RM participants still have the chance to reduce residual risk if the level is insufficient. In the scheme of DIN EN ISO 14971 comprehensiveness is not fully ascertained until entering risk control. We therefore prefer ISO 31000 in setting up an explicit formal risk identification step.

The structure of the MBR identification process is driven by software design, not by participants, allowing RM steps to be segmented clearly. All RM events are constantly decoupled from the model through a continuous software layer. RM participants are never to look at or change the model directly, but only through software tools. The inputs to each step are requested by RM design and entries can be traced back to participants. If RM events prompt information display or documentation, software tools will generate all documents on demand from the model. As long as an application programming interface (API) is provided, risk identification and analysis may be carried out by the panelists with any RM technique desired. APIs enable a formalised execution of the identification step in the panel while the software layer ensures the consecutive execution of the RM stages.

4.4 Comparable Panel Results

Different information gathering techniques will always produce different documents, a fact that is not changed by a model-based approach. Mode and motive of the experts' decisions are woven into the procedural protocol and it is hard to undo that fabric. However, what can (and necessarily must) be done by MBR is to detach the decisions from the documentation, as they are changes to the model itself. So, the vectorised model is not influenced by the methods used; hence, there is a free choice of information gathering techniques.

Resuming the example from section 2.4, the selector is now not forced any longer to choose between approximating a skill set or task. A model-based approach brings the synthesis of "hard" fact data that is stored in the model structure itself and "soft" meta and description data that offers information about the evolutionary history of both the component and its risk assessment. Any changes to the actual product or its lifecycle "mutate the model genome"; any generated document will automatically carry the change. The meta data makes changes traceable. Depending on the utility, the software layer could emphasise or withhold that information from the user. Writing information back, very specific descriptions and protocols can be saved in raw text in the database and be linked with the concerning element's unique identifier, keeping the model lightweight.

Furthermore, combining model and database allows us take advantage of the component and composite structure diagrams to compare possible treatments in input/output testing. While each component or even part or property of it stays traceable, interchangeable elements can be compared regarding to their impact on risk.

4.5 Human factor

Human bias is of negative impact to RM processes if it stays unidentified or produces a gap between a participant's capacity and willingness to perform. The latter can occur to the biased person as well as someone else whose disposition to contribute is affected by bias-driven behavior. A smart task design in MBR could reduce that impact by separating generation and evaluation of RM material in the panel from its reorganisation and display, helping participants to examine the current task without the inhibiting consequences to their or others' roles as stakeholders. Engineers could be more prone to accept changes to their designs and medics more open to discuss hints to application errors coming from medical laymen. Predetermined visualization obscures the origins of the risk identification data from experts, which should lead to a more objective view to complete data sets. For example, it is not relevant to the process of identifying critical characteristics if such a one is derived from field data or as a theoretic formation stemming from another RM process. Overall, an integration of all stakeholders through unified visualization and ubiquitous access should diminish the bias or subjective thinking as it will serve database information to experts with clear and limited assignments.

The same mechanisms within MBR that help balance human bias can be used to integrate the different professional mindsets of stakeholders into an interdisciplinary RM process. The high level of formalisation we propose for the modelling syntax should assist participants in understanding what fellow panelists from other backgrounds want to communicate, while the possibility for raw descriptions assures each expert can express his thoughts as detailed as desired. Nevertheless, not all connotations can be saved in the procedure, as non-document-based RM still is text-based. For that reason it is still important to choose the RM techniques wisely according to the mindsets and work history of the participants.

Eventually, MBR will not eliminate all circumstantial effects on the RM process, but its ability to separate automated workflow from task design can support and enable RM to achieve better results, where we need the special faculties of human minds, as it can spare humans paper work and factor out human distortion wherever a computer can do the better job. We do not think of an MBR software layer as a way to replace human experts, but rather a front desk assisting them and letting them focus on their expert work.

4.6 Incompatible work environment

Because all documents would be generated in real time and at the interface requesting them, MBR natively supports all kind of API connecting it to the software already common to the different work environments. The latest versions of UML bring new features with the XML Metadata Interchange (XMI) specification, for the first time allowing porting of certain aspects not only on model level, but on code level between different modelling environments. Exchanging properties of model components in XMI will make non-trivial transformations dispensable and thus creating APIs easier.

As the model evolves through RM iterations, the transformation within the software layer needs to stay persistently linear, meaning all vectorisation (RM process -> model) and visualization (model -> RM process) must be reversible. It must be ensured that different, but congruent entries stemming from different mindsets automatically trigger the identical change to the model as well as that identical changes triggered by different participants would result in identical documentation and visualization.

4.7 Poor risk treatment

MBR can encourage decision makers to call for better risk treatment in two main ways: by immediately lowering investments exchanging expensive man-hours and back office for computing time and by making information about treatment alternatives easier accessible and thus offering better estimates on ROI.

The drastic reduction on red tape and document management does not only express itself in lower cost, but also in time savings, which again will allow more servicing points alongside process chain with less side effects on latter. More servicing points stand for more iterations, so the model can be compared at different developmental stages; showing impact on ROI ahead of time. More time at hand also means that there is margin for running tests on alternative treatment options, at least for the most basic test of control and treatment. All this simplifies and safeguards the decisions about which risks need treatment or not and lead to the realisation of further required risk treatment.

5 Conclusion

The demands that the production of modern medical devices holds for RM can no longer be met by document-based approaches. The concurrence of higher complexity with shorter innovation cycles finds them more and more on the edge of operability. Besides, document-based RM does already not fulfill the requirements in the fields of comprehensive risk identification, predictability, interoperability of techniques, multidisciplinary integration of stakeholders and their work environments or RM enforcement throughout all stages. Our approach to combine modelling of the product life cycle with database supported RM procedures has potential to improve on these conditions. In the next step, software prototypes and trials against document-based RM will approve the assumptions.

Model-based system engineering is not the holy grail of production system engineering and MBR will not spare anyone the effort and expense of a well-designed RM process carried out by expert human beings. But in the best case, it can combine the strengths of human faculties and computing power and extract more comprehensive and numerically better assessable results from RM. Utilising this advantage throughout the product life cycle is viable path to safer medical devices.

References

Chan S, Larsen G N (2010) A framework for supplier-supply chain risk management: Tradespace factors to achieve risk reduction — Return on investment. In Technologies for Homeland Security (HST): 29-34. IEEE

Delligatti L (2013) SysML Distilled: A Brief Guide to the Systems Modeling Language. Addison Wesley

DIN EN ISO 4135:2001   Anaesthetic and respiratory equipment – Vocabulary

Ganeshkumar P (2002) Basic principles of risk management for medical device design. White Paper.   Wipro   Technologies   https://www.wipro.com/documents/resource-center/library/Whitepaper_Medical_Devices_Basic_Principles_of_Risk_Management_for_Medi cal_Device_Design.pdf Accessed 4 November 2015

Grubisic V, Ogliari A, Gidel T (2011) Recommendations for risk identification method selection according to product design and project management maturity, product innovation degree and project team. In: Culley SJ, Hicks BJ, McAloone TC, Howard, TJ, Cantamessa M, editors. Proceedings of the 18th International Conference on Engineering Design (ICED 11) August 15-18; Copenhagen.

Hall D (2011) Making risk assessments more comparable and repeatable. J. Systems Engineering 14:2, pp. 173–179

IEC/ISO 31010 (2009-11) Risk management - Risk assessment techniques.

INCOSE (2007) Systems Engineering Vision 2020. INCOSE-TP-2004-004-02. http://oldsite.incose.org/ProductsPubs/pdf/SEVision2020_20071003_v2_03.pdf Accessed 4 November 2015

ISO 31000:2009(E) (2009-11-15) Risk management — Principles and guidelines

MedTech Europe (2014) The European Medical Technology Industry in Figures http://www.eucomed.org/medical-technology/facts-figures Accessed 4 November 2015

Murray J (2012) Model Based Systems Engineering (MBSE) Media Study - Julia Almond-Murray http://www.syse.pdx.edu/systems/program/portfolios/julia/MBSE.pdf Accessed 4 November 2015

Oehmen J, Mohammad B, Warren S, Muhammad A (2010) Risk management in product design: current state, conceptual model and future research. In: Florida Institute of Technology, University of Dayton (eds) Proceedings of the ASME 2010 International Design Engineering Technical Conference & Computers and Information in Engineering Conference.

Pammolli F, Riccaboni M, Oglialoro C, Magazzini L, Baio G, Salerno N (2005): Medical Devices Competitiveness and Impact on Public Health Expenditure. https://mpra.ub.uni-muenchen.de/16021/ Accessed 4 November 2015

Radermacher K, Zimolong M, Stockheim G (2004) Analysing reliability of surgical planning and navigation systems. In: Lemke H U, Vannier M W(eds.), International Congress Series, Volume 1268, pp.824-829, as found in: Schmitt, R, Zentis T (2011) New approach for risk analysis and management in medical engineering. In: Proceedings Reliability and Maintainability Symposium (RAMS), p.1, 24-27 Jan.

Rakitin R (2006) Coping with defective software in medical devices. J. Computer 39:4, pp.40-45

Redmill F (2002) Risk Analysis-A Subjective Process. J. Engineering Management 12:2, pp.91-96,

Schmitt R, Zentis T (2011) New approach for risk analysis and management in medical engineering. In: Proceedings Reliability and Maintainability Symposium (RAMS), pp.1-6, 24-27 Jan.

Škec S, Štorga M, Marjanović D (2013) Mapping Risk Analysis Methods on Product Development Process. In: Fernandes A A, Natal Jorge R M, Patrício L, Medeiros A (Eds) International Conference on Integration of Design, Engineering & Management for Innovation, (IDEMi 2013). Porto, Portugal, 4-6 September

WHO (2010) Trends in medical technology and expected impact on public health. WH0/HSS/EHT/DIM/10.7 http://apps.who.int/medicinedocs/en/d/Js17702en/ Accessed 4 November 2015

Zentis T, Czech A, Prefi T, Schmitt R (2011) Technisches Risikomanagement in produzierenden Unternehmen. Apprimus

Zentis T, Schmitt R (2011) Technical risk management for an ensured and efficient product development on the example of medical engineering. p.2

Proving the Absence of Software-Induced Memory Corruption

Daniel Kästner, Christian Ferdinand

AbsInt GmbH

Saarbrücken, Germany[1]

Abstract *Software-induced memory corruptions can be caused by stack overflows, run-time errors such as invalid pointer accesses or buffer overflows, and data races. They can trigger software crashes, invalidate separation mechanisms in mixed-criticality software, and are a frequent cause of errors in multi-core applications. In contrast to hardware faults, software-induced memory corruptions are always systematic errors, and hence it is possible to formally prove their absence. Abstract interpretation is a formal method for static program analysis which supports formal soundness proofs (it can be proven that no error is missed) and which scales. This article gives an overview of abstract interpretation and its application to prove the absence of stack overflow, run-time errors, and data races, and reports on practical experience with the tools StackAnalyzer and Astrée.*

1 Introduction

A failure of a safety-critical system may cause high costs or even endanger human beings. With the unbroken trend towards growing software size in embedded systems more and more safety-critical functionality is implemented in software. Preventing software-induced system failures becomes an increasingly important task. Memory corruption, i.e. the unintentional modification of memory locations, is a particular dangerous class of errors. It can be caused by transient, intermittent, or permanent hardware faults, and it can be caused by malfunctioning software. Dealing with random hardware faults is well-understood whereas software-induced memory corruption often is not addressed in a systematic way. Programming errors causing memory corruption are hard to locate because cause and effect may seem completely unrelated and because they are typically hard to repro-

---

[1] AbsInt GmbH, Science Park 1, 66123 Saarbrücken, Germany

duce. Memory corruption errors can cause erroneous and erratic program behavior, software crashes, they can invalidate separation mechanisms in mixed-criticality software, and are a frequent cause of failures in multi-core applications. The accidents caused by the unintended acceleration of the 2005 Toyota Camry illustrate the potential consequences of memory corruption: the expert witness' report commissioned by the Oklahoma court in 2013 lists various software-induced memory corruption errors found in the code and identifies a stack overflow as most probable failure cause (TRA 2013, Barr 2013).

One potential source of memory corruption errors is miscompilation, i.e., programming errors induced by a buggy compiler. With the advent of formally verified compilers, e.g., the CompCert compiler (Bedin Franca 2012, Leroy et al. 2016), it becomes possible to prove that the generated code behaves exactly as specified by the semantics of the source program. This leaves the three main types of software-induced memory corruption:

1. Resource bound violations, in particular stack overflows
2. Run-time errors, e.g., invalid pointer dereference and manipulation, accesses to uninitialized variables, array-out-of-bounds accesses, etc.
3. Data races in concurrent program execution

Contemporary safety norms – including DO-178B, DO-178C, IEC-61508, ISO-26262, and EN-50128 – require to identify potential hazards and to demonstrate that the software does not violate the relevant safety goals. In all of them demonstrating the absence of memory corruption errors is a verification goal which is mostly formulated indirectly by addressing stack usage, run-time errors (e.g., division by zero, invalid pointer accesses, arithmetic overflows), corruption of content, synchronization mechanisms, and freedom of interference. In the case of DO-178B, e.g., these properties typically are covered by the *accuracy and consistency* verification objective (ABS 2015).

The absence of such errors does not belong to the program properties typically covered by functional requirements; they are safety-relevant quality objectives, often called non-functional requirements. This distinction is relevant, since functional program properties can be addressed by automatic and model-based testing, or by model checking and theorem proving. The non-functional program properties cannot be directly mapped to test cases, e.g. a test case for stimulating the worst-case stack usage typically is not known. Identifying a safe end-of-test criterion is an unsolved problem since failures usually occur in corner cases and full test coverage – which for these properties would require full control and data coverage – cannot be achieved.

Abstract interpretation is a formal methodology for static program analysis which is well suited to analyze non-functional software properties. It supports formal soundness proofs (it can be proven that no error is missed) and scales to real-life industry applications. Abstract interpretation-based static analyzers provide full control and data coverage and allow conclusions to be drawn that are valid for all program runs with all inputs. Such conclusions may be that no timing or space constraints are violated, or that run-time errors or data races are absent: the absence of these errors can be guaranteed. Nowadays, abstract interpretation-based static analyzers that can detect stack overflows and violations of timing constraints (Souyris et al. 2005) and that can prove the absence of run-time errors (Delmas and Souyris 2007), are widely used in industry. From a methodological point of view, abstract interpretation-based static analyses can be seen as equivalent to testing with full data and control coverage. They do not require access to the physical target hardware, can be easily integrated in continuous verification frameworks and model-based development environments, and they allow developers to detect run-time errors as well as timing and space bugs in early product stages. For validating non-functional program properties they define the state-of-the-art technology (Kästner and Ferdinand 2011).

In this article we will give an overview of the theory of abstract interpretation and its application to prove the absence of stack overflows, run-time errors, and data races. We report on practical experience with the tools StackAnalyzer and Astrée, and outline their automatic tool qualification strategy.

2 Abstract Interpretation

The theory of abstract interpretation is a mathematically rigorous formalism providing a semantics-based methodology for static program analysis. It is one of the formal verification methods suggested by the DO-333 supplement of DO-178C (DO-333 2011). The semantics of a programming language is a formal description of the behaviour of programs. The most precise semantics is the so-called concrete semantics, describing closely the actual execution of the program. Yet in general, the concrete semantics is not computable. Even under the assumption that the program terminates, it is too detailed to allow for efficient computations. The solution is to introduce an abstract semantics that approximates the concrete semantics of the program and is efficiently computable. This abstract semantics can be chosen as the basis for a static analysis. Compared to an analysis of the concrete semantics, the analysis result may be less precise but the computation may be significantly faster. By skilful definition of the abstract semantics, a suitable trade-off between precision and efficiency can be obtained.

Abstract interpretation supports formal correctness proofs: it can be proven that an analysis will terminate and that it computes an over-approximation of the concrete semantics, i.e., that the analysis results are *sound*. An analysis is called sound when the results hold for every possible program execution and every possible input scenario.

- For worst-case stack usage analysis soundness means that the reported stack height is never below the actual stack usage in any concrete execution. Over-estimations might occur.
- For run-time error analysis soundness means that the analysis never omits to signal an error that can appear in some execution environment. If the analyser does not report any potential error, definitely no run-time error can occur. When a potential error is reported, the analyser cannot exclude that there is a concrete program execution triggering the error. If there is no such execution, the notification about the potential error is a false alarm.

3 Stack Overflows

In embedded systems, the run-time stack (often just called "the stack") typically is the only dynamically allocated memory area. Therefore in this article we do not address overflows of heap-allocated storage but only focus on stack overflows as source of resource-related memory corruption.

The stack is used during program execution to keep track of the currently active procedures and facilitate the evaluation of expressions. Each active procedure is represented by an activation record, also called stack frame or procedure frame, which holds all the state information needed for execution. The stack size at a given program point depends on the program path executed and the sizes of the frames of all currently active functions.

In a multi-tasking system, in general each task and each interrupt service routine (ISR) can be assigned their own stack. The maximal stack usage then results from adding the stack maxima of all relevant tasks and ISRs at a critical instant, i.e. the worst-case interruption scenario. Operating systems for safety-critical systems typically use static-priority scheduling strategies. The overall worst-case stack usage then can be determined from the priorities of tasks and the interrupt hierarchy. In the case of OSEK (OSEK 2005) the necessary information can be derived from the OS configuration (Kästner and Ferdinand 2014).

3.1. StackAnalyzer

StackAnalyzer computes a safe upper bound of the maximal stack usage by whole-program static analysis at the executable code level. Since it is based on abstract interpretation it can be formally proven that the maximal stack usage of each task will never be underestimated. From the per-task stack maxima the system-level stack maximum can be determined from the priorities of the tasks and the interrupt hierarchy. StackAnalyzer approximates the semantics of the machine code of the microprocessor by using an abstract model of the processor architecture. The abstract model does not need to cover the entire state of the microprocessor, only the parts affecting the stack are needed. The hardware state relevant for worst-case stack analysis includes the processor registers and the memory cells. For a naive analysis only the stack pointer register is needed, but for precise results it is important to perform an elaborate value analysis on the contents of processor register and memory cells.

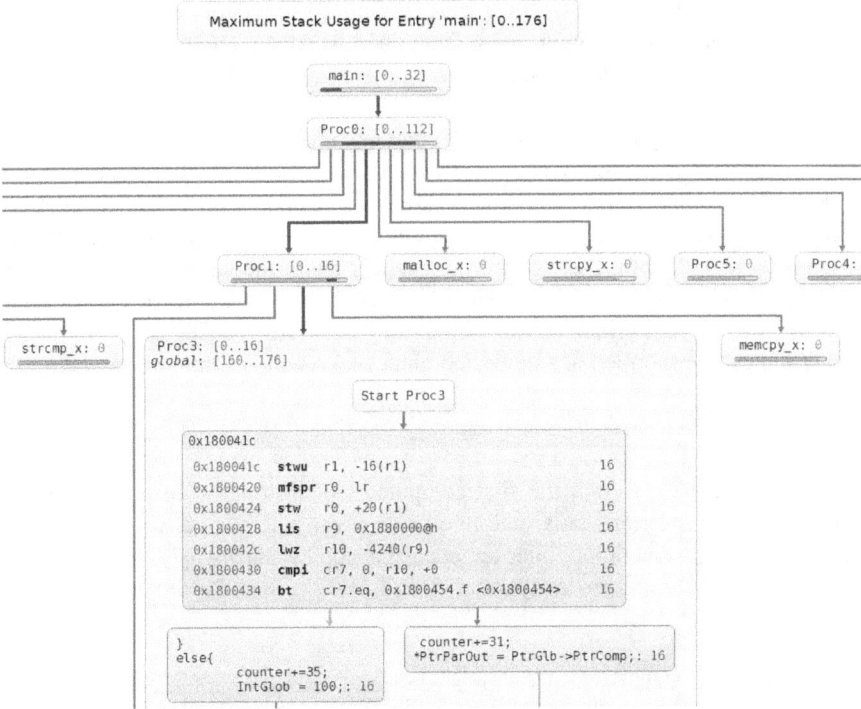

Fig. 1. StackAnalyzer analysis results and call graph visualization

In the following we will give an overview of the structure and analysis phases of the tool StackAnalyzer (Ferdinand et al. 2007, Kästner and Ferdinand 2014).

Decoding Stage

The input for the decoding phase is the fully linked binary executable that contains the task to be analyzed. The instruction decoder identifies the machine instructions and reconstructs the control-flow graph. To ensure safety of later analysis results, the reconstructed CFG itself must be safe, i.e., all possible paths that can occur during execution of the program must be represented.

Finding the target addresses of absolute and PC-relative calls and branches is straightforward, but determining target addresses computed from register contents can become difficult. Examples of such computed addresses are, e.g., indirect calls via function pointers or the implementation of high-level programming language constructs like switch tables. Uncertainties may lead to over-approximations of the actual control flow of the analyzed task which reduces analysis precision.

To deal with this, StackAnalyzer uses specialized decoders that are adapted to certain code generators and/or compilers: They usually recognize branches to a previously stored return address, and know the typical compiler-generated patterns of branches via switch tables.

Yet non-trivial applications may still contain some computed calls and branches (in hand-written assembly code) that cannot be resolved by the decoder; these unresolved computed calls and branches are documented by appropriate messages and require user annotations. Such annotations may list the possible targets of computed calls and branches, or tell the decoder about the address and format of an array of function pointers or a switch table used in the computed call or branch. The annotations are written in the formal language AIS (Kästner and Ferdinand 2014) and can be supplied in a dedicated input file. As an example the annotation

```
instruction "main" + 1 computed calls "cma", "cmo";
```

informs StackAnalyzer that the first computed call instruction of function main can invoke the functions cma and cmo. Astrée can automatically generate AIS specifications for function pointer targets as needed by StackAnalyzer, enabling a fully automatic stack usage analysis (cf. Sec. 4.1).

Value Analysis. Value analysis aims at statically determining the contents of the registers and memory cells at each program point and for each execution context. Disambiguating contexts is important to achieve high analysis precision. StackAnalyzer computes an abstract information for every pair of a program point p and a possible calling context of p. The results of the value analysis are used to predict the addresses of data accesses, computed calls and branches, and to find infeasible paths caused by conditions that always evaluate to true, or always evaluate to false in a specific context. By concentrating on the value of the stack pointer during value analysis, the analysis can determine how the stack increases and decreases along the various control-flow paths. This information can be used to derive the maximum stack usage of the entire task.

Loop Bound Analysis. As a part of the value analysis a so-called loop bound analysis is performed. It uses the results of value analysis to determine lower and upper bounds for the number of iterations of loops. Knowing such bounds can improve the precision of value analysis, e.g., if the loop iterates through an array.

For stack analysis the loop bound analysis can be restricted to loops that increase or decrease the stack in every iteration. If the iteration bound of a loop cannot be determined automatically, users can provide it in the AIS file or as source code annotations.

Iteration between Decoding and Value Analysis. The target of an indirect call or branch depends on the value of some register. Sometimes, an explicit value is written to the register some time before the call so that the call target can in principle be determined automatically. Unfortunately, the setting of the register often does not occur immediately before the call, but some time earlier, e.g., before a loop containing the call, so that matching of a simple code pattern is not sufficient to find the target address. The executable reader, which should follow function calls to decode the reachable instruction sequences, is not aware of the register values. These values are determined by value analysis, which however cannot construct control-flow graphs. The solution of the problem is to iterate between decoding and value analysis. If value analysis has found some register values needed by the decoder to resolve some indirect calls or branches, then the decoder is run again so that it can use these register values. Since this changes the control-flow graph, the value analyzer is run again to take care of the changes. If it finds more register values or the information about the values already found changes, a third iteration is performed, etc.

Visualization. The graphical user interface of StackAnalyzer provides different views of the result, including visualizations of the call graph and control flow graph, and enables the information contained in the executable to be browsed in a user-friendly way (cf. Figure 1).

4 Run-Time Errors and Data Races

Run-time errors are errors that occur during run-time of the software. In this section we will focus on run-time errors which correspond to undefined or unspecified behavior with respect to the semantics of the programming language. This class of errors is of particular interest for the programming language C since it includes many common problems which cannot be detected by the compiler or prevented during run-time. Examples are arithmetic exceptions (e.g. divide by zero), overflows, and validity of addresses for pointers or array bound errors. The C99 standard provides a list of unspecified and undefined behaviors in Section J of ISO/IEC 9899:1999 (E). Errors induced by concurrent execution, e.g., data races and inconsistent thread synchronization, are not addressed by the C99 standard.

Depending on their effects run-time errors can be grouped in two categories. The first category of run-time errors is related to conditions in which the source semantics is undefined. After such a run-time error the actual execution will do something unknown. Examples are invalid array or pointer accesses which might corrupt memory and destroy the data integrity of the program. It can even happen that the program code is dynamically modified resulting in erratic behavior, or that the program crashes with segmentation faults or bus errors. Further examples of errors from that class are integer division by zero, floating-point overflows and invalid operations without mathematical meaning which might cause the program to be stopped by an interrupt.

The second category of run-time errors is due to unspecified but implementation-defined behavior; here it is predictable what will happen after the error has occurred. Examples are integer overflows or invalid shifts for which the actual computations are quite different from the expected mathematical meaning.

4.1. Astrée

Astrée signals all potential runtime errors and further critical program defects. It is sound, i.e., if no errors are signalled, this means there are no errors from the class of errors under investigation – the absence of errors has been proved. It reports program defects caused by unspecified and undefined behaviours according to the C norm (ISO/IEC 9899:1999 (E)), program defects caused by invalid concurrent behaviour, violations of user-specified programming guidelines, and computes program properties relevant for functional safety. Users are notified about:

- integer/floating-point division by zero
- out-of-bounds array indexing
- erroneous pointer manipulation and dereferencing (null, uninitialized, and dangling pointers)
- data races (read/write or write/write concurrent accesses by two threads to the same memory location without proper mutex locking)
- inconsistent locking (lock/unlock problems)
- invalid calls to operating system services (e.g. calls to the OSEK service TerminateTask() on a task with unreleased resources)
- integer and floating-point arithmetic overflows
- read accesses to uninitialized variables
- code unreachable under all circumstances
- violations of optional user-defined assertions to prove additional runtime properties, e.g., to guarantee that output variables are within the expected value ranges
- violations of coding rules (MISRA 2004, MISRA 2012) and code metric thresholds. The supported code metrics include the statically computable HIS metrics (HIS 2008), e.g., comment density, and cyclomatic complexity.
- non-terminating loops

Floating-point computations are handled precisely and safely by taking all potential rounding errors into account. Furthermore Astrée computes data and control flow reports containing a detailed listing of accesses to global and static variables sorted by functions or variables, and caller/called relationships between functions. The analyser can also report each potentially shared variable, the list of asynchronous tasks accessing it, and the types of the accesses (read, write, read/write).

The C99 standard does not fully specify data type sizes, endianness nor alignment which can vary with different targets or compilers. Astrée is informed about these target settings by a dedicated configuration file and takes the specified properties into account.

Workflow

In the following we will use the term *alarm* to denote a notification about a potential run-time error. While Astrée finds all potential run-time errors, it may err on the other side and produce false alarms. For industrial use producing the fewest possible number of false alarms is an important goal. Only with zero alarms the absence of run-time errors is automatically proven. The design of the analyser aims at reaching the zero false alarm objective (Blanchet et al. 2002, Blanchet et al. 2003, Cousot et al. 2005), which was accomplished for the first time on large industrial applications at the end of November 2003. For keeping the initial number of false alarms low, a high analysis precision is mandatory. To achieve high precision Astrée provides a variety of predefined abstract domains, including the following ones:

- The interval domain approximates variable values by intervals.
- The octagon domain (Miné 2006) covers relations of the form $x \pm y \leq c$ for variables x and y and constants c.
- Floating-point computations are precisely modelled while keeping track of possible rounding errors (Blanchet et al. 2003).
- The memory domain empowers Astrée to exactly analyse pointer arithmetic and union manipulations. It also supports a type-safe analysis of absolute memory addresses.
- The clock domain has been specifically developed for synchronous control programs and supports relating variable values to the system clock (Cousot et al. 2007).
- With the filter domain (Feret 2004) digital filters can be precisely approximated.

Any remaining alarm has to be manually checked by the developers – and this manual effort should be as low as possible. Astrée explicitly supports investigating alarms in order to understand the reasons for them to occur. Alarm contexts can be interactively explored, the computed value ranges of variables can be displayed for each different context, the call graph is visualized, etc. (cf. Figure 2).

If there is a true error it has to be fixed. A false alarm can possibly be eliminated by a suitable parameterization of Astrée: If the error cannot occur due to certain preconditions which are not known to Astrée, they can be made available to Astrée via dedicated directives. These annotations make the side conditions explicit which have to be satisfied for a correct program execution. If the false alarm is caused by insufficient analysis precision, steering directives are available that allow to locally tune the analysis precision to eliminate the false alarm. As an example the ASTREE_unroll directive can be used to enforce disambiguating every iteration of one specific loop. The key feature is that Astrée is fully parametric with respect to the abstract domains. Abstract domains can be parameterized to tune the precision of the analysis for individual program constructs or program points (Mauborgne and Rival 2005). This means that in one analysis run important program parts can be analysed very precisely while less relevant parts can be analysed very quickly – without compromising system safety.

Fig. 2 Astrée analysis results

All directives can be specified in a formal language AAL (Kästner and Pohland 2015) and stored in a dedicated file. An AAL annotation consists of an Astrée directive and a path specifying the program point to insert the directive at. The path is specified in a robust way by exploiting the program's syntactical structure without relying on line number information. E.g., the annotation

```
__ASTREE_annotation((main {+1 loop} insert before:
                 __ASTREE_unroll((3)))));
```

inserts the directive `__ASTREE_unroll((3))` immediately before the first loop in function `main`. The AAL language is a prerequisite for supporting model-based code generators. It makes it possible to separate the annotations from the source code, so that when the code is regenerated, all previously generated annotations from structurally unchanged code parts are still valid, even if the line numbers change. In cases where there are structural changes of the code, Astrée provides a mechanism to detect whether annotations are still placed at the intended location (Kästner and Pohland 2015).

In the case some alarms cannot be eliminated by increasing the analyser precision they can be classified and commented by using the `__ASTREE_comment` directive. This can be done in a convenient way by using the Astrée result viewer which will automatically generate the corresponding AAL annotation. The available classifications are *uncommented, true, true (high), true (low), true (not a defect), false* or *undecided*.

Handling Concurrency

Whereas previous Astrée versions have been limited to sequential C software, Astrée has been extended by a novel low-level concurrent semantics (Miné 2012) which provides a scalable abstraction covering all possible thread interleaving. The interleaving semantics enables Astrée, in addition to the classes of run-time errors found in sequential programs, to report data races, i.e., read/write or write/write concurrent accesses by two threads to the same memory location without proper mutex locking and lock/unlock problems, i.e., inconsistent synchronization. After a data race, the analysis continues by considering the values stemming from all interleaving. In addition to the range of each variable at each program point, Astrée reports the set of shared variables it discovers, together with the set of threads accessing these variables, the kinds of operations performed (reads or writes), and their range of values.

To cover all potential execution interleaving Astrée analyzes separately each thread and collects abstract versions of the effects they have on the shared memory. Threads are reanalyzed iteratively, taking into account such global effects, until stabilization, at which point a sound over-approximation of all possible behaviors has been found. Thread priorities are exploited to reduce the amount of spurious interleaving considered in the abstraction and to achieve a more precise analysis. Astrée includes a built-in notion of mutual exclusion locks, on top of which actual synchronization mechanisms offered by operating systems can be modeled (such as POSIX mutexes or semaphores); program-enforced mutual exclusion is also exploited by Astrée to reduce spurious interleaving. When these features are insufficient to match the concurrency semantics of the analyzed program, Astrée reverts to unrestricted preemption, which ensures a sound analysis coverage for all concurrency models, including execution on multi-core processors. In particular, Astrée is not limited to collaborative threads nor discrete sets of preemption points.

Programs to be analyzed are seldom run in isolation; they interact with an environment. In order to soundly report all run-time errors, Astrée must take the effect of the environment into account. In the simplest case the software runs directly on the hardware, in which case the environment is limited to a set of volatile variables, i.e., program variables that can be modified by the environment concurrently, and for which a range can be provided to Astrée by formal directives. More often, the program is run on top of an operating system, which it can access through function calls to a system library. When analyzing a program using a library, one possible solution is to include the source code of the library with the program. This is not always convenient (if the library is complex), nor possible, if the library source is not available, or not fully written in C, or ultimately relies on kernel services (e.g., for system libraries). An alternative is to provide a *stub* implementation, i.e., to write, for each library function, a specification of its possible effect on the program. Astrée provides stub libraries for the ARINC 653 standard, the OSEK/AUTOSAR standards, and for POSIX threads. More details on these models are available in (Miné and Delmas 2015).

OSEK/AUTOSAR programs consist in a set of tasks and a set of interrupts, which are all modeled as Astrée threads. The standard proposes several conformance classes, with support for increasingly complex features (such as extended tasks, fully preemptive scheduling, multiple task activation, etc.). The model proposed in Astrée supports the most general class, which guarantees that all programs can be soundly analyzed. A particularity of OSEK is that system resources, including tasks, are not created dynamically at program startup; instead they are hardcoded in the system: a specific tool reads a configuration file in OIL format describing these resources and generates a dedicated version of the system to be linked against the application. Astrée supports a similar workflow by providing an OSEK stub library, and a tool to automatically convert OIL configurations into the required C data structures.

5 Practical Experience

In the following we will summarize practical experience with StackAnalyzer and Astrée. Both have been used in numerous safety-critical industry projects from various domains, including aerospace, automotive, and nuclear energy subject to certification according to various safety standards. They can be operated in stand-alone mode, and can be seamlessly integrated in other development tools based on a dedicated XML-based exchange format XTC[1]. E.g., there is a tool coupling between StackAnalyzer and the model-based code generator Esterel SCADE (Ferdinand et al. 2008), StackAnalyzer and Astrée are coupled with the model-based code generator dSPACE TargetLink (Kästner, Rustemeier et al. 2014) and the model-based testing tool BTC EmbeddedTester (Kästner, Brockmeyer et al. 2014). The couplings make it possible to detect stack problems and run-time errors early in the development process. Results are reported back to the modelling level, annotations can be automatically generated from the model level without any user interaction being required.

5.1 StackAnalyzer

StackAnalyzer depends on the instruction set architecture and is available for a wide range of targets, including ARM, Infineon C16x, Infineon TriCore/Aurix, TI C28x, TI C33, Fujitsu FR81S, M68k, PowerPC, and V850.

Experience shows that even for large applications precise stack bounds can be calculated within short computation time. The computation time is mainly influenced by the program structure and the task size; no significant variations between different instruction set architectures could be observed. In Table 1 stack analysis results are shown for some tasks from different industry sectors and for different target architectures. The tasks have been arbitrarily chosen among the largest available tasks of industrial production software. Colum *Arch* denotes the target architecture, column *Industry* the application area (E: electronics, AU: automotive, AE: aerospace). In column *Task Size* the code size of the task under analysis is shown in kilobytes; column $Stack^{max}$ shows the computed maximal stack size in bytes. The analysis time is given in column *Analysis Time*.

Table 1. Stack analysis results

Arch	Industry	TaskSize [KB]	$Stack^{max}$[B]	Analysis Time
ARM	E	24.67	2184	8s
ARM	AU	124.86	396	1m 12s
M68020	AE	48.34	34752	1m 50s
M68020	E	0.96	128	1s
PowerPC	AE	681.63	67728	2m 52s
PowerPC	AU	82.94	1312	12m 25s
C16x	AU	93.55	168u/56s	2m 31s
C16x	AE	25.84	394u/32s	7s

---

[1] http://www.absint.com/xtc/

Some processors like the Infineon C16x, TriCore, and Aurix have separate user and system stacks; StackAnalyzer computes separate maxima for both the user and the system stack. In the table above 'u' denotes the user stack while 's' denotes the system stack.

5.2 Astrée

We first summarize results for the sequential version of Astrée and then describe experiments with the new interleaving semantics.

One of the examined software projects from the avionics industry comprises 132,000 lines of C code including macros and contains approximately 10,000 global and static variables (Blanchet et al. 2003). The first run of the sequential version of Astrée reported 1200 false alarms; after adapting Astrée the number of false alarms could be reduced to 11. The analysis duration was 1h 50 min on a PC with 2.4 GHz and 1GB RAM.

The article (Delmas and Souyris 2007) gives a detailed overview of the analysis process for an Airbus avionics project. The software project consists of 200,000 lines of pre-processed C code, performs many floating-point computations and contains digital filters. The duration of the sequential analysis for the entire program is approximately 6 hours on a 2.6 GHz PC with 16 GB RAM. At the beginning, the number of false alarms was 467 and could be reduced to zero in the end.

The following table shows experiments for the concurrent version of Astrée on avionics case studies from (Miné and Delmas 2015). It shows the size of the code (in lines), the selectivity (percentage of lines proved correctly), the analysis time and memory consumption.

Table 2. Results obtained with concurrency-aware Astrée analysis

Size	Selectivity	Analysis Time	Memory
2.1 M	99.94%	24 h	27 GB
1.9 M	99.56%	154 h	18 GB
2.2 M	99.52%	160 h	23 GB
31.8 K	97.28%	50 min	0.6 GB
33.1 K	97.18%	35 h	2.5 GB

We can see that a sound analysis which finds all potential run-time errors and data races can be successfully finished for software projects with several million lines of code in acceptable time on standard PC hardware. We can also see that the selectivity is very high, which is a very important aspect to minimize the time to investigate the reported findings, and to fix any defects detected.

6 Tool Qualification

To provide high confidence in the correct functioning of a tool it is necessary to demonstrate that the tool works correctly in the operational context of its users. This is a common requirement of most current safety standards. The correct functioning of a tool might be affected by the OS version, system libraries installed, software patch levels, etc. Moreover, depending on the user's development process structure and tool landscape the probability for detecting tool errors may vary. The tool qualification process can be automated by dedicated *Qualification Support Kits* (QSKs) as shipped as a part of a software tool.

The AbsInt QSKs for StackAnalyzer and Astrée are centered round an automatic validation suite and essentially consists of two parts, a report package and a test package. The report package consists of two different documents: the Tool Operational Requirements (TOR) report and the Verification Test Plan (VTP). The TOR lists the tool functions and technical features which are stated as low-level requirements to the tool behavior under normal operating conditions. The VTP defines the test cases demonstrating the correct functioning of all specified requirements from the TOR. The test package contains an extensible set of test cases with a scripting system to automatically execute them and generate reports about the results.

7 Conclusion

Contemporary safety norms – including DO-178B, DO-178C, IEC-61508, ISO-26262, and EN-50128 – require to identify potential hazards and to demonstrate that the software does not violate the relevant safety goals. In all of them demonstrating the absence of memory corruption errors is a verification goal. Software-induced memory corruptions can be caused by stack overflows, run-time errors like invalid pointer accesses or buffer overflows, and data races. They can trigger software crashes, invalidate separation mechanisms in mixed-criticality software, and are a frequent cause of errors in multi-core applications. Tools based on abstract interpretation can perform static program analysis of embedded applications. Their results are determined without the need to change the code and hold for all program runs with arbitrary inputs. Especially for non-functional program properties they are highly attractive, since they provide full data and control coverage and can be seamlessly integrated in the development process. The theory of abstract interpretation provides a formal methodology for semantics-based static analysis of dynamic program properties. It allows provably sound over-approximations of the worst-case stack usage to be determined and the absence of run-time errors and data races to be proven. In this chapter, we have presented two example tools: StackAnalyzer calculates safe upper bounds on the maximum stack usage of tasks and can prove the absence of stack overflows. Astrée can be used to prove the absence of run-time errors and data races in C programs. It can be specialized to the software under analysis, achieves very high precision, and automatically supports embedded OS norms. StackAnalyzer and Astrée have been successfully used as verification tools for certification according to contemporary safety standards. They are used in numerous safety-critical industry projects from various domains, including aerospace, automotive, and nuclear energy and have been proven in industrial practice. The tool qualification process can be automatized to a large extend by dedicated Qualification Support Kits.

Acknowledgments The work presented in this paper has been supported by the German BMBF project FORTISSIMO.

References

ABS (2015) Safety Manual for aiT, Astrée, StackAnalyzer. AbsInt GmbH, 2015.
Barr M (2013) Bookout v. Toyota, 2005 Camry software Analysis by Michael Barr, http://www.safetyresearch.net/Library/BarrSlides_FINAL_SCRUBBED.pdf
Blanchet B, Cousot P, Cousot R, Feret J, Mauborgne L, Miné A, Monniaux D, and Rival X (2002) Design and Implementation of a Special-Purpose Static Program Analyzer for Safety-Critical Real-Time Embedded Software, invited chapter in T. Mogensen, D. Schmidt, and I. Sudborough, editors, The Essence of Computation: Complexity, Analysis, Transformation. Essays Dedicated to Neil D. Jones, LNCS 2566, pages 85–108. Springer, Oct. 2002.
Blanchet B, Cousot P, Cousot R, Feret J, Mauborgne L, Miné A, Monniaux D, and Rival X (2003) A Static Analyzer for Large Safety-Critical Software. In Proceedings of the ACM SIGPLAN 2003 Conference on Programming Language Design and Implementation (PLDI'03), pages 196–207, San Diego, 2003. ACM Press.
Bedin Franca R, Blazy S, Favre-Felix D, Leroy X, Pantel M, and Souyris J (2012) Formally verified optimizing compilation in ACG-based flight control software. In ERTS 2012: Embedded Real Time Software and Systems.
Cousot P, Cousot R, Feret J, Mauborgne L, Miné A, Monniaux D, and Rival X (2005) The ASTRÉE analyzer. In M. Sagiv, editor, Proc. of the European Symposium on Programming (ESOP'05), volume 3444 of Lecture Notes in Computer Science, pages 21–30. Springer, 2005.
Cousot P, Cousot R, Feret J, Miné A, Mauborgne L, Monniaux D, and Rival X (2007) Varieties of Static Analyzers: A Comparison with ASTRÉ. In First Joint IEEE/IFIP Symposium on Theoretical Aspects of Software Engineering, TASE 2007, pages 3–20. IEEE Computer Society, 2007.
Delmas D, and Souyris J (2007) ASTRÉE: from Research to Industry. In Proc. 14th International Static Analysis Symposium (SAS2007), number 4634 in LNCS, pages 437–451, 2007.
Dunn M (2013) Toyota's killer firmware: bad design and its consequences. EDN Network, October 2013. Available at http://www.edn.com/design/automotive/4423428/Toyota-s-killer-firmware–Bad-design-and-its-consequences.
Ferdinand C, Heckmann R, and Franzen B (2007) Static Memory and Timing Analysis of Embedded Systems Code. In P. Groot, editor, Proceedings of the 3rd European Symposium on Verification and Validation of Software Systems (VVSS 2007), vol. 07-04 of TUE Computer Science Reports, pages 153–163, 2007.
Ferdinand C, Heckmann R, Le Sergent T, Lopes D, Martin B, Fornari X, Martin F (2008) Combining a High-Level Design Tool for Safety-Critical Systems with a Tool for WCET Analysis on Executables. 4th European Congress ERTS - Embedded Real Time Software, Toulose, 2008.
Feret J (2004) Static Analysis of Digital Filters. In European Symposium on Programming (ESOP'04), number 2986 in LNCS, pages 33–48. Springer-Verlag, 2004.

HIS (2008) Hersteller Initiative Software (HIS) HIS Source Code Metriken v1.3.1, 01.04.2008.

Kästner D, Brockmeyer U, Pister M, Nenova S, Bienmüller T, Dereani A, Ferdinand C (2014) Combining Model-based Analysis and Testing. Embedded Real Time Software and Systems Congress ERTS², Toulouse, 2014.

Kästner D, Ferdinand C (2011) Efficient Verification of Non-Functional Safety Properties by Abstract Interpretation: Timing, Stack Consumption, and Absence of Runtime Errors. In Proceedings of the 29th International System Safety Conference ISSC2011, Las Vegas, 2011.

Kästner D, Ferdinand C (2014) Proving the Absence of Stack Overflows. In SAFECOMP '14: Proceedings of the 33th International Conference on Computer Safety, Reliability and Security (SAFECOMP), Florence, 2014. Springer LNCS 8666, Springer, Heidelberg.

Kästner D, Pohland J (2015) Program Analysis on Evolving Software. In Matthieu Roy, editor, CARS 2015 – Critical Automotive applications: Robustness & Safety, Paris, France, September 2015, https://hal.archives-ouvertes.fr/hal-01192985/.

Kästner D, Rustemeier C, Kiffmeier U, Fleischer D, Nenova S, Heckmann R, Schlickling M, Ferdinand C (2014) Model-Driven Code Generation and Analysis. SAE World Congress 2014, available at http://papers.sae.org/2014-01-0217/.

Leroy X, Blazy S, Kästner D, Schommer B, Pister M, Ferdinand C (2016) CompCert – A Formally Verified Optimizing Compiler. Embedded Real Time Software and Systems Congress ERTS², Toulouse. To appear.

Mauborgne L, Rival X (2005). Trace Partitioning in Abstract Interpretation Based Static Analyzers. In 14th European Symposium on Programming ESOP'05, no. 3444 in LNCS, pages 5–20, 2005.

Miné A (2006) The Octagon Abstract Domain. Higher-Order and Symbolic Computation, 19(1): 31 – 100, 2006.

Miné A (2012) Static analysis of run-time errors in embedded real-time parallel C programs. Logical Methods in Computer Science (LMCS), 8(26):63, Mar. 2012.

Miné A, Mauborgne M, Rival X, Feret J, Cousot P, Kästner D, Wilhelm S, Ferdinand C (2016) Taking Static Analysis to the Next Level: Proving the Absence of Run-Time Errors and Data Races with Astrée. Embedded Real Time Software and Systems Congress ERTS², Toulouse, 2016. To appear.

Miné A, Delmas D (2015) Towards an Industrial Use of Sound Static Analysis for the Verification of Concurrent Embedded Avionics Software. Proc. of the 15th International Conference on Embedded Software (EMSOFT'15), Amsterdam, Oct. 2015.

MISRA (2004) MISRA Limited. MISRA-C:2004 Guidelines for the use of the C language in critical systems. ISBN 978-0-9524156-4-0, October 2004.

MISRA (2012) MISRA Limited. MISRA-C:2012 Guidelines for the use of the C language in critical systems. ISBN 978-1-906400-11-8, March 2013.

OSEK (2005) OSEK/VDX. OSEK/VDX Operating System. Version 2.2.3., 2005.

POSIX (1995) IEEE Computer Society and the Open Group. Portable operating system interface (POSIX) – Application program interface (API) amendment 2: Threads extension (C language). Technical report, ANSI/IEEE Std. 1003.1c-1995, 1995.

DO-333 (2011) Radio Technical Commission for Aeronautics. Formal Methods Supplement to DO-178C and DO-278A, 2011.

Souyris S, Pavec E L, Himbert G, Jégu V, Borios G, and Heckmann R (2005) Computing the Worst Case Execution Time of an Avionics Program by Abstract Interpretation. In Proceedings of the 5th International Workshop on Worst-case Execution Time (WCET '05), Mallorca, pages 21 – 24, 2005.

TRA (2013) Transcript of Morning Trial Proceedings had on the 14th day of October, 2013 Before the Honorable Patricia G. Parrish, District Judge, Case No. CJ-2008-7969, http://www.safetyresearch.net/Library/Bookout_v_Toyota_Barr_REDACTED.pdf

The Role of Standardisation and Guidance in the Development of Sub-sea Glider Technologies

John Hoddinott, Edward Horabin, Luke Hankins

BMT Isis Ltd

Fareham, UK

Abstract *A sub-sea glider is a type of autonomous underwater vehicle propelled by changes in buoyancy and is typically used for oceanographic research and monitoring. The BRIDGES (Bringing together Research and Industry for the Development of Glider Environmental Services) project is funded under the European Commission Horizon 2020 programme and aims to develop tools to further understanding, monitoring and sustainable exploitation of the marine environment. BRIDGES has the objective to improve sub-sea glider technology, system integration, operational management and standardisation. This paper summarises the research undertaken by BMT Isis as part of the BRIDGES consortium into understanding the safety benefits and applicability of standardisation to the design, manufacture and operation of gliders. The paper explores to what extent standardisation and guidance should be introduced into what is a relatively new and evolving domain. It also investigates what opportunities there are to transfer approaches from other domains; maximising the potential safety benefits whilst encouraging innovation.*

1 Introduction

1.1 The BRIDGES Project

The BRIDGES (Bringing together Research and Industry for the Development of Glider Environmental Services) project is funded under the European Commission Horizon 2020 programme and aims to develop tools to further understanding, monitoring and sustainable exploitation of the marine environment. The four-year project has the objective to improve sub-sea glider technology, system integration, operational management and standardisation.

BRIDGES aims to provide new opportunities for offshore industries, such as oil and gas, sea mining, marine research and environmental monitoring. This new tool consisting of ocean gliders – that are robust, cost-effective, re-locatable, versatile and easily-deployable – will support autonomous long-term in-situ exploration of the deep ocean over a wide range of spatial and temporal scales. Two Explorer gliders will be developed that are built on the successful unique European underwater glider, the Sea Explorer. During BRIDGES the Sea Explorer will be modularized and adapted to more diverse operations. The two glider designs will be optimised for missions at 2,400 metres ("Deep") and 5,000 metres ("Ultra Deep") sea depth. Multiple near-shore and deep sea trials and demonstrations are planned, with a forecast market introduction of the Explorer gliders of 2020.

The BRIDGES consortium is composed of 19 public and private partners from 7 EU countries and 2 associated countries, covering renowned scientific institutes, industrial groups and innovative Small to Medium Sized Enterprises (SMEs).

1.2 Sub-sea Gliders

A sub-sea glider is a type of autonomous underwater vehicle (AUV) which is propelled by changes in buoyancy and achieves lateral motion by the flow of water over its wings. The resulting 'saw-tooth' profile through the water allows measurements to be made at a range of depths. This energy-efficient technique allows gliders to travel thousands of miles and operate for months at a time. Sub-sea gliders are currently used for oceanographic research and monitoring but have the potential to operate in the oil & gas and offshore mining industries.

The buoyancy of a sub-sea glider is typically altered by one of four methods, pumping ballast water into and out of a tank in the glider, using a compressed air buoyancy engine, pumping oil into and out of an external bladder, or relying on the thermal gradient between water at the surface and at depth. Challenges for ultra-deep operations include pumping ballast water out against phenomenal external pressure at depths of up to 5,000 metres, and the compressibility of the hull, which will reduce the glider's buoyancy at depth.

The pitch, and hence the glide angle, of the glider can be controlled by adjusting the position of the vessel's centre of gravity, usually by moving a mass fore and aft within the hull. The glider's heading can be altered using a conventional rudder on a fin.

A hybrid propulsion system, incorporating a motor and propeller in addition to the glider's wings, can increase the speed and manoeuvrability of a glider but the additional power and weight requirements can reduce the overall range and endurance.

Communication between operators and the gliders is typically by satellite modem, as the short antenna height when surfaced limits the range of terrestrial radio communications. Operator 'tracking' of gliders is based on simulation models, as the gliders cannot communicate when submerged and they may stay underwater for up to 24 hours at a time. An on-board GPS sensor allows gliders to get a positional fix when surfaced, and an inertial guidance system is used to maintain positional awareness underwater.

The sensor package on a sub-sea glider may include acoustic (active and passive) instruments, particle measurement devices and dissolved oxygen sensors, as well as on-board chemistry analysis labs. For long-duration missions, the sensor power budget may only be a few watts, shared between the instruments. Data from the sensors may be transmitted whilst the glider is surfaced and/or stored on-board for download and analysis when the glider is recovered at the end of its mission.

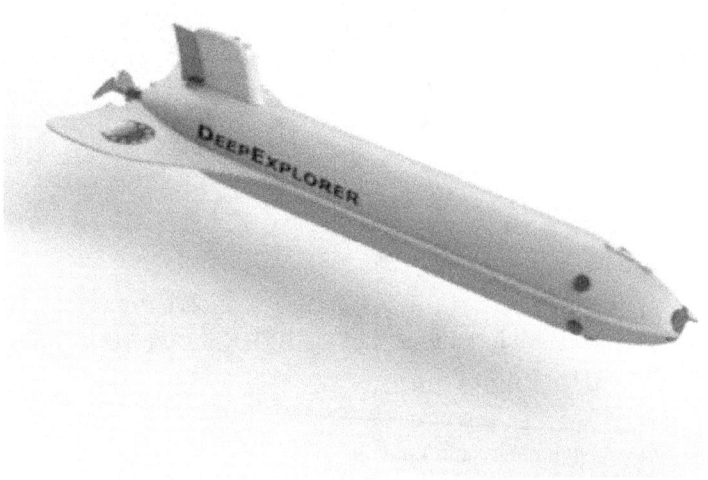

Fig. 1. Concept for BRIDGES Deep Explorer Glider

1.3 Typical Sub-sea Glider design specifications

Design specifications for gliders vary greatly dependant on their planned use, the myriad of operating environments and research objectives means that often gliders are constructed to very different specifications.

Typically, weights, dimensions and load outs will be designed to meet the expected conditions, although on average figures can be expected to be within the following ranges:

Weight - Between 50 and 200 Kilograms
Length - Between 1 and 3 meters
Diameter - Between 0.25 to 0.5 meters
Velocity - Up to 1.5 meters per second

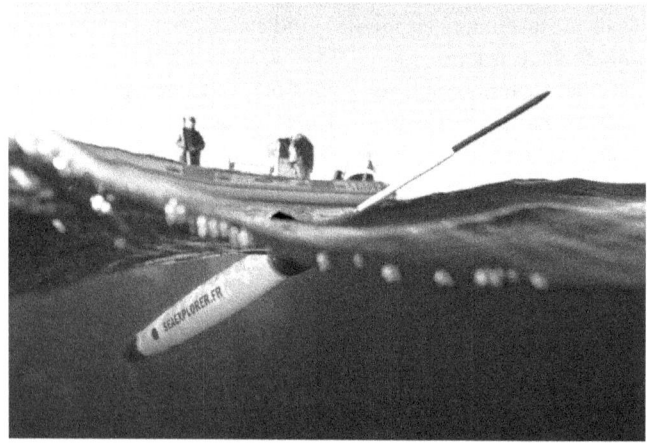

Fig. 2. Deployment of SeaExplorer

2. Understanding Risks

Broadly speaking, the gliders produced as part of the BRIDGES project will present risks during four lifecycle stages:

1. Construction;
2. Configuration, Maintenance and Launch;
3. Operation;
4. End of Life.

Each of these stages pose different challenges that will need to be met through appropriate guidelines, regulations and controls. Construction methods and hazards posed to industry during construction falls outside the scope of the BRIDGES project.

Through appropriate use of common hazard identification techniques it has been possible to create a short list of hazards that exist within the remaining three lifecycle stages (Tables 1 through 3).

The configuration, maintenance and launch periods are those periods where personnel may be directly interfacing with the glider, be that onboard ship or in workshops, during which certain elements of the system pose hazards to the personnel.

Table 1. Configuration & Maintenance Hazards

Hazard Title	Hazard Description	Safety Risks	Env. Risks
Manual Handling	Glider weight is above guidance for safe manual handling.	✓	
Moving Parts	Various moving parts (specifically propellers) may pose hazards to personnel configuring and maintaining the unit.	✓	
Hot and Cold	Depending on usage, parts of the glider may become excessively hot or cold and pose contact hazards.	✓	
Contamination	Depending on usage, the unit may become contaminated with dangerous chemicals that may pose a hazard to personnel working with the device.	✓	✓
High Pressure Discharge	Accidental pump activation on surface would result in a high pressure discharge in the proximity of personnel.	✓	
Sea State	Attempts at launching and recovering in poor sea states.	✓	
Electrical	Gliders contain high capacity batteries	✓	
Fire	The batteries in use combust, leading to a fire that is hard to control and put out.	✓	✓

During the operational phase, the glider will be acting autonomously and may be out of contact for extended periods of time. The glider itself will be required to make decisions to help mitigate dangers it encounters.

Table 2. Operational Hazards

Hazard Title	Hazard Description	Safety Risks	Env. Risks
Fire	The batteries in use combust, causing a fire and damaging the glider.		✓
Entanglement	Glider gets entangled during operations, limiting its ability to continue its mission and potentially (in the case of entanglements with fishing nets and similar) causing risks to members of the public operating these devices.	✓	

Hazard Title	Hazard Description	Safety Risks	Env. Risks
Collision	Collision with Manned Vessels, Members of the Public, Fixed Infrastructure or the Sea Bed. Potentially causing damage/injury to either the glider or the object with which it collides.	✓	✓
Navigational Obstruction	The glider presents a navigational obstruction to passing vessels. Due to a lack of manoeuvrability of the glider it is likely that this would result in the vessel in question having to alter course and potentially conduct unsafe navigational actions.	✓	
Discharge	During normal operations and fault conditions the glider discharges elements that are hazardous to the environment.		✓
Inadvertent Handling by Public	Due to repeated surfacing and the risk of being washed ashore, members of the general public may come into direct, unsupervised contact with hazardous elements of the glider.	✓	
Harmful Interaction with Environment	Depending on location, the glider could damage local flora and fauna.		✓
Contamination of Glider	Some ecological factors such as oil spills could coat the glider and limit its ability to perform its mission. It could also transfer that contamination to other locations.		✓
Surface in Wrong Location	Given stronger than expected current flows, the glider may surface outside of the expected areas and be out of range for contact.		✓
Security / Hacking	Due to the isolated nature of the glider it is possible that outside malicious interaction could be conducted.	✓	

End-of-life hazards occur during decommissioning, disposal or total loss of the glider sub-systems or the glider itself.

Table 3. End of Life Hazards

Hazard Title	Hazard Description	Safety Risks	Env. Risks
Loss of Glider	Any number of system failures in the glider could result in mission failures, and potentially (depending on failure type) environmental damage.	✓	✓
Disposal	Some elements of glider construction may prove environmentally damaging following disposal of the glider system.		✓

3 The Role of Standards in BRIDGES

Standards can be categorised into a number of generic types and serve a variety of purposes. The principle types of standards commonly applied to technical projects include:

- Standard definition – used to formally establish terminology;
- Standard units – commonly applied metrics for physical measurements;
- Standard specification – an unambiguous statement of requirements for a physical item, material, system or service;
- Standard test method – defines an approach to perform an unbiased and repeatable measurement of physical properties or performance;
- Standard practice – a common set of instructions for performance of an operation.

The application of standards provides a range of benefits that encompass:

- Statement of what is known to be common industry practice;
- Definition of an acceptable means to comply with relevant regulations;
- Ensuring commonality of approach to facilitate integration of elements within a system;
- Interoperability of systems with each other and the interaction of systems that may come in to conflict;
- Product commercialisation;
- Ongoing system upgrade, enhancement and technology insertion;
- Consistency in the conduct and performance of design, manufacture, operation, maintenance and disposal.

4 Standards and Regulatory Approaches from other Domains

A review was undertaken into a number of related domains in order to understand and characterise the approach taken to regulation and standards. The review findings are summarised in the following subsections. The domains chosen were considered relevant to the BRIDGES project because:

- Aspects of the regulation and standards are directly applicable to BRIDGES;
- The domain displays key technological features that were analogous to sub-sea glider technologies;
- The domain addresses analogous risk issues to those identified as applicable to sub-sea glider operations.

At the end of this section, a table summarises the characteristics of the regulations and standards within the various domains.

4.1 Maritime

The maritime industry is regulated through a hierarchal arrangement with the International Maritime Organisation (IMO) providing the global standard-setting authority for the safety, security and environmental performance of international shipping. IMO conventions cover all aspects of international shipping from ship design, construction, equipment, manning, operation and disposal to ensure safety, environmental protection, energy efficiency and security of international shipping. The IMO applies the Formal Safety Assessment (FSA) (IMO, 2002) methodology to the development of its conventions. FSA is a structured and systematic methodology that utilises risk analysis and cost benefit assessment to derive regulation that achieves a balance of risk reduction and cost.

The IMO's principle conventions include:

- The International Convention for the Safety of Life at Sea (SOLAS) (IMO, 1974) is an international maritime safety treaty. Its primary objective is to specify minimum standards for ocean-going vessels, and it ensures that ships' flag states remain responsible for minimum safety standards in construction, equipment and operation.
- The International Convention for the Prevention of Pollution from Ships (MARPOL) (IMO, 1973) was developed by the IMO and is the main international convention covering prevention of pollution of the marine environment by ships from operational or accidental causes.
- The International Convention on Standards of Training, Certification and Watchkeeping for Seafarers (STCW) (IMO, 1978) establishes basic requirements for training, certification and watchkeeping for seafarers on an international level.
- The International Ship and Port Facility Security (ISPS) Code (IMO, 2004) prescribes responsibilities to governments, shipping companies, shipboard personnel, and port/facility personnel to identify security threats and take preventative measures against security incidents.

Commercial vessels are registered or licensed by the 'flag state' who have the authority and responsibility to enforce regulations over vessels registered under its flag.

Fig. 3. Commerical vessel operating under flag state authority

Marine equipment can only be installed on board ships flying the flag of an EU country, Norway, Iceland and other flag states if it is marked with the Marine Equipment Directive 96/98/EC (MED) (European Union, 1996) mark of conformity, also known as the "wheelmark".

Classification societies are non-governmental organisations that establish and maintain prescriptive technical standards for the construction and operation of ships and offshore structures.

The maritime regulations and standards detailed in this section are generally only applicable to manned surface vessels and therefore will not apply directly to sub-sea gliders or any other form of marine autonomous vessel. The UK-based Maritime Autonomous Systems (MAS) Regulatory Working Group (RWG) (Society of Maritime Industries, 2015) is an organisation that is leading the development of a best practice regulatory framework for MAS that will be submitted to the Maritime and Coastguard Agency (MCA). This may ultimately lead to the MCA making recommendations for changes to the International Maritime Organisation (IMO) conventions to accommodate MAS.

4.2 Offshore Oil and Gas

Offshore regulation of the Oil and Gas Industry is spread over three authorities in the UK, comprising:

- Health and Safety Executive, with responsibilities for human safety;
- Department for Environment and Climate Change (DECC) with responsibility for environmental compliance and leak containment;
- The MCA with responsibility for spill clean-up at sea.

The HSE regulates offshore safety using a goal-driven safety case regime focused around regulatory expectations. Operators have the opportunity to be innovative and to achieve the required high levels of safety by adopting practices that meet the particular circumstances outlined in the regulatory standards. This approach fosters innovation and continuous improvement in operational and technological integrity. This approach is supported through mechanisms in place for independent, third party verification in the crucial areas of well design and integrity of safety critical equipment.

In contrast, the offshore environmental regulation regime is based on the implementation of EU regulation. This aspect of the regulation is largely focused on preventing or minimising any leakage of hydrocarbons during normal operations. Consequently it is relatively prescriptive compared to the safety regime, with less scope or encouragement for innovation in approach.

The oil and gas industry is arguably one of the first domains to exploit the potential of unmanned systems. Remotely Operated underwater Vehicles (ROVs) represent a type of unmanned system that have been in regular service since the mid-1960s, undertaking a variety of operations predominantly in exploration, installation and maintenance. NORSOK, the Norwegian petroleum industry body, has developed standards for ROV operations. NORSOK U-102 ROV services (NORSOK, 2012) has been produced containing information and typical requirements, deliveries and documentation expected from operators of ROVs. It also contains requirements for ROVs and for other services which have similarities to the ROVs and the way they are operated, including AUVs, remotely operated tools (ROTs), remotely operated towed vehicles (ROTVs) and dredging machines.

Fig. 4. Remotely Operated underwater Vehicles (ROVs)

4.3 Defence Unmanned Maritime Systems

Unmanned Maritime Systems (UMS) is one of the agreed 22 priority areas with the potential to become successful as joint European research work in the European Defence Research & Technology (EDRT) strategy. As a result, several research projects with participation from many European Union (EU) nations in the area of UMS came together in the European Defence Agency UMS programme.

UMS has the objective to enhance European capabilities in a number of naval applications by means of several research projects related to unmanned systems. Unmanned vehicles in particular are expected to be an integrated part of modern fleets. Given the expeditionary nature of modern European naval operations it is necessary to address interoperability issues.

It was recognised that national or international rules, regulations and legislation governing safe operation of unmanned maritime vehicles at sea are virtually non-existent. Common understanding of minimum safety procedures and a joint view on rules and regulations among European Navies would enhance interoperability in future joint maritime operations and training. To establish a foundation for achieving interoperability, a forum was created to address all regulations, legislation and safety issues related to design and operations of UMS - the European Defence Agency (EDA) Safety and Regulations for European Unmanned Maritime Systems (SARUMS) forum.

The objective of SARUMS is to provide European Navies with a best practice safety framework for UMS that recognises their operational usage, legal status and the needs of Navies. The philosophy behind this guidance will be based on the management of risk as well as applicable rules and regulations. The group is currently developing a document for this purpose titled "Best practice guide for UMS handling, operations, design and regulations". A significant improvement in interoperability and standardisation in design and operation of UMS is expected if nations decide to adopt this guidance document.

4.4 Unmanned Civil Aviation

The civil aviation industry has adopted unmanned technologies and has seen a proliferation of Unmanned Aerial Systems (UAS), particularly at the smaller end of the spectrum where they are used in a variety of survey, photography and monitoring applications as well as for recreational use. The vast majority of these systems are remotely piloted with limited autonomous operation capability.

In the UK, regulation of civilian airspace is the responsibility of the Civil Aviation Authority (CAA) through the application of the Air Navigation Order (ANO), CAP 393 (CAA, 2015). Safety of smaller UAS is principally controlled through the ANO Articles 166 and 167 which apply to what the CAA define as 'Small Unmanned Aircraft'. UAS are therefore considered aircraft but the CAA effectively allows derogation from the vast majority of requirements that would apply to larger manned craft, so long as the UAS meets a set of defined criteria and articles 166 and 167 of the ANO are adhered to in relation to the responsibilities of the 'Remote Pilot'. The 'Remote Pilot' is required, amongst other obligations, to:

- Ensure the system is airworthy and safe to fly;
- Maintain visual contact with the craft through Visual Line of Sight (VLOS) operations so as to avoid collision;
- Not permit overflight of persons, structure or vehicles;
- Liaise with Air Traffic Control (ATC) when operating in controlled airspace.

Fig. 5. Unmanned autonomous quadcopter

For larger UAS, or operations outside of those prescribed for Small Unmanned Aircraft in the ANO, the CAA requires airworthiness to be assessed and the safety of operations to be justified through a formal Safety Case submission. Relatively few UAS have been approved for operation through this route, principally due to the lack of robust sense and avoid technologies - a requirement to maintain separation from other airspace users. ASTRAEA (Autonomous Systems Technology Related Airborne Evaluation & Assessment) is a UK industry-led consortium focusing on the technologies, systems, facilities, procedures and regulations that will allow autonomous vehicles to operate safely and routinely in civil airspace over the United Kingdom (ASTRAEA, 2015). ASTREA is one of the few civilian programmes to have successfully trialled operation of an autonomous aircraft outside of restricted airspace.

The CAA has published guidance for unmanned aircraft system operations in UK airspace as CAP 722 (CAA, 2015). This guidance covers aspects such as Approvals, Regulatory Policy, Airworthiness and Operations. Reference is made in the ANO to acting 'reasonably' which a court would likely interpret as meaning standard practice, custom or guidelines have been followed. In this way, these published guidelines effectively become part of the regulatory requirement.

Meteorological balloons are another aspect of aviation that provides parallels with autonomous sub-sea glider applications. These balloons are unguided and will typically operate at very high altitudes, above the majority of other airspace users. They are required to traverse through operational airspace during ascent and parachute-controlled decent phases where the risk of collision with other air users exists. This risk is controlled through the requirement of the operator to:

- Obtain permission from the CAA to operate;
- Apply for a NOTAM (Notice to Airmen) to ensure other airspace users are aware of the operation;
- Liaise with Air Traffic Control (ATC).

4.5 Space

Due to the high level of platform autonomy, restrictive electrical power budgets, limited contact with ground-based controllers and the extreme physical environment in which they operate, spacecraft share many similarities with sub-sea gliders. As the launch costs for spacecraft are so great (in the order of $10,000 per kg of payload inserted into Low Earth Orbit), very high platform and subsystem reliability are critical to the success of a mission.

In order to ensure that a completed spacecraft will be reliable enough to complete its mission, testing is conducted on individual components and subsystems in addition to the integrated spacecraft. Due to the high manufacturing costs for spacecraft, testing is often conducted on the real flight articles, rather than dedicated test prototypes. So, for instance, the actual completed spacecraft will be subjected to the extremes of vacuum, high and low temperatures, vibration, noise and shock that it will encounter during launch and operation.

Space agencies such as the US National Aeronautics and Space Administration (NASA), the European Cooperation for Space Standardization (ECSS), the European Space Agency (ESA) and the Japan Aerospace Exploration Agency (JAXA) have developed a number of technical standards to de-risk the design and manufacture of satellites and space probes. Amongst many others, these include:

- The use of design tools, such as NASA-STD-(I)-0007 - NASA Computer-aided Design Interoperability (NASA, 2009);
- Manufacturing techniques, such as NASA-STD-5006 - General Welding Requirements for Aerospace Materials (NASA, 2015); and
- Testing and inspection procedures, such as NASA-STD-7002 - Payload Test Requirements (NASA, 2004).

These standards may be generic (non-prescriptive), entirely prescriptive requiring the use of specified techniques and templates, or a combination of the two.

Fig. 6. Long range satellite

4.6 Other Generic Standards

Other standards that were reviewed for potential relevance to the BRIDGES project included:

- IEC 61508 (International Electrotechnical Commission, 2010) which is concerned with the functional safety achieved by safety-related systems that are primarily implemented in electrical and/or electronic and/or programmable electronic (E/E/PE) technologies. It covers:

  - A risk based approach to determine safety integrity requirements of E/E/PE safety-related systems;
  - A safety lifecycle model as the technical framework for the activities necessary for ensuring functional safety is achieved;
  - System aspects to include: hardware and software sub-systems; and failure mechanisms (random hardware and systematic);
  - Preventing failures and controlling consequences;
  - The techniques and measures that are necessary to achieve the required safety integrity.

- ISO 9001 (International Organization for Standardization, 2015), a certified quality management system for organisations that want to consistently provide products and services that meet the needs of their customers and other relevant stakeholders. ISO 9001 is based on eight quality management principles and divided into several sections. The most relevant sections to BRIDGES are likely to include Product Realization and Measurement, Analysis and Improvement.

Table 4. Characterisation of Standards and Regulatory Approaches

Aspect	Characterisation				
	Maritime	**Oil and Gas**	**Defence**	**Civil Aviation**	**Space**
Precedence	International conventions enacted through national regulatory bodies and non-governmental classification	European Law enacted through national regulatory bodies with split of responsibility for safety, environmental compliance and spill clean-up. Industry driven standards developed in unregulated areas (e.g. ROVs)	User/industry driven standards	International conventions enacted through the national regulatory body	User/industry driven standards
Prescription	Risk based approach to the development of typically prescriptive requirements	Goal-driven safety regime and prescriptive environmental requirements	Non-prescriptive guidance	Highly prescriptive regulation for standard recognized operations, risk based approach to address safety of non-standard aircraft and operations	Highly prescriptive standards reflecting the potentially high consequence of mission failure
Regulation or Standards	Regulation supported by nominated standards	Regulation supported by nominated standards	Standards providing description of best practice approach	Regulation supported by nominated standards	Prescriptive standards
Depth of Detail	Detailed coverage of design, build and operational aspects	Detailed coverage of design, build and operational aspects	Detailed coverage of operational aspects, limited coverage of design and build where industry best practice is yet to be established	International conventions enacted through the national regulatory body	Detailed coverage of design, build and operational aspects

5 Defining Standards Requirements

5.1 Standardisation Approach

The use of autonomous sub-sea gliders is still a relatively new area and research is still being conducted to find the most appropriate combination of design elements to make these devices as effective as possible. This is currently leading to creative ideas being tried and tested in the field and whilst regulations and standards will always play a role in the development and use of these devices, it is important that these rules do not become too restrictive and thus stifle innovation in the still evolving domain.

It is known that a number of regulations (especially those relating to environmental protection) will have prescriptive elements that still apply to sub-sea gliders, but in most areas the existing regulations will not be applicable to such a vessel. Where this is the case, innovation and new ideas can be developed by using a more goal-based approach to reduce risks to an acceptable level through standards that are applied.

Regulations and guidance are currently limited; this is in part due to the lack of regulatory recognition of the risks in this arena. This is leading to an industry-driven approach to regulation, similar to the industry-driven guidance issued by operators of ROVs in the oil and gas fields. In the long term it might be expected that the data and experience gained from sub-sea glider use will enable the regulations and standards to deal with detailed features of glider design. However, in the current climate with the existing lack of historical data on glider use, standards will, by necessity, be relatively high level.

5.2 Aspects to be Standardised

Whilst standardisation may become useful in all areas of BRIDGES, it is important to focus on those areas that may lead to safety-related risks. The following paragraphs describe aspects to be considered for standardisation within the BRIDGES project.

Materials. The sub-sea gliders will be operating in varied environments and will interact with a number of different operating conditions. It is important that the materials used in construction are known to be able to handle such environmental hazards as extreme heat (thermal vents) and extreme cold (artic waters) whilst remaining strong enough to handle shocks and impacts that may occur during missions and transport operations.

Construction methods. The modular design of the glider system will allow for great versatility in the operation of the BRIDGES gliders, but this comes at a cost of needing to ensure that all manufacturers are working to the same standards and using similar construction methods. A failure to achieve this may result in parts with radically different lifespans or, in extreme cases, modules that simply cannot interact in the way they are meant to.

Scientific performance. The BRIDGES gliders will be designed to allow for accurate and useful data retrieval from areas that have been previously inaccessible or simply too costly to access. To allow the retrieved data to be as useful and as scientifically valid as possible, it is essential that there are methods in place to coordinate the retrieval of this data. Due to the conditions that the gliders operate in, the system will need regular calibration and assessment to ensure that they remain functioning as accurately as required.

Navigation. The maritime community operates under a number of regulations and standards that control how vessels at sea operate and navigate. This system ensures that the worldwide community of marine operators can interact safely at sea and accurately navigate to their destinations. The gliders will, at times, be interacting with this community and, as an ocean-going vessel, may well be required to meet some of the guidelines and standards that other marine vessels act under.

In addition, due to sub-sea gliders being autonomous craft capable of acting independently for long periods of time over great distances, it is vital that the navigational system used by gliders is fit for purpose.

Communication. The communication systems between both the glider and base-station and between the glider's various systems will inevitably form a complex system with many elements and possible complications. Bandwidth for transmission of mission results and data will be limited at times, so care will be needed to ensure that vital data is prioritised for transmission, with lower priority data being stored for retrieval at the end of the mission.

Contamination / environmental pollution. Due to the wide range of areas that the glider may operate in, it is possible that contamination may occur to the glider itself; with the long-range abilities of gliders it is feasible that contamination could also be transferred into other areas of operation. Regulations already exist for managing the transfer of ballast water, the effects of anti-fouling systems and environmental guidelines for arctic water travel; all of these regulations may bear some important lessons for glider use.

Power systems. The largest factor upon a glider's feasible range is the suitability of the power systems. The glider's power supply will need to be able to maintain both the buoyancy and manoeuvring systems, and the various sensor payloads that are installed. Power systems will need to be rated to handle the stress and stains that occur at significant depths, depths which may also affect battery performance and longevity.

Potentially a combination of miniaturisation, power budgeting techniques and power usage data could result in a system that can manage its own power requirements to ensure mission success.

Collision Avoidance. The International Regulations for Preventing Collisions at Sea (The COLREGs) requirements do not directly apply to sub-sea gliders, but must be considered in terms of how other users would recognise and interact with a glider operating at, or near, the sea surface.

Emergency / fault. Collisions, fire, equipment failure and other factors could result in emergency situations for the glider, potentially removing the glider's ability to continue its mission. In situations such as these it will be important that the responses that occur are known in advance. Minimum equipment lists, redundancy and emergency location devices may play roles in the glider configurations, and intelligent response systems to known possible faults may define how the glider reacts in situations that are unrecoverable.

Manoeuvrability. The very design of gliders limits their manoeuvrability more than most waterborne craft. The combination of low power usage, buoyancy-based travel mechanics and low movement speed limits the gliders to slow turns and slow dive / ascents. Whilst on the surface, the glider's movement options are limited to its ability to dive and therefore becomes a passive object when surfaced, potentially causing navigational issues for other sea going vessels.

Hydrodynamics. The gliders will be designed with the aim of presenting a suitably hydrodynamic body, enabling the glider to make best use of battery power. The modular nature of the science bays may present threats to that hydrodynamic profile, and certain sensors that penetrate the hull may reduce the effectiveness of the rudder and propeller. Standards could help guide the manufacturers of the science bays into what protrusions would affect performance.

Markings / colours. Current ocean-going vessels operate under a strict set of guidelines concerning visibility, typically this is achieved through the use of lighting systems on board. Sub-sea gliders are unlikely to have the capacity to host lighting systems and will instead be required to achieve visibility through other means. The overall colour scheme of the glider, and additional written markings could be used to alert the general marine population of both the presence of the device, and any potential dangers that the device presents. This does also however, create potential issues involving unwanted attention from unauthorised users (see security).

Pressure hull / penetrations. Pressure hulls are a regular feature of many aquatic vehicles, and there are many construction and testing requirements that help ensure the safety of these hulls. In the case of the gliders, the risks associated with the pressure hull failing are unlikely to affect the safety of persons, however, the integrity of these hulls still affects the ability of the system to complete its mission as a failure of the hull would likely be catastrophic for the glider.

Security. The systems included in autonomous craft typically concentrate on the goal of mission success and focus on the physical limitations of the equipment and the environment it will be operating in. Malicious or unauthorised usage of these devices are a very real possibility that should also be considered.

Reliability. The glider systems will need to operate autonomously for extended periods of time, in environments that have, at times, been unexplored. Any equipment failure could be catastrophic and yet it will be difficult to fully assess component lifespans when the components may never have been used before, or at least have not been used in conditions as extreme as expected.

Launch and recovery. Due to the expected weight of the glider systems, suitable systems will be required for launch and recovery operations. Depending on what these systems entail, regulations such as the Lifting Operations and Lifting Equipment Regulations 1998 (LOLER) may be applicable.

Units, Terminology and language. Due to the BRIDGES project being funded by the EU it seems likely that the units used through construction will be SI units. However, due to numerous failures of large scale projects due to inconsistent units it is still vital that the units used are stated and standardised throughout the glider lifecycle. The use of a common set of terminologies and common language(s) will also aid in the reduction of errors during production and use.

Transportation. The complete glider systems will involve varied and potentially dangerous elements. Some transportation providers may be reluctant to transport the gliders without suitable controls. Specifically there are currently concerns for air transportation, where the transport of high powered batteries is heavily controlled and may be too restrictive for the transport of gliders by this route.

Testing. Testing procedures are vital in every scientific field and ensuring that a device will operate correctly and return accurate and usable data is essential. Testing guidance and procedures should be an integral part of the systems that support glider operation.

Satellite communications. The gliders will be required to transmit and receive data during missions and will need to communicate that data when not necessarily in vicinity of the base stations that launched the units. Satellite communication techniques are a valuable method for facilitating communications, but they do not come without complications. The gliders may surface at unexpected times and in unplanned areas, making booking of satellite communication slots problematic.

Legal Aspects. The gliders will be required to meet numerous legal requirements, including certifications, approvals before being able to be effectively used in certain territorial waters. For example, whilst it is envisaged that the gliders will primarily operate in international waters, the nature of sea currents and certain close-to-shore survey missions will require that each country of operation's own regulations are followed, either for a purposeful launch in the area or simply that the glider drifts accidently into the waters in question.

6 Summary

BRIDGES is a long term project spanning a number of years and there is still much research and design work to be conducted. The approaches and topics discussed in this paper will form the basis of the ongoing research into standards and guidance, and whilst this will not result in a complete set of standards, it will guide the way to the eventual creation of such a set of guidance.

There are a number of IMO conventions and other marine regulations that may bear relevance/importance to this project, however the bulk of the standards and guidance that exist do not cater for sub-sea gliders. Despite this, there are still lessons to be learnt from the ways in which the existing regulations have been developed and these lessons need not only come from marine industries.

The drive for innovation and creative design work has led to the conclusion that the regulations and standards that are applied to sub-sea gliders should be as non-prescriptive as possible, with goal based approaches supporting construction methods. Due to there currently not being an international oversight committee or organisation, industry will be required to push forward standards and guidance, which in turn will allow those standards to evolve naturally over time, without limiting innovation or the growth of the market.

Acknowledgments BMT Isis Ltd would like to acknowledge the work of the other BRIDGES consortium partners and thank them for their contribution to the preliminary deep glider design which has made it possible to develop this paper.

References

ASTRAEA, 2015, Current Projects, Available from: http://astraea.aero/current-projects-2/. [Accessed 25 September 2015]

BRIDGES, 2015, Temporary Website for the H2020 BRIDGES Project, Available from: http://www.bridges-h2020.eu

Civil Aviation Authority (2015), CAP 393 Air Navigation - The Order and the Regulations, The Stationery Office

Civil Aviation Authority (2015), CAP 722 Unmanned Aircraft System Operations in UK Airspace - Guidance

European Aviation Safety Agency, 2015, Available from https://www.easa.europa.eu [Accessed 25 September 2015]

European Union (1996), Council Directive 96/98/EC of 20 December 1996 (Marine Equipment Directive (MED))

International Civil Aviation Organization, 2015, Available from http://www.icao.int

International Association of Classification Societies, 2011, Classification Societies - What, Why and How?, Available from: http://www.iacs.org.uk/document/public/explained/Class_WhatWhy&How.PDF [Accessed 25 September 2015]

International Electrotechnical Commission (2010), IEC 61508:2010 Functional safety of electrical/electronic/programmable electronic safety-related systems

International Maritime Organization (1973), International Convention for the Prevention of Pollution from Ships (MARPOL)

International Maritime Organization (1974), International Convention for the Safety of Life at Sea (SOLAS)

International Maritime Organization (1978, with major revisions in 1995 and 2010), International Convention on Standards of Training, Certification and Watchkeeping for Seafarers (STCW)

International Maritime Organization (2002), Guidelines for Formal Safety Assessment (FSA) for use in the IMO Rule-making Process, MSC/Circ. 1023

International Maritime Organization (2004), International Ship and Port Facility Security (ISPS) Code

International Maritime Organization, 2015, Available from: http://www.imo.org [Accessed 25 September 2015].

International Organization for Standardization (2015), ISO 9001:2015 Quality Management Systems

NASA (2004), NASA-STD-7002 - Payload Test Requirements

NASA (2009), NASA-STD-(I)-0007 - NASA Computer-aided Design Interoperability

NASA (2015), NASA-STD-5006 - General Welding Requirements for Aerospace Materials

NORSOK (2012), U-102 Remotely Operated Vehicle (ROV) Services, Edition 2

Society of Maritime Industries, 2015, MAS Regulatory Working Group, Available from: http://www.maritimeindustries.org/MAS-Regulatory-Working-Group- [Accessed 25 September 2015].

UK Civil Aviation Authority, 2015, Available from: http://www.caa.co.uk [Accessed 25 September 2015]

Improving the testability of high integrity FPGAs

Matthew Noonan[1]

Resource Group

Worcester, UK

Abstract *FPGA usage within high integrity systems is becoming both more popular and more complex. One of the challenges of putting an FPGA in a high integrity system is the cost of verifying its correct operation, and this is made significantly more difficult by the increasing complexity of FPGA applications.*
For a typical DO-254 Level A aerospace FPGA application, at least 50% of the overall effort and engineering budget is spent on verification activities. As design decisions are set in stone early in the development process, it is common to discover unexpected verification problems when it is too late to do anything about it.
This paper seeks to explore and quantify the effect of various architectural and design structures on their 'testability' for FPGA systems in high integrity applications, as well as identifying test based mitigations for common problems. Using anonymised data from Resource Group's high integrity customers, a study of varied design structures has been analysed into a set of testability rules and a summary of their respective impact on the effort of verification.

1   Definition of testability in relation to FPGAs

To address testability, it must first be defined what testability is in relation to FPGAs (Field Programmable Gate Arrays). From a software point of view testability is:

"The degree to which a system or component facilitates the establishment of test criteria and the performance of tests to determine whether these criteria have been met" (IEEE 1990)

At a broad level this definition applies to FPGAs just as well as traditional software targets. There will always be differences in how testability is judged be-

---

[1] Embedded Systems & Solutions, Resource Group

tween software and firmware and to define this we need to look at the specific requirements needed at FPGA test.

2   Importance of testability in high integrity systems

The requirements of FPGA test for high integrity systems is easily defined from the global standards that are applied in high integrity industries. A set of common test requirements from standards such as IEC 61508 (IEC 2010), DO-254 (RTCA 2000) and EN 50128 (BSI 2011) can be summarised as:

1. Evidence of compliance to functional requirements (positive proof of requirements)
2. Evidence of robustness to unplanned influences (negative/robustness testing)
3. Traceability evidence of requirements coverage at test (Full requirements coverage)
4. Coverage of all code elements through test execution (Code coverage – variable by SIL/DAL)

These four objectives have many sub-elements to them but together represent the high level objectives of requirements testing for high integrity systems. These objectives are the same for software and firmware but whilst software has learnt testability lessons over a long period of time, FPGA applications are slow to implement the same lessons despite being potentially more testable than pure software (Thomson 2015).

In 2014 the National Microelectronics Institute (NMI) conducted an FPGA usage study within the UK and EIRE. This survey sought to identify what devices are used, by whom, and what problems are being faced. A brief summary of the results shows that RTL verification is acknowledged as a significant issue:

Table 1 – RTL Verification results from NMI FPGA usage survey 2014

Application type	Number of survey reponders	Average amount of project time spent on RTL verification
All applications	125	21%
High Integrity applications only	21	29%

RTL verification was also identified in this survey as the 2nd biggest challenge[1] that the projects face.

3 Increased use of and complexity or FPGAs

As identified above, FPGA applications are slow to apply the lessons of high integrity software and this absence of testability awareness is a growing problem due to the increased usage of FPGAs in high integrity applications. This is supported by a higher usage of FPGAs in place of Application Specific Integrated Circuits (ASIC) technology in industries like space where FPGAs are offering a faster time to market and an increasing radiation tolerance (Leon 2013) as well as greater applications for FPGA technology (Leong 2008).

There is also little debate that the complexity of FPGAs has increased over the past few years (Moretti 2004). This is further indicated by the fact that FPGA capacity has increased by a factor of 10,000 in the past 30 years (Trimberger 2015) and that even high integrity applications are making use of the higher capacity variants.

4 Introduction to the data

Resource Group' embedded systems & solutions division provides safety critical FPGA applications support to customers in industries such as aerospace, defence, power and rail. Subjective analysis of the projects for these many customers and industries identifies a range of factors influencing the effectiveness, effort and cost of FPGA verification. The raw data analysed below has been anonymised but can be summarised as follows:

Table 2 - Resource Group project summary

Project type	Number of projects	Total size (VHDL[2])
IEC 61508 SIL 2	3	22,500 lines of code
EN 50128 SIL 4	1	30,000 lines of code
DO-254 DAL B	2	52,000 lines of code
DO-254 DAL A	4	33,000 lines of code

---

[1] The survey showed 52% of responders identifying Timing Analysis and Closure as a challenge and 48% of responders saying that RTL verification was a challenge.

[2] VHDL is VHSIC Hardware Description Language, VHSIC being Very High Speed Integrated Circuit. This is one of the 2 common hardware description languages used for FPGA development and test, the other being Verilog.

The breadth of projects here provides a good basis for identifying common trends affecting testability however the volume of projects within each category is low. This may show a need for more analysis in future where there are specific requirements for each group.

The sample data showed a good spread of major architectural decisions: basic logic; pipelined logic; multi-layered logic; highly-coupled logic and high single-dependency logic. However, due to such small sample numbers in each of these categories no specific conclusions have been drawn against these variables.

Initial analysis of testability influences was performed by looking at the project costs against baseline metrics for such tasks. Where anomalies were identified these were categorised as either: Requirements issues, Test environment issues, Estimating error (i.e. baseline estimates were incorrect) or other issue. This analysis was done on a 'function' level wherever possible and at a whole project level where not. As an example, an FPGA based project may have logic built from many functions: comms, algorithm, housekeeping, memory control, etc.

Any function or project identified as having Requirements or Test environment issues was then considered with others to identify common trends of problems throughout the industries represented.

The impact of each issue was to be subjectively assessed against:

- Testing effectiveness: How much did the issue limit, hinder or degrade the effectiveness of the testing (ignoring effort)?
- Testing effort: How much did the issue increase the effort of testing to achieve the same result if the issue were not present?
- Testing cost: For most scenarios this was proportional to any increase in testing effort however, some testability issues requires the application of different and more expensive tools.

Early conclusions showed that the most common influences are shared with software projects, most prominently the need for well defined, unique and concise requirements. In addition, choices over the architecture of the FPGA, as with software architecture had a large influence over both effectiveness and effort. As these elements of testability are shared between FPGA and software development, use of software studies such as Baudry et al. (2005) and Boehm (1984) will show the impact to tests with wider sample data than is presented here.

Due to the sensitive nature of customer data, specific strengths and weaknesses can only be presented here as general conclusions and lessons. The data will not be presented publicly with criticism of any specific customer's designs. However, the subjective analysis of the data does identify FPGA specific lessons and these can be summarised under four major subjects: Unit level design and test; Integration level design and test; Test interfaces and; Simulation time.

4.1. Unit level design and test

High integrity software will commonly break requirements into high level and low level. The low level requirements are then proven through unit level testing with a focus on robustness and code coverage. High level requirements testing is performed later and only has a focus on functional testing. By performing these activities separately, it is easier to reach the objectives of each.

Typically we have seen that FPGA projects try to accomplish all their requirement testing objectives in a single level of test. This leads to a greater level of effort to achieve code coverage objectives and can greatly reduce the effectiveness of the testing activities. Both of these effects stem from a 'middle level' requirements approach where a single set of requirements are made to cover functionality and robustness. These requirements invariably lead to requirements that are too high to justify all elements of the code (leading to code coverage shortfalls) and too low to allow quick and simple testing of functionality (degrading the effectiveness of early testing, causing more faults to be found only at formal test later in the development).

As projects using FPGAs increase in complexity and responsibility, the need to approach the process with two levels of design and test will become more pertinent. The individual business case will need to be case-by-case but the benefits of including a low-level-requirements and high-level-requirements phase include de-risking code coverage objectives and quicker and easier test development and maintenance.

4.2. Integration level design and test

For some, the simple or cohesive nature of their design will warrant that only a single requirements and test phase is adopted. For these projects there are still major testability lessons that should be applied. The following points have been observed as common areas for improvement:

4.2.1. Consider partitioning

Adding partitions from the requirements through the FPGA architecture will greatly increase the maintainability of tests into the future. One of the greatest testability impacts identified from the data set is that testing at a single level for the whole FPGA gives tests a high co-dependency. In particular driving conditions for tests proving output functionality will be dependent on stable and correct working of the input and process logic. As such, any change can have significant impact to both the number of tests affected and the amount of effort needed to accommodate change. **Fig 1** shows a highly coupled architecture. Each unit can represent a module of code that is performing a specific function e.g. signal validation or mathematical functions. It can be seen that any change or test failure in 'input side' units would have an impact to many if not all 'downstream' units. This impact can increase the effort to test in the first instance and will also increase the number of tests that need to be reworked every time there is a change within the FPGA.

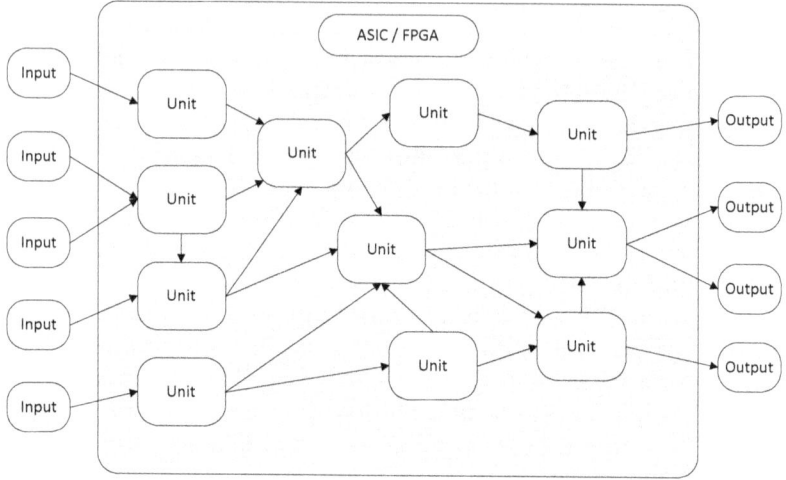

Fig 1. A highly coupled architecture

By adding partitioning as shown in **Fig 2**, the architecture of the FPGA forces low coupling and increases cohesion. This limits the impact of change to 'paths' through the FPGA and provides clear boundaries meaning that changes can be limited in impact.

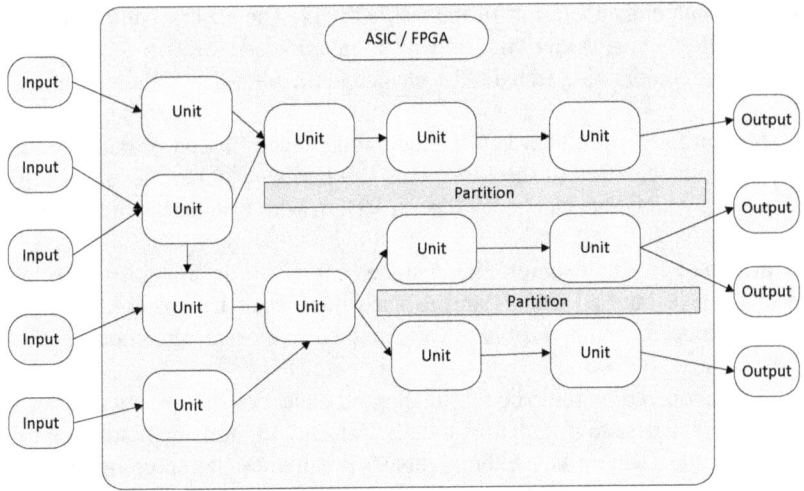

Fig 2. A partitioned architecture

4.2.2. Visibility of signals is key

In proving compliance to requirements there is a need to consider more than just the outside pins of the FPGA. There is a clear need to monitor internal signals when checking the correct operation of requirements, in particular when testing fault tolerances which would have little or no impact on external signals.

A common problem within the test dataset that has increased both testing effort and cost has been managing the visibility of internal signals for testing purposes. This can be classed into two grades of problem: difficult or impossible.

Difficult: As an example a tester has a requirement to check the output of decision logic or input condition but the data is question is obfuscated within a packed data array. The tester must resolve to analysing bits 3, 8 and 12 of a 64 bit data packet (as an example) to identify if the requirement has been met. This approach has several negative impacts to the testability:

- The risk of human error is high – The bit positions are not meaningful and must be determined by reference to large volumes of table data that themselves are prone to error.
- The effort of peer review is high – As with test authorship, the dependence on external table data makes the effort higher and more prone to error.

- The maintenance effort of the test is high – The arbitrary allocation of bit positions means that this test will be invalidated by either a change in the requirement being tested OR a change in the bit order of the data array.

A solution to this problem is to include good encapsulation of data during design and implementation of the code. This is a particularly pertinent lesson given the ease with which data can be handled in VHDL with little or no encapsulation.

Impossible: In this example the designer has given no thought to the testers need to observe internal signals and as such then tester has no ability to access them. It then becomes impossible to validate these requirements rendering the test activity of very little value.

This scenario can be remedied through good education and practices of designers to provide access to all internal signals. An easy to implement solution to this is shown in **Fig 3** where key internal signals are mapped to spare output pins to give the tester greater visibility.

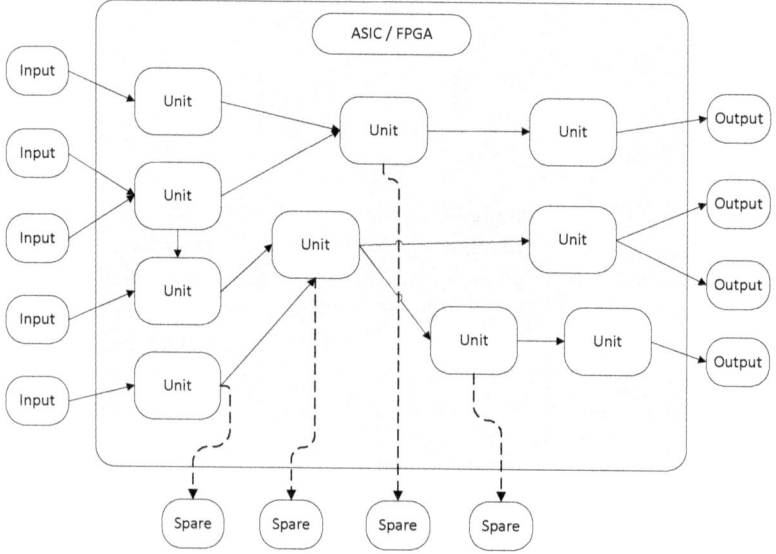

Fig 3. An easy method to increase signal visibility

4.3. Test interfaces

A more permanent solution to signal visibility is to include a dedicated test interface as part of the system architecture or even as standard across all FPGA products as illustrated by **Fig 4**.

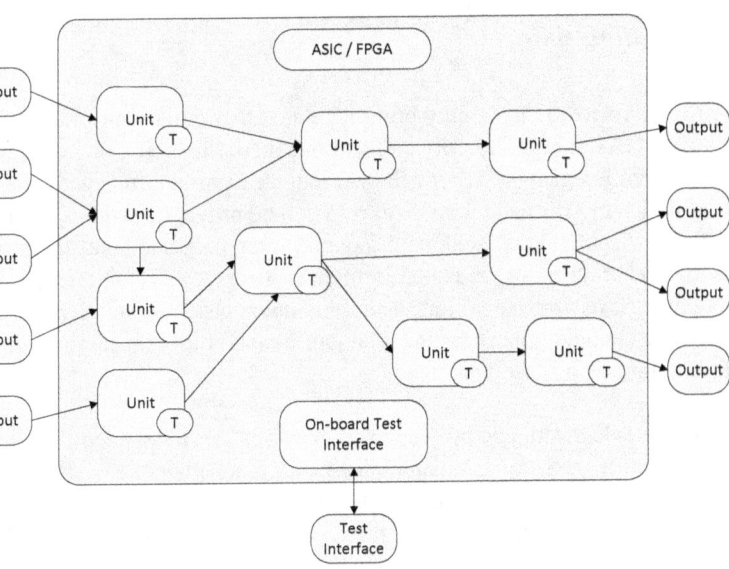

Where all elements marked (T) give visibility of signals to the on-board test interface

Fig 4. A formal test interface architecture

By standardising on a test interface it becomes a simple and repeatable task to interrogate internal signals at test. However, this approach requires a lot of considerations for high integrity applications:

Maintenance: Developing a test interface for use across all products is an excellent way of increasing testability if this is done in a consistent manner. This functionality will need maintenance into the future.

Be required: To enable a test interface to be included within a high integrity application it will need explicit requirements to justify its use. These can likely be a common set of requirements used on every project but these themselves will require maintenance.

Bigger/Slower: A permanent test interface may not be practical if the FPGA is pushed for space or at the limits of speed. A test interface will take up logic space itself and, depending on its design and integration may also affect the timings of your primary functionality.

4.4. Simulation time

The last FPGA specific testability point is related to simulation time. As the complexity of FPGAs in high integrity applications increases, test simulation time also increases. High integrity systems will also require coverage runs which take more time that RTL-only runs and gate-level runs can be orders of magnitude higher.

Of the projects studied within this paper, over half had total RTL only (no coverage) execution times measured in months if run serially on a single standard PC simulator. Two of these projects had this same metric measured at over a year.

There are some simple measures that can be taken to help limit the impact of simulation time on project deadlines:

Simulator selection: The project toolset choices can have a ten-fold impact on the simulation time. Some simulators are faster than others on a small scale and most offer more advanced versions that will simulate faster than the base or free versions. It is worthwhile looking at the costs of the more advanced tools versus the costs of project delays when planning projects.

Simulator hardware: Most current simulation tools are processor usage heavy and single threaded. It is worthwhile planning what machines are going to perform the simulations so that they are built for the task.

Test size: Most simulators give the option to stitch code coverage metrics together as a post-process activity. Given this there is no reason to run all the work through the same machine in series. Breaking tests down to the smallest sensible blocks helps with this and also aids test maintenance in the long term.

Time killers: There are some design oversights that can add enormous amounts of simulation time with no useful outcome. Specifically these are lengthy start up routines and fixed time delays within the design. When testers are executing many combinations of scenarios through the same functions, it can add a fixed and pointless overhead to every test case if the test has to go through an FPGA start up or fixed time delay. These can be easily avoided by a small amount of designer awareness that allows these long length steps to be configured or skipped by testers if so desired. The ways to accomplish this are varied but can include configurable constants to shorten time or enable/disable start up routines etc. On a similar point, clock multipliers used throughout the design are better if consistently derived from global constants. This approach allows testers to shorten these multipliers and 'speed-up' the global time within the FPGA for tests that don't directly involve these times.

5   Conclusion of testability impact on cost

In the same way that it is incredibly difficult to accurately estimate the cost of development and verification of a high integrity project, it is also difficult to say the impact of each testability decision on individual projects.

If we were to discuss a hypothetical 'average' FPGA project then we could assume factors like ~20,000 lines of VHDL, Level A/SIL 4 development and a moderate level of control complexity. A project of this nature could easily have a FPGA development and verification budget of around £1 million. This 'average' project could also expect to spend roughly half of this budget on development activities (Requirements, architecture, designs and code and their reviews) and the other half on verification activities (Lower level test, integration level tests and their reviews).

Taking this hypothetical example as a benchmark then the testability rules presented above can be divided into 'improvements' in that they will typically improve on the average process and reduce the overall budget of this project or 'precautions' in that they will protect against high risk areas of effort and cost runaway that are frequently seen in high integrity testing phases.

Whilst each project will be unique based on its architecture, processes, application and budget, generally the ideas presented above can be split as follows:

Improvements
- Unit level designs and tests
- Increased encapsulation
- Test specific interfaces
- Simulator selection
- Simulator hardware

Precautions
- Increased partitioning
- Adequate visibility of signals for test
- Avoidance of time killers
- Reduced test size

By looking at where these actions have been taken in different projects within the data set, it is possible to justify up to 15% saving in verification effort from the 'average' by implementation of the improvement actions.

In a similar way, analysis of projects where the precautions group have not been implemented show up to a 100% increase in verification effort and cost compared to the 'average' project.

Due to the small data set and the need to anonymise and abstract the data it is not possible to define specific costs savings against individual testability actions. This could be accomplished through an industry wide buy-in for improvement and sharing of project data to increase the data set and allow future analysis to be more accurate and conclusive and to allow industry members to estimate there actions with confidence through methods like Monte Carlo analysis against the baseline data set.

Acknowledgements

Resource Group would like to thank their clients for their support in this study, it is our combined hope that studies such as this can lead to more efficient and effective knowledge of verification within our industries.

References

Amos, B. 2015, 'NMI FPGA Usage Survey 2014',
  http://www.testandverification.com/VerificationFutures/2015/Doug_Amos.pdf?ea3bdf
  viewed 21 October 2015
Baudry, B and Le Traon, Y. 2005, 'Measuring design testability of a uml class diagram. Information and Software Technology', 47(13):859–879.
Boehm, B W. IEEE Software 1.1 (Jan/Feb 1984): 75-88
CENELEC, BSI GEL/9/1, 2011, 'EN 50128: Railway applications. Communication, signalling and processing systems. Software for railway control and protection systems', BSI
IEC, 2010, 'IEC 61508: Functional Safety of Electrical/Electronic/Programmable Electronic Safety-related Systems', IEC
IEEE 1990, IEEE standard glossary of software engineering terminology, IEEE255
Leon, A. F. 2013, 'Trends and patterns of ASIC and FPGA use in European space missions', ESA
Leong, P 2008, 'Recent trends in FPGA architectures and applications', 4th IEEE International Symposium on Electronic Design, Test & Applications
Moretti, G 2004, 'FPGAs break down the complexity gridlock', Electronics Weekly, 2165, pp. 20-22, Business Source Complete, EBSCOhost, viewed 21 October 2015.
RTCA SC-180, EUROCAE WG-46, 2000, 'DO-254: Design Assurance Guidance For Airborne Electronic Hardware', RTCA/EUROCAE
Thomson J. R. 2015, High Integrity Systems and Safety Management in Hazardous Industries, Butterworth-Heinemann, Oxford. 38p-41p
Trimberger, S 2015, 'Three ages of FPGAs: A retrospective on the first thirty years of FPGA technology', Proceedings of the IEEE (Vol. 103, No. 3, March 2015)

Time for a New Approach to Accident Investigation?

Graham Braithwaite

Cranfield University[1]

Cranfield, UK

Abstract *The investigation of aircraft accidents is an important source of learning for the aviation industry and beyond. Whilst protected by international standards and European and national legislation, and emulated across other sectors, aircraft accident investigation faces a number of challenges. These include: technical challenges surrounding complexity, concerns about the protection of sensitive evidence, maintaining investigator currency and the maintenance of investigator competency. Faced with such challenges, is it time that aircraft accident investigation benefited from a change in approach or has it evolved to meet the challenges that have surfaced along the way?*

1 Introduction

The investigation of aircraft accidents as a source of learning is as old as man's first attempts to fly. Accidents were frequent and the most precious resource – in those days' aircraft rather than people – were difficult to replace. As Grahame-White (1912) observed over a century ago, 'Bitter lessons were learned; but the industry profited by them. The margin of safety of machines was increased, not according to theory – which has sometimes proved a dangerous thing in connection with flying – but in accordance with the lessons learned in the actual navigation of the air'.

---

[1] Director of Transport Systems, Professor of Safety and Accident Investigation, Cranfield University, Cranfield, UK

In 1915, Captain G B Cockburn was appointed to the independent post of 'Inspector of Accidents' within the Royal Flying Corps, a role which continues to the present day in the form of both the UK Air Accidents Investigation Branch (AAIB) (for civil aircraft) and the Defence Accident Investigation Branch (Defence AIB) (for military aircraft). Over this century of investigation, great advances have been made through the detailed and forensic examination of accidents and their causes. So too, the emphasis of investigations has allowed the development of a deeper understanding of the underlying or systemic causes of accidents.

With the UK AAIB celebrating the centenary of a permanent professional aircraft accident investigator in 2015, it seem propitious to ask whether the tried and test approach to accident investigation remains effective.

2 The Purpose of Investigation

In the event of an aircraft accident, the initial response is rightly focused on rescuing those who may be injured or recovering the bodies of those killed. Soon after this, inquiries will commence from a number of different perspectives, such as law enforcement, regulatory compliance, safety investigation and the allocation of financial liability. Where an act of deliberate criminality is suspected, it is likely that a police investigation would continue to take precedence, but in most circumstances, accidents would be handed over to a safety investigation agency (SIA) to take primacy. This does not prevent other forms of inquiry which may wish to have access to similar sources of evidence and have aims which may be incompatible with the safety investigation agency. Following a fatal accident involving a transport category aircraft, it would be commonplace for the following parties to make inquiries:

- Law enforcement
- Safety investigation agency
- Occupational health and safety regulator
- Medical examiner / Coroner / Procurator Fiscal
- Aviation safety regulator
- Loss adjuster on behalf of insurer
- Aircraft manufacturer
- Component manufacturers
- Aircraft operator
- Independent investigators on behalf of others e.g. families

Not all of the agencies would investigate each time – for example in the UK, the aviation safety regulator would only investigate if it considered that a major breach of regulation had occurred and the occupational health and safety regulator, the Health and Safety Executive would generally defer to the AAIB on matters of aviation safety.

For aviation, the purpose of investigation is clearly defined at the international level and then enacted through legislation at the national and, in the case of member states, the European level.

2.1 The International Scene

The International Civil Aviation Organisation (ICAO) first prepared the International Standard and Recommended Practices (SARPS) for the investigation of civil aviation accidents in 1951 as *Annex 13 to the Convention on Civil Aviation.* Whilst the original Annex was named 'Aircraft Accident Inquiry' this was changed in 1975 to 'Aircraft Accident Investigation' as in many States, the word 'inquiry' had judicial connotations (ICAO, 2015). There have been ten editions of *Annex 13* plus minor revisions in between and the most recent version was published in 2010 (ICAO, 2010)

ICAO clearly states 'the sole objective of the investigation of an accident or a serious incident shall be the prevention of accidents and incidents. It is not the purpose of this activity to apportion blame or liability'. SARPS are specifications that Member States are expected to ether conform with or file a 'notice of difference' with the ICAO Council. They are not legally binding and therefore need State-level implementation through relevant legislation to be enforceable. As such, compliant Member States have national legislation that empowers its safety investigation agency (SIA) to do its job to the same core principle e.g. *The Civil Aviation (Investigation of Air Accidents and Incidents) Regulations 1996* for the UK.

Annex 13 details the way in which an aircraft accident investigation should be conducted, the rights and responsibilities of participants, reporting format and perhaps most importantly, the protection of evidence. Its power is in its widespread adoption and in particular, acceptance of its aim.

However, while all ICAO member states are obliged under *Annex 13* to conduct an investigation into a serious incident or accident (or delegate it to another competent state), in practice, the depth and quality of an investigation is somewhat at the mercy of where in the world it occurs. All countries are not created equal in terms of their investigation capability or indeed philosophy. Consequently, the investigations that have done the most to advance the discipline are simply because of *where* the event occurred rather than *what* causes lay behind it.

2.2 The European Scene

Transport accident investigation within Europe has been supported through a variety of Directives and, in the case of aviation, a subsequent European Council Regulation. Directive 94/56/EC was aimed at aviation; 04/49/EC at rail; and directives 1999/35/EC, 2002/59/EC and amended via 2009/18/EC were aimed at marine. The directives emphasized the importance of not-for-blame investigation in improving safety, for example 94/56/EC (European Union, 1994) notes... 'the expeditious holding of technical investigations of civil aviation accidents and incidents improves air safety in helping to prevent the occurrence of such accidents and incidents.' It adds that, and in accordance with ICAO Annex 13, that '...the sole aim of the technical investigation is to draw lessons which could prevent future accidents and incidents and... ...therefore the analysis of the occurrence, the conclusions and the safety recommendations are not designed to apportion blame or liability.'

In 2010, the European Council enacted regulation governing the investigation and prevention of accidents and incidents in civil aviation (European Union, 2010) known as 'EU996/10'. This regulation replaced the earlier Directive 94/56/EC with a binding obligation on EU Member States that remained compatible with the SARPS in Annex 13. Particular emphasis was placed upon three aspects:

- independence of investigation;
- protection of sensitive safety information; and
- information for victims' families

In the case of **independence of investigation**, the Regulation states

'The safety investigation of accidents and incidents should be conducted by or under the control of an independent safety investigation authority in order to avoid any conflict of interest and any possible external interference in the determination of the causes of the occurrences being investigated.' (European Union, 2010)

The aviation industry has been concerned for many years about the apparent increase in criminalization of aircraft accidents (See FSF et al. 2010) and with it the risk of jeopardizing the learning culture fostered through a not-for-blame approach. Quinn (2007) notes, '...when considering the chilling impact the threat of prosecution can and does have on safety investigations, it becomes clear that the future of aviation safety depends on unhindered communication between investigators, witnesses and those involved in accidents'. As Michaelides-Mateou and Mateou (2010) observe, 'This view undoubtedly results in the fear of prosecution which impedes safety and casts doubts upon the integrity of the technical investigation.'

The EU Regulation needed to accommodate the different legal approaches across Europe and clarify a route whereby regulators and specifically, the European Aviation Safety Agency, had a route to participate in an investigation. As the Union is not a 'State' in ICAO terms, but yet acts on behalf of member states in many aviation regulatory matters, a way needed to found for its inclusion in the process '...without affecting the independent status of the investigation.'

The **protection of sensitive safety information** is of particular interest because of the concern over access by judicial authorities of evidence that is either considered privileged or has been collected for a different purpose.

EU 996/10 states that the following records shall not be made available or used for purposes other than safety investigation:

a) all statements taken from persons by the safety investigation authority in the course of the safety investigation;

b) records revealing the identity of persons who have given evidence in the context of the safety investigation;

c) information collected by the safety investigation authority which is of a particularly sensitive and personal nature, including information concerning the health of individuals;

d) material subsequently produced during the course of the investigation such as notes, drafts, opinions written by the investigators, opinions expressed in the analysis of information, including flight recorder information;

e) information and evidence provided by investigators from other Member States or third countries in accordance with the international standards and recommended practices, where so requested by their safety investigation authority;

f) drafts of preliminary or final reports or interim statements;

g) cockpit voice and image recordings and their transcripts, as well as voice recordings inside air traffic control units, ensuring also that information not relevant to the safety investigation, particularly information with a bearing on personal privacy, shall be appropriately protected, without prejudice to paragraph 3[1].

The primary example would be the cockpit voice recorder (CVR) which is 'accepted' by the flight crew on the premise that it would only be used for the not-for-blame investigation of a serious incident or accident and not as a source of evidence in litigation. EU996/10 specifically identifies that 'Flight data recorder recordings shall not be made available or used for purposes other than those of the safety investigation, airworthiness or maintenance purposes, except when such records are de-identified or disclosed under secure procedures.'

With regard to **information for victims' families**, it is essential to remember that the audience for an accident investigation goes way beyond the technical. Whilst those who design, maintain, operate, regulate and train staff to operate aircraft are the main recipients of formal recommendations, there is a much broader group of people with an interest in understanding what happened, why it happened and what will be done to prevent it from happening again. This will also include those seeking either restitution or those who are either fiscally or criminally responsible. Such interests may be disparate and hard to service, but nevertheless remain important.

---

[1] Paragraph 3 acknowledges that the administration of justice or the authority competent to decide on the disclosure of records according to national law may decide that the benefits of the disclosure of the records for any other purposes permitted by law outweigh the adverse domestic and international impact that such action may have on that or any future safety investigation.

In the USA, the Aviation Disaster Family Assistance Act (1996) was enacted following several major accidents in which '…air carriers, local responders and federal agencies did not provide en effective coordinated effort to meet the needs of family members and survivors' (NTSB, 2008) The Act made it clear that the relevant agencies and operator had a responsibility to support and inform those affected by an accident which may number in the hundreds. EU 996/10 obliges Member States to '…establish a civil aviation accident emergency plan at national level. Such an emergency plan shall also cover assistance to the victims of civil aviation accidents and their relatives.' In terms of the investigation (European Union, 2010),

> The safety investigation authority in charge shall be authorised to inform victims and their relatives or their associations or make public any information on the factual observations, the proceedings of the safety investigation, possibly preliminary reports or conclusions and/or safety recommendations, provided that it does not compromise the objectives of the safety investigation and fully complies with applicable legislation on the protection of personal data.

> Before making public the information referred to in paragraph 4, the safety investigation authority in charge shall forward that information to the victims and their relatives or their associations in a way which does not compromise the objectives of the safety investigation.

3 The Evolution of Investigation

The International Civil Aviation Organisation (ICAO) describe three eras of aviation safety, characterized as:

- The technical era (early 1900s until late 1960s)
- The human factors era (early 1970s until mid 1990s)
- The organizational era (mid 1990s to the present day)

So too, investigation has been characterized a similar way, although this may in fact reflect the emphasis in *investigation* rather than changes in *causation*. In other words, as investigation has evolved and adopted new methods to capture and analyse evidence, so the ability of investigations to uncover underlying factors has changed. The causes *per se* may not have changed radically but as technologies such as flight data recorders and cockpit voice recorders collect evidence that hitherto had been inaccessible and as safety thinking has evolved to better understand complex systems and the role of organisational factors, so too findings of investigations have changed. Stoop and Dekker (2012) discuss three phases in the development of safety investigations: '…a technological phase in which the notion of technical failure was dominant'; a second phase which '..started in the nineties when independence from state interference was deemed necessary'; and a third phase which '…characterizes the transition from accident investigation into safety investigation". The authors contend that this represented the shift from event analysis to systems analysis.

However, back in 1993, ICAO Human Factors Digest No. 10 *Human Factors, Management and Organization* (ICAO, 1993) addressed '...the influence of management factors in aviation safety, from the perspective of organizational accidents' whilst acknowledging that the role of management factors in accident prevention had been highlighted in the literature '...forty or more years ago'. The document provided a case study or an organizational accident – the loss of a DC10-30 aircraft on a special sightseeing flight over the Antarctic in 1979. Notably, although the accident was subject to an *Annex 13* investigation (Office of Accident Investigation, 1979), it was in fact a Royal Commission (Mahon, 1981) that examined the deeper, systemic issues within the airline management and regulatory oversight. At the time, his findings were unpopular and he was accused of exceeding his brief when he described the evidence presented before him by the airline as '...an orchestrated litany of lies'. The report was finally accepted by the New Zealand Parliament in 1999.

Flight 901

Air New Zealand flight 901 collided with Mount Erebus on 28th November 1979 after descending to 1,500 feet whilst on a sightseeing flight over Antarctica. 257 people died. The accident investigation focused on the decision of the flight crew to descend in an area of overcast cloud in contravention of minimum safe altitude rules. However, the causes of the accident were complex and subsequent inquiries determined that an optical illusion associated with polar conditions, an unreported correction to an error in navigation waypoints, the pre-flight briefing, experience of previous flights and the management of polar flights were some of a large number of contributory factors.

Just over a decade on from the Royal Commission, those investigation agencies that looked into organizational factors were considered to be pioneering (e.g. the Australian Bureau of Air Safety Investigation and Canadian Transport Safety Board). The use of Reason's (1990) Organisational Accident approach in the investigation of a fatal accident in Young, New South Wales (BASI, 1994) was one such investigation. Barry Sargeant, the Investigator-in-charge of the accident noted, "It really was a new way of looking at the information we received in the process of the investigation. This revealed the complex relationship between individuals associated with the occurrence, and the design and characteristics of the system within which those individuals operated." (CASA, 2001) BASI's success with the approach led to its wider adoption by ICAO.

The approach adopted by the aviation sector has been replicated elsewhere. For example, the Marine Accident Investigation Branch (MAIB) of the UK Department for Transport was formed in 1989 (following the capsize of MV Herald of

Free Enterprise in 1987 with the loss of 193 souls) and the Rail Accident Investigation Branch (RAIB) was formed in 2003 (following the Southall, Ladbroke Grove, Hatfield and Great Heck accidents in 1997, 1999, 2000 and 2001 respectively). The principle of not-for-blame investigation to improve safety is matched across and none of them have regulatory responsibility for enforcement.

However, the loss of MV Herald of Free Enterprise was also seminal moment in terms of the legal profession's move towards the establishment of 'corporate manslaughter'. Whilst such a prosecution was unsuccessful in the Herald case, it was pivotal in what ultimately became the Corporate Manslaughter and Corporate Homicide Act 2007. Under the act, '…an organisation is guilty of the offence if the way in which its activities are managed or organised causes a death and amounts to a gross breach of a relevant duty of care to the deceased. A substantial part of the breach must have been in the way activities were managed by senior management.' (Ministry of Justice, 2007) Hence, just as system safety advocates had pushed for investigators to go deeper into organizational and management factors in their investigations, so too do the prosecutors.

Outside the UK, other countries' investigation capabilities map into other sectors. For example, the US National Transportation Safety Board (NTSB) covers aviation, marine, railroad, road and pipeline accidents; the Canadian Transport Safety Bureau (TSB) covers marine, pipeline, rail and air accidents; and the Australian Transport Safety (ATSB) Bureau covers aviation, rail, marine and space accidents. The Dutch Safety Board (DSB) is unique in that it '…can launch investigations into incidents (disasters, accidents and near-accidents) in the various transport sectors and in the fields of defence, industry and trade, health care, nature and environment, and crisis management and relief. (DSB, 2015) Such developments are primarily the result of the trust that has built up in the way aviation investigates occurrence as Smart (2004) observed:

> The fundamental reason why the aviation model has been used in the other modes is because it has been able to establish public and industry trust its ability to conduct thorough and objective investigation into the circumstances of aircraft accidents. This trust extends to a confidence that the process will swiftly address the public safety issues arising from any accident while at the same time meeting the needs of survivors and bereaved families by keeping them updated on the progress of the investigation.

4 Challenges for Investigation

The integrity of aircraft accident investigation is a key facet of its success. In other words, as the industry and its customers have benefited from the not-for-blame approach, so trust has increased and investigations have been able to go deeper. This notwithstanding, threats do exist for aircraft accident investigation in the areas of complexity of investigation; access to, preservation and protection of evidence; investigator competency and currency; and societal demands.

4.1 Complexity

Aircraft and the systems that support them have grown in capability and, as a by-product, often complexity. Even those who operate and maintain aircraft often have an incomplete understanding of what goes on within an aircraft's flight management system – a condition known as 'system opacity' (Campbell and Bagshaw, 2002). For investigators trying to understand faults within such complex aircraft, especially if significantly damaged, they become highly reliant on manufacturers or are confronted with an absence of evidence. As Roed-Larsen and Stoop (2012) note 'in many cases the level of physical evidence is missing in the factfinding phase because systems operate virtual without physical components. They may fail at a functional level rather than due to mechanical failure caused by mechanical overloads, or operating outside the designed performance envelope' (Vaughan, 1999; Leveson, 2004 cited in Roed-Larsen and Stoop 2012).

For different reasons, penetrating the deeper systemic factors such as organizational risk controls and management actions remains difficult, especially in entities that are multinational. For example, the accident involving Fairchild Metro III aircraft which crashed at Cork Airport in 2011 was described as ...' the most challenging of the more than 500 Investigations that have been completed by the Unit since its formation in 1994 (AAIU, 2015). In particular the relationship between ticket seller, aircraft operator and the various regulators was difficult to unravel. As Reason noted (Strauch 2002), 'The achievements of accident investigators are all the more remarkable when one considers the snares, traps and pitfalls that lie in their path'.

The task of accident investigation is also frequently complex and therefore costly in nature, and despite the addition of recorder technology to aid in the task, this trend appears to continue. Aside from the material loss and human consequence of an accident, the investigation is likely to involve multiple parties from many countries over a prolonged period. Specialist agencies will likely be included to conduct detailed testing and analysis, all of which must be closely witnessed in order to maintain a chain of evidence and therefore the integrity of the process. Although ICAO sets investigation agencies the target of one year for the publication of a final investigation report, many investigations have exceeded this. The complexity of an investigation or indeed the apparent absence of 'an answer' can stretch even the best-resourced agencies. The investigation into the loss of Swissair flight 111 which crashed off Nova Scotia in 1998 lasted 5 years (TSB, 2003); the investigation into the loss of TWA flight 800 (NTSB, 2000) which crashed off Long Island in 1996 took 4 years; and in the case of Air France flight 447, it took investigators 2 years to find the Digital Flight Data Recorder (DFDR), cockpit voice recorder (CVR) and main wreckage field and a further year to complete the investigation.

As accidents have reduced in frequency and are perceived by some to be reactive in nature, the concept of investigating near misses and incidents is compelling. They represent a greater volume of data with live and generally willing witnesses. However, the availability of resources and the willingness to investigate is not the same as when an accident occurs. Certain national investigation agencies have initiated proactive investigations on the basis of a series of near misses, but these are not always popular by those under 'investigation'. A lack of a complete causal sequence and philosophical disagreement as to whether the type of incidents are a good indicator of the types of accident that are likely to occur limit the effectiveness of this approach. The concept seems straightforward, but the application is rarely so.

4.2 Evidence

Without evidence, an investigator's view is nothing more than an opinion. Aircraft accident investigators therefore draw upon physical, documentary and human evidence to test their hypotheses of cause. Some of this evidence is sensitive and therefore should only be used for improving flight safety. Notwithstanding the attempts of Annex 13 and EU 996/10 to protect evidence, the principle is only as good as a court's judgment.

In 2013, an AS332 Super Puma helicopter crashed on approach to Sumburgh Airport in the Shetland Islands resulting in four fatalities. In accordance with Annex 13 and EU 996/10, the UK Air Accidents Investigation Branch (AAIB) conducted a not-for-blame investigation. In 2015, and before the AAIB's final report was published, Lord Jones (Scottish Courts and Tribunals, 2015) responded to a petition to the court by the Lord Advocate regarding request from the Procurator Fiscal to the AAIB to make available the combined voice and flight data recorder (CVFDR) for use in his investigation. The AAIB had '…refused to do so in the absence of an order from the court, under reference to the provisions of paragraphs 1 and 2 of article 14 of Regulation EU 996/2010.' It was the view of the AAIB that such disclosure what contrary to the EU Regulation.

In June 2015, Lord Jones granted an order releasing the CVFDR to the Crown Office and Procurator Fiscal Service, subject to seven conditions that were designed to address concerns raised during proceedings. Jones' judged '…the Lord Advocate's investigation into the circumstances of the death of each of those who perished in this case is both in the public interest and in the interests of justice'. Further, he stated 'the cockpit voice recording and the flight data recording which the Lord Advocate seeks to recover will provide relevant, accurate and reliable evidence which will enable SARG *(CAA Safety and Regulation Group)* to provide an expert opinion of value to assist him in his investigation of the circumstances of the death of the four passengers whose lives were lost, and his decision whether and, if so, against whom to launch a prosecution.'

The ruling was made on the basis of the facts of the individual case and Jones' argued that it met the '…tests laid down by the 1996 Regulations and the EU regulation'. In doing so it did not create precedent for future cases or the routine disclosure of cockpit voice recordings.

In a separate development, precedence seems to have been set regarding the admissibility of AAIB reports in legal proceedings with the case of Rogers and Hoyle in 2015. The case was brought on behalf of the dependents of Mr Rogers – a passenger who had been killed in a vintage Tiger-Moth accident in 2011 – who wished to append the AAIB report in support of civil litigation against the surviving pilot. The English Court of Appeal affirmed the High Court's decision that accident reports prepared by the AAIB are admissible in evidence in civil proceedings.

In response, Belsham et al. (2015) observed

'…Although we would expect such applications to be only rarely, if ever successful, any order for disclosure of these records is potentially toxic to the effectiveness of accident investigation, which (as is recognised by the very regulations governing accident investigation) relies on a deliberate lack of transparency to promote safety and not drive it underground.'

In contrast Healy-Pratt and Stewart (2015) of Stewarts Law who were representing the dependents of the deceased cite the family as saying

'We are very pleased that the Court of Appeal has seen how helpful the work of the AAIB can be to assist with the facts of an air disaster. We are also comforted by the fact that this will assist not only our own case but also future victims of air crashes. Our aviation lawyers, Stewarts Law, will now be able to continue with our civil claim as we seek justice for the loss of Orlando in such tragic circumstances. '

A further challenge related to evidence comes in terms of developing an accurate understanding of human and organizational behaviour, where physical evidence is often limited and where witness evidence if often absent, in accurate or incomplete. For example, although the French investigation agency (BEA) considered the performance of the crew in detail following the loss of an Air France A330 over the Atlantic in 2009 (BEA, 2012), they were limited in their understanding to the evidence provided by the digital flight data recorded (DFDR) and cockpit voice recorder (CVR) which captured actions involving the aircraft and conversation only. In other words, they had no additional information with which to determine why the crew did what they did or as Dekker (2011) describes it 'why taking or not taking action made sense to them'.

Strauch (2015) notes that although '…the objectives of accident investigation are similar to those of empirical research', investigations typically rely on a sample size of one accident to describe a causal sequence. Investigators therefore rely on a considerable amount of data (evidence) from a variety of sources to reach their conclusions. This can become especially difficult when trying to find evidence for deeper systemic causes such as the influence of safety culture. Indeed,

Strauch (2015) argues that safety culture cannot and should not be examined in accident investigations for a variety of reasons, but also acknowledges:

'...the considerable data that accident investigations typically collect can better describe a company's actual practices in system operations than could be obtained from most direct assessments of safety culture.'

4.3 Investigator competency and currency

ICAO (2003) argues that the task of aircraft accident investigation should only be undertaken by 'qualified investigators'. However, few such qualifications exist and there is no professional accreditation beyond that given by certain organisations (such as the Australian Transport Safety Bureau) to their own staff. ICAO does recognize within it Manual of Aircraft Accident and Incident Investigation: *Part 1 Organization and Planning* (2000):

'In addition to technical skills, an accident investigator requires certain personal attributes. These include integrity and impartiality in the recording of facts, logic and perseverance in pursuing inquiries, often under difficult or trying conditions and tact in dealing with a wide range of people who have been involved in the traumatic experience of an aircraft accident.'

As each accident is largely unpredictable, maintaining currency for investigators is a challenge. Whilst some agencies are able to send investigators to more frequent general aviation accidents or serious incidents involving transport category aircraft, others do not have the resources to do so, or (thankfully) do not experience many accidents within their territory.

The European Safety Reliability and Data Association (ESReDA) Working Group on Accident Investigation identified (Dechy et al. 2012):

'few investigators and LFE (Learning from Experience) experts have general training in conducting AI (Accident Investigation), and even less on the systemic and organisational approaches, to the extent that we could wonder about the weaknesses of the analysts and their approaches as a contributing factor to the repeated failures of the LFE... ...In addition, few oganisations (except for some investigation boards and industrialists) adopted the socio-technical and organisational paradigms in their approaches and methods. Thus, the technical dominant paradigm with the engineering culture and the human error paradigm remain very dominant despite many scientific statements since the eighties and nineties.'

Whilst myriad techniques exist within the academic literature to help to investigate and analyse accidents, comparatively few make it to the front line, especially in terms of the investigation of the 'softer' systemic issues such as human and organizational influences. As the Energy Institute (2008) remarked:

'Whilst many... ...businesses investigate and analyse both incidents and accidents – whether with major hazards or occupational potential – human and organisational factors aspects are rarely addressed enough. This is particularly true for non-engineering aspects of HSE's priority human factors issues, such as supervision and organisational culture. The problem is compounded by the volume of tools available to investigate and analyse incidents, many of which have some good points; however, none of them presents an ideal solution.'

Underwood and Waterson (2013) found that 'the different analysis approaches taken by the researcher and practitioner communities suggest that a research–practice gap exists in the domain of SAA (systemic accident analysis).' Indeed, anecdotal feedback from investigators demonstrates the impracticality of many tools that look good as an academic paper, but which have never been applied by those stood 'knee deep in mud, metal and body parts'.

The challenge of maintaining competency may yet increase due to the relative infrequency of events, changes in technology, decreasing financial resources and increasing complexity. A focus on training that goes beyond establishing an initial level of competence and instead provides a continuous process of knowledge and skill development to maintain competency is required. This will require cooperation between agencies e.g. through networks such as ENCASIA – the European Network of Civil Aviation Safety Investigation Authorities (European Commission, 2015) or DASIF (the Defence Aviation Safety Investigators Forum) and innovation on the part of training provider such as through the use of simulation to develop and practice skills.

4.4 Social demands

The tolerance for uncertainty in the investigation of aircraft accidents is limited, partly as a consequence of the many successes that, against apparently impossible odds, investigators have achieved. Many investigations have achieved near celebrity-status thanks to mass media coverage and countless documentaries. However, this creates an expectation that the answer will always be found, even when in practice, the evidence may not support a definitive conclusion.

A recent example was the publication of the AAIB report into the Police helicopter which crashed onto the Clutha Bar in Glasgow in 2013 (AAIB, 2015). The investigation was hampered by a lack of DFDR and CVR data and an absence of witnesses, but Scottish First Minister Nicola Sturgeon described it as 'deeply disappointing' that after a two-year investigation, the report did not reach a clearer conclusion (BBC News, 2015). She then raised the expectations of the families that the subsequent Fatal Accident Inquiry '...can help the families get the answers they seek' without identifying what additional evidence was likely to become available.

The insatiable desire for information is both served and fueled by a growing formal and social media network. The appetite was no more clearly expressed than following the disappearance of Malaysian flight MH370 when the story remained a front-page story on news websites for many weeks.

When something goes wrong involving aviation, there is an expectation that this industry will be able to identify the problem in a timely fashion and put in place corrective actions; 'don't know' is not an acceptable answer.

5 An Opportunity for Investigation?

Stoop and Dekker (2012) question whether safety investigations are obsolete due to '...their reactive nature and the lack of learning potential they provide', but later conclude that such investigations are at least partially pro-active and provide '...indispensable feedback... ...in order to develop new knowledge and insights in the performance of complex systems'.

In 2013, Eurocontrol – the European Organisation for the Safety of Air Navigation published a white paper entitled 'From Safety-I to Safety-II' to advocate a shift in safety management '...from ensuring that 'as few things as possible go wrong' to ensuring that 'as many things as possible go right'. (Eurocontrol, 2013)

'The new paradigm also means that the priorities of safety management must change. Instead of limiting investigations and learning from incidents, a SMS should allocate some resources to look at the events that go right and try to learn from them.'

The challenge is to apply the not-for-blame, methodical, evidence-driven approach that has been championed in aircraft accident investigation to the challenge of building organizational resilience through understanding how a complex systems performs well. This will require the development of new techniques and the justification of additional resources in an area where performance is generally negatively reported. In other words, how will organisations justify additional expenditure on investigation where to the outside world at least, there is apparently nothing to investigate.

6 Conclusion

Over a century of aircraft accident investigation, the emphasis has evolved greatly as new sources of evidence and techniques for analysis have been developed. However, whilst such evolution has overcome a range of challenges, so to expectations have grown against a backdrop of increasing technical complexity and the desire to go deeper into the less tangible organizational and cultural factors. Investigation has maintained its relevance by changing, but perhaps the greatest potential lies in extending the highly successful approach into understand how complex systems succeed and not just fail?

References

AAIB – Aircraft Accidents Investigation Branch (2015) Report on the accident to Eurocopter (Deutschland) EC135 T2+, G-SPAO Glasgow City Centre, Scotland on 29 November 2013. Aircraft Accident Report 3/2015. Department for Transport, Farnborough, UK

AAIU – Aircraft Accident Investigation Unit (2015) Publication of AAIU Final Report No. 2014-001 Accident to SA 227-BC Metro III EC-ITP at Cork Airport on 10 February 2011. http://www.aaiu.ie/sites/default/files/report-attachments/Final%20PRESS%20RELEASE%202014-001_0.pdf Press Release Accessed 8 November 2015

BASI - Bureau of Air Safety Investigation (1994) Investigation Report 9301743, Piper PA31-350 Chieftain VH-NDU Young, NSW 11 June 1993. Canberra, Australia

BBC News (2015) Reaction: Clutha Helicopter Crash Report http://www.bbc.co.uk/news/uk-scotland-glasgow-west-34615349 Accessed 18th November

BEA - Bureau d'Enquêtes et d'Analyses pour la sécurité de l'aviation civile (2012) Final Report on the accident on 1st June 2009 to the Airbus A330-203 registered F-GZCP operated by Air France flight AF 447 Rio de Janeiro – Paris. Ministère de l'Écologie, du Développement durable, des Transports et du Logement. France

Belsham G, McInnes D, Turner W and J Hickland (2015) Aviation Safety In The Balance? http://incelaw.com/en/documents/pdf_library/strands/aviation/safety-in-the-balance.pdf Accessed 8 November 2015.

Campbell R D and Bagshaw, M (2002) Human Performance and Limitations 3rd ed. Ashgate, Aldershot, UK

CASA - Civil Aviation Safety Authority (2001) Breaking New Ground: One Man's Reflection. Flight Safety Australia March-April. Canberra, Australia

Dechy N, Die Y, Funnemark E, Roed-Larsen S, Stoop J, Valvisto T and A L V Arellano (2012) Results and lessons learned from the ESReDA's Accident Investigation Working Group Introducing article to "Safety Science" special issue on "Industrial Events Investigation" Safety Science 50 (2012) 1380-1391

Dekker S (2011) Drift Into Failure: From Hunting Broken Components to Understanding Complex Systems. Ashgate, Aldershot, UK

Dutch Safety Board (2015) http://www.onderzoeksraad.nl/en/ Accessed 8 November 2015.

Energy Institute (2008) Guidance on Investigating and Analysing Human and Organisational Factors Aspects of Incidents and Accidents. Energy Institute, Lodon, UK

Eurocontrol (2013) From Safety-I to Safety-II: A White Paper. Downloaded from https://www.eurocontrol.int/sites/default/files/content/documents/nm/safety/safety_whitepaper_sept_2013-web.pdf

European Commission (2015) European Network of Civil Aviation Safety Investigation Authorities homepage - http://ec.europa.eu/transport/modes/air/encasia/activities/index_en.htm Accessed 17 November 2015.

European Union (1994) Council Directive 94/56/EC of 21 November 1994 establishing the fundamental principles governing the investigation of civil aviation accidents and incidents. Brussels.

European Union (2010) Regulation (EU) No 996/2010 of The European Parliament and of The Council of 20 October 2010 on the investigation and prevention of accidents and incidents in civil aviation and repealing Directive 94/56/EC. https://www.gov.uk/government/uploads/system/uploads/attachment_data/file/384433/Regulation_996_2010_of_20_October_2010_accident_investigation.pdf Accessed 8 November 2015

FSF, RAeS, ANAE, CANSO, ERA, IFATCA, PAMA and ISASI (2010) Joint Resolution Regarding Criminalization of Aviation Accidents. http://flightsafety.org/files/resolution_01-12-10.pdf Accessed 15 October 2011

Grahame-White, C (1912) Aviation. Collins, London, UK

Healy-Pratt J and S Stewart (2015) Rogers v Hoyle: Legal victory for claimants in UK aviation claims. Press Release http://www.stewartslaw.com/rogers-v-hoyle-legal-victory-for-claimants-in-uk-aviation-claims.aspx Accessed 8 November 2015

ICAO – International Civil Aviation Organisation (1993) Human Factors Digest No. 10 Human Factors, Management and Organization. Circular 247-AN/148 Montreal, Canada

ICAO - International Civil Aviation Organization (2003) Training Guidelines for Aircraft Accident Investigators. Cir 298 AN/172. Montreal, Canada

ICAO – International Civil Aviation Organisation (2010) Annex 13 to the Convention on International Civil Aviation. Aircraft Accident and Incident Investigation. 10th ed. Montreal, Canada

ICAO – International Civil Aviation Organisation (2015) The Postal History of ICAO: Annex 13 - Aircraft Accident and Incident Investigation. http://www.icao.int/secretariat/postalhistory/annex_13_aircraft_accident_and_incident_investigation.htm Accessed 8 November 2015

Mahon, P Justice (1981) Report of the Royal Commission into the Crash on Mount Erebus, Antarctica, of DC10 Aircraft Operated by Air New Zealand Limited. Wellington, New Zealand

Michaelides-Mateou S, Mateou A (2010) Flying in the Face of Criminalization: The Safety Implications of Prosecuting Aviation Professionals for Accidents. Ashgate, Aldershot, UK.

Ministry of Justice (2007) A guide to the Corporate Manslaughter and Corporate Homicide Act 2007. Crown Copyright.

NTSB – National Transportation Safety Board (2000) Report into the In-flight Breakup Over the Atlantic Ocean of Trans World Airlines Flight 800, Boeing 747-141, N93119. Report AAR-00-03 NTSB Washington

Office of Accident Investigation (1979) Aircraft Accident Report No. 79-139. Wellington, New Zealand

Quinn, K P (2007) Battling Accident Criminalization. Aerosafety World. Journal of the Flight Safety Foundation, Arlington, Virginia

Reason, J. (1990) Human Error. Cambridge University Press.

Reason, J (1991) Identifying the latent causes of aircraft accidents before and after the event. International Society of Air Safety Investigators 22nd International Annual Seminar, Canberra, 4-7 November 1991, pp. 1-13

Scottish Courts and Tribunals (2015) Outer House, Court of Session Judgment CSOH 80 P628/14 - Opinion of Lord Jones in the cause Frank Mulholland QC, the Lord Advocate (Petitioner) for An order in terms of Regulation 18 of the Civil Aviation (Investigation of Air Accidents and Incidents) Regulations 1996. https://www.scotcourts.gov.uk/search-judgments/judgment?id=0452dda6-8980-69d2-b500-ff0000d74aa7 Accessed 8 November 2015

Smart K (2004) Credible Investigation of Air Accidents. Journal of Hazardous Materials 111 (2004) 111-114

Stoop J and S Dekker (2012) Are Safety Investigations Pro-active? Safety Science 50 (6), 1422-1430

Strauch B (2002) Foreword by Reason J in Investigating Human Error Ashgate, Aldershot UK.

Alexander C, Ishikawa S, Silverstein M et al (1977) A pattern language: towns, buildings, construction. Oxford University Press

TSB - Transportation Safety Board of Canada (2003) Aviation Investigation Report A98H0003

AUTHOR INDEX

www.ingramcontent.com/pod-product-compliance
Lightning Source LLC
Chambersburg PA
CBHW051437170526
45166CB00001B/16